AF822429

Astrophysics and Space Science Proceedings

The Impact of HST on European Astronomy

F. Duccio Macchetto
Editor

Space Telescope Science Institute (STScI), Baltimore, MD 21218, USA

Springer

Editor
Dr. F. Duccio Macchetto
Space Telescope Science
Institute (STScI)
3700 San Martin Dr.
Baltimore, MD 21218
USA
macchetto@stsci.edu

ISSN 1570-6591 e-ISSN 1570-6605
ISBN 978-90-481-3399-4 e-ISBN 978-90-481-3400-7
DOI 10.1007/978-90-481-3400-7
Springer Dordrecht Heidelberg London New York

Library of Congress Control Number: 2009942446

© Springer Science+Business Media B.V. 2010
No part of this work may be reproduced, stored in a retrieval system, or transmitted in any form or by any means, electronic, mechanical, photocopying, microfilming, recording or otherwise, without written permission from the Publisher, with the exception of any material supplied specifically for the purpose of being entered and executed on a computer system, for exclusive use by the purchaser of the work.

Cover design: eStudio Calamar S.L.

Printed on acid-free paper

Springer is part of Springer Science+Business Media (www.springer.com)

Foreword

Remembrance of Things Past

It scarcely seems credible that it was almost exactly thirty years ago that I first met Duccio Macchetto at the first meeting of the newly formed Science Working Group of what was then called the Space Telescope project. We were there in slightly different roles, Duccio as the project scientist for the Faint Object Camera and I as an interdisciplinary scientist. Henk van de Hulst was also there as the official representative of ESO. The approval of the project was the end result of a great deal of lobbying and politicking both in the USA and Europe, the European contribution proving essential to the approval process in the USA. Those interested in the nitty-gritty of the process should read Robert Smith's outstanding history of the Hubble Space Telescope.

We should have realized early on that we were in for a rough time. At that first meeting of the Science Working Group I remember vividly NASA Headquarters telling us that the Space Telescope was a success-oriented programme that would cost M$ 680. Well, we could live with the cost-tag, but we should have had concerns about the expression "success-oriented". This meant that everything should turn out exactly as planned, the project would be carried out within the projected time-scale and budget and the telescope would be launched in 1983. Well, the rest is history.

We learned a lot of useful jargon along the way. During one particularly worrisome period, the project manager announced with confidence "We have no problems; we have the following concerns" — all spoken with a completely straight face. During the difficult years up to about 1983, the programme ran through a number of project managers. The root cause was the same — the project could not be carried through with the projected budget and the blame for this fell on the project managers. Duccio and I were bemused by the fact that each new project manager was introduced as the person who had large numbers of successful space projects under their belt and was exactly the right person to solve all the Space Telescope's funding and management problems. They never lasted more than about two or three meetings before they were replaced by another outstanding manager with the same credentials who would soon suffer the same fate. Eventually, when the project was about to run out of money really seriously, a realistic budget for the project was agreed and Joe Rothenberg did a splendid job of bringing the whole mission to a successful conclusion.

The Europeans were in an interesting position since we only had a 15% stake in the project, but many of us had argued successfully that we had to be in the Space Telescope project as of right rather than as guests. Probably the most important thing the three of us had to do was to remind our American colleagues that this really was an international project and that Europe had a lot to contribute to the project in many different ways. It was interesting to contrast the ESA approach with that of our US colleagues who were largely University based, but supported by major aerospace companies. The professionalism of Duccio and his ESA colleagues shone through while all the agonizing was centred on the many problems which the US instruments had to overcome and building the telescope within a reasonable budget.

Despite all the agonies, the telescope was launched in 1990 and then the spherical aberration problem with the primary mirror surfaced. This was the mirror which had completely blown our minds when we were shown the amazing interferometer fringes in the early 1980s. The late Bill Fastie, one of the two optics specialists on the Science Working Group, claimed it was the best large mirror that had ever been polished. He was right, but the problem was that the mirror had been tested with the same set-up which had been used to polish it. The interesting aspect of the story was that the Science Working Group could not actually get their hands dirty in participating in the tests. They simply had to accept what they were shown by the contractors.

The very positive side of this story was the subsequent crash programme to rescue the mission and restore the full scientific capability of the telescope. The Space Telescope Science Institute and NASA worked together in an extraordinary endeavour which resulted in the successful refurbishment mission which took place in 1993. Again, the rest is history. Like the rest of the Science Working Group, I eventually receive my guaranteed time observations in 1996. My student Philip Best did a splendid job in analyzing our wonderful images of large redshift radio galaxies. As he remarked in the colloquium he gave describing his excellent analysis of the data, he was only four years old when I had submitted the original proposal in 1977.

What is really remarkable is the extraordinary change in astronomical perspective since the project was approved in 1977. It was a really brave decision to adopt CCD detectors for the Wide Field-Planetary Camera when there was not really any working CCD camera on any ground-based telescope. One of the great dramas of the construction programme was the remarkable ingenuity displayed by the late Jim Westphal, Jim Gunn and the team which constantly came up with innovative solutions to the many problems of implementing CCD cameras in space.

The one really key lesson that we learned from the subsequent programme of observations was undoubtedly that the science which was undertaken by the HST was far more imaginative, creative and exciting than anything which the original proposers had imagined. That is what this ESLAB Symposium celebrated. The European contribution was essential to the success of the programme and brought with it a different astronomical perspective which I believe was very important scientifically, as well as technically and politically.

Equally amazingly, Duccio has seen the programme all the way through from 1977 to the present Symposium thirty years later. During this time, Duccio's job has been to look after European interests in the construction, commissioning and operational phase of the Hubble Space Telescope. He has done this quite exceptionally well and it will be difficult to think of European involvement in the programme without him. I can even forgive him the time he nearly poisoned the whole Science Working Group in Paris by feeding them raw oysters.

I was sorry I could not be at the Symposium, because of a wrecked right Achilles tendon, and share even more of my memories of these dramatic and ultimately superbly successful days. Memories will fade and we will only remember the fabulous success of the HST and Europe's contribution to it. But there was a huge amount of human lifeblood and dedicated effort involved. Nobody epitomizes that spirit more than Duccio. His legacy is the superb quality of the science presented at this Symposium.

Cambridge
September 2007

Malcolm Longair

Preface

Since its launch in 1990 the Hubble Space Telescope has made major contributions to all areas of astronomy and astrophysics. These range from the study of nearby planets, the processes of star and planet formation, the stellar and interstellar components of galaxies, the discovery that most, if not all, galactic nuclei harbor a massive black hole that profoundly affects their evolution, to the realisation that the universe as a whole is undergoing acceleration as a result of a yet unknown form of "dark energy".

European astronomers have made significant contributions to projects that have led to unprecedented progress in our understanding. However, the impact of HST goes beyond these direct contributions and continues to shape the key questions that need to be addressed not only with HST but with other space and ground based observatories. In addition Hubble has had an important role on the performance and productivity of several European facilities, such as the VLT or XMM-Newton, and in the near future it will benefit from the synergy with Herschel and ALMA.

The primary aim of this symposium, the 41st in ESLAB series of the European Space Agency, was to review the key contribution that HST has made in all areas of astronomy and emphasise their impact on European astronomical research. The symposium took place from 29 May to 1 June 2007 at the European Space Research and Technology Centre in Noordwijk, the Netherlands, and this book contains the proceedings of the oral and poster papers that were presented and discussed during the meeting.

The publication of this book took longer than anticipated, since obtaining the individual contributions from some very busy colleagues proved harder than we thought. However, we gladly decided to wait for them, as the purpose of this volume is to offer a broad perspective of the advancements made possible by the HST over its almost two decades of operation, rather than portraying just the most recent discoveries. We are fortunate that this volume sees the light right at the time when a new HST, fully refurbished and upgraded after Servicing Mission 4, becomes available to the European and world-wide astronomical community to answer some of the many still open questions.

Noordwijk
May 2009

Duccio Macchetto & Guido De Marchi

Contents

Contributors

the ACS Coma Cluster Treasury Survey Team

the ACS IDT Team

Alessandra Aloisi Space Telescope Division of the European Space Agency, Space Telescope Science Institute, Baltimore, MD 21218, USA, aloisi@stsci.edu

Hakim Atek Institut d'Astrophysique de Paris, Paris, France

D.J. Axon Department of Physics, Rochester Institute of Technology, Rochester, NY 14623, USA, djasps@rit.edu; Centre for Astrophysics Research, Science & Technology Research Institute, University of Hertfordshire, Hatfield AL10 9AB, UK

Ruymán Azzollini Instituto de Astrofísica de Canarias, 38205 La Laguna, S/C de Tenerife, Spain

Francesca Bacciotti INAF, Osservatorio Astrofisico di Arcetri, Largo E. Fermi 5, 50125 Firenze, Italy, fran@arcetri.astro.it

R. Bacon Université de Lyon 1, CRAL, Observatoire de Lyon, 9 av. Charles André, 69230 Saint-Genis Laval, France

Barbara Balmaverde INAF, Osservatorio Astronomico di Torino, Strada Osservatorio 20, 10025 Pino Torinese, Italy, balmaverde@oato.inaf.it

John E. Beckman Instituto de Astrofísica de Canarias, c/Vía Láctea s/n, 38205 La Laguna, S/C de Tenerife, Spain, jeb@iac.es; Centro Superior de Investigaciones Científicas, Serrano, 117, 28006 Madrid, Spain

Torsten Böker European Space Agency, Dept. RSSD, Keplerlaan 1, 2200 AG Noordwijk, Netherlands, tboeker@rssd.esa.int

Bernhard Brandl Sterrewacht Leiden, Leiden University, Leiden, The Netherlands

Wolfgang Brandner Max-Planck-Institut für Astronomie, Heidelberg, Germany

Paul Bristow ESO, Karl-Schwarzschild-Str. 2, 85748 Garching, Germany, bristowp@eso.org

Alessandro Capetti INAF, Osservatorio Astronomico di Torino, Strada Osservatorio 20, 10025 Pino Torinese, Italy, capetti@oato.inaf.it

M. Cappellari Sub-Department of Astrophysics, University of Oxford, Denys Wilkinson Building, Keble Road, Oxford OX1 3RH, UK

Karina I. Caputi Institute for Astronomy, Swiss Federal Institute of Technology (ETH Hönggerberg), Wolfgang-Pauli-Strasse 16, 8093, Zurich, Switzerland, caputi@phys.ethz.ch

Patrizia Caraveo INAF–IASF, Via Bassini 15, 20133 Milan, Italy, pat@iasf-milano.inaf.it

David Carter Astrophysics Research Institute, Liverpool John Moores University, Twelve Quays House, Birkenhead, CH41 1LD, UK, dxc@astro.livjm.ac.uk

V. Charmandaris LERMA, Paris Observatory, Paris, France; University of Crete, Heraklion, Greece, vassilis@physics.uoc.gr

Marco Chiaberge Space Telescope Science Institute, 3700 San Martin Dr., Baltimore, MD 21210, USA, marcoc@stsci.edu; INAF, Istituto di Radioastronomia, Via P. Gobetti 101, 40129 Bologna, Italy

You-Hua Chu Department of Astronomy, University of Illinois, Urbana, USA

R. Cid Fernandes Depto. de Física-CFM, Universidade Federal de Santa Catarina, Florianópolis, SC, Brazil

Jean-François Claeskens Institut d'Astrophysique et de Géophysique, Université de Liège, Allée du 6 Août, 17, B5C, 4000 Liège, Belgium, claesken@astro.ulg.ac.be

Gisella Clementini INAF, Osservatorio Astronomico di Bologna, via Zamboni 33, 40126 Bologna, Italy, gisella.clementini@obao.inaf.it

Lodovico Coccato Max-Planck-Institut fuer extraterrestrische Physik, Garching bei Muenchen, Germany

Deirdre Coffey INAF, Osservatorio Astrofisico di Arcetri, Largo E. Fermi 5, 50125 Firenze, Italy, dac@arcetri.astro.it

Enrico Maria Corsini Università degli Studi di Padova, Dipartimento di Astronomia, Vicolo dell'Osservatorio, 3, 35122 Padova, Italy

COSMOS Team

Paul A. Crowther Dept of Physics & Astronomy, University of Sheffield, Hounsfield Road, Sheffield, S3 7RH, UK, Paul.Crowther@sheffield.ac.uk

Michael Dahlem Australia Telescope National Facility, Narrabri, NSW, Australia, Michael.Dahlem@csiro.au

Elena Dalla Bontà Università degli Studi di Padova, Dipartimento di Astronomia, Vicolo dell'Osservatorio, 3, 35122 Padova, Italy, elena.dallabonta@unipd.it

L. Danese Astrophysics Sector, SISSA/ISAS, Via Beirut 2-4, 34014 Trieste, Italy, danese@sissa.it

R.L. Davies Sub-Department of Astrophysics, University of Oxford, Denys Wilkinson Building, Keble Road, Oxford OX1 3RH, UK

L. Deharveng Marseille Observatory, Marseille, France, Lise.Deharveng@oamp.fr

Ralf-Jürgen Dettmar Astronomical Institute, Ruhr-University Bochum, Bochum, Germany, dettmar@astro.rub.de

Andrew E. Dolphin Raytheon Corporation, Waltham, USA

Jochen Eislöffel Thüringer Landessternwarte Tautenburg, Sternwarte 5, 07778 Tautenburg, Germany, jochen@tls-tautenburg.de

E. Emsellem Université de Lyon 1, CRAL, Observatoire de Lyon, 9 av. Charles André, 69230 Saint-Genis Laval, France

J. Falcón-Barroso European Space and Technology Centre, Keplerlaan 1, 2200 AG Noordwijk, The Netherlands, jfalcon@rssd.esa.int

Annette M.N. Ferguson Institute for Astronomy, University of Edinburgh, Blackford Hill, Edinburgh, Scotland EH9 3HJ, UK, ferguson@roe.ac.uk

Laura Ferrarese Herzberg Institute of Astrophysics, Victoria, Canada

Francesco R. Ferraro Dipartimento di Astronomia, Via Ranzani 1, 40127 Bologna, Italy, francesco.ferraro3@unibo.it

Robert A.E. Fosbury ST-ECF, Garching bei München, Germany, rfosbury@eso.org

Filippo Fraternali Department of Astronomy, University of Bologna, Bologna, Italy, filippo.fraternali@unibo.it

R.M. González Delgado Instituto de Astrofísica de Andalucía (CSIC), Granada, Spain

Dimitrios A. Gouliermis Max-Planck-Institut für Astronomie, Heidelberg, Germany, dgoulier@mpia-hd.mpg.de

G.L. Granato INAF, Osservatorio Astronomico di Padova, Vicolo dell'Osservatorio 5, 35122 Padova, Italy, gianluigi.granato@oapd.inaf.it; Astrophysics Sector, SISSA/ISAS, Via Beirut 2-4, 34014 Trieste, Italy

the GRAPES Team

Richard Grijs Department of Physics & Astronomy, University of Sheffield, Hicks Building, Hounsfield Road, Sheffield S3 7RH, UK,

R.deGrijs@sheffield.ac.uk; National Astronomical Observatories, Chinese Academy of Sciences, 20A Datun Road, Chaoyang District, Beijing 100012, China

Robert A. Gruendl Department of Astronomy, University of Illinois, Urbana, USA

Leonel Gutiérrez Albores Instituto de Astrofísica de Canarias, c/Vía Láctea s/n, 38205 La Laguna, Tenerife, Spain, leonel@iac.es; Universidad Nacional Autónoma de México, Ensenada, Mexico

Michael Hauser Space Telescope Science Institute, 3700 San Martin Drive, Baltimore, MD 21218, USA, hauser@stsci.edu

Matthew Hayes AlbaNova U. Centre, Stockholm Observatory, Stockholm, Sweden

Eva Hennekemper Max-Planck-Institut für Astronomie, Heidelberg, Germany

Thomas Henning Max-Planck-Institut für Astronomie, Heidelberg, Germany

M. Heydari-Malayeri LERMA, Paris Observatory, Paris, France, m.heydari@obspm.fr

Helmut Jenkner Space Telescope Science Institute, 3700 San Martin Drive, Baltimore, MD 21218, USA, jenkner@stsci.edu; European Space Agency, Research and Scientific Support Department, Noordwijk, The Netherlands

Andrés Jordán European Southern Observatory, Karl-Schwarzschild-Str. 2, 85748 Garching bei München, Germany, ajordan@eso.org

Inga Kamp Space Telescope Science Division of ESA, STScI, Baltimore, MD 21218, USA; Kapteyn Astronomical Institute, PO Box 800, 9700 AV Groningen, The Netherlands, kamp@astro.rug.nl

Florian Kerber ESO, Karl-Schwarzschild-Str. 2, 85748 Garching, Germany, fkerber@eso.org

Jean-Paul Kneib Laboratoire d'Astrophysique de Marseille, CNRS–Université de Provence, Technopole de Marseille-Etoile, 38, rue Frédéric Joliot-Curie, 13388 Marseille cedex 13, France, jean-paul.kneib@oamp.fr

D. Krajnović Sub-Department of Astrophysics, University of Oxford, Denys Wilkinson Building, Keble Road, Oxford OX1 3RH, UK

Daniel Kunth Institut d'Astrophysique de Paris, Paris, France, kunth@iap.fr

H. Kuntschner Space Telescope European Coordinating Facility, European Southern Observatory, Karl-Schwarzschild-Str. 2, 85748 Garching, Germany

A. Lapi Astrophysics Sector, SISSA/ISAS, Via Beirut 2-4, 34014 Trieste, Italy, lapi@sissa.it

Søren S. Larsen Astronomical Institute, Utrecht University, Princetonplein 5, 3584 CC, Utrecht, The Netherlands, larsen@astro.uu.nl

Claus Leitherer STScI, Baltimore, USA, leitherer@stsci.edu

Daniel J. Lennon Isaac Newton Group of Telescopes, La Laguna, Tenerife, Spain, djl@ing.iac.es; Instituto de Astrofisica de Canarias, La Laguna, Tenerife, Spain

Géraldine Letawe Département AGO, Université de Liège, allée du 6 août 17, 4000 Liège, Belgium

Yannick Letawe Département AGO, Université de Liège, allée du 6 août 17, 4000 Liège, Belgium, yletawe@ulg.ac.be

Duccio Macchetto Space Science Department of the European Space Agency, Space Telescope Science Institute, Baltimore, MD 21218, USA, macchetto@stsci.edu

Lucas Macri National Optical Astronomy Observatory, 950 N Cherry Ave, Tucson, AZ, USA, lmacri@noao.edu

Piero Madau Department of Astronomy and Astrophysics, University of California, Santa Cruz, CA 95064, USA

Pierre Magain Département AGO, Université de Liège, allée du 6 août 17, 4000 Liège, Belgium

Sangeeta Malhotra School of Earth and Space Exploration, Arizona State University, Tempe, USA

J. Mao Astrophysics Sector, SISSA/ISAS, Via Beirut 2-4, 34014 Trieste, Italy, mao@sissa.it

Guido Marchi ESA, Space Science Department, Keplerlaan 1, 2200 AG Noordwijk, Netherlands, gdemarchi@rssd.esa.int

Roeland Marel Space Telescope Science Institute, Baltimore, MD, USA, marel@stsci.edu

F. Martins MPE, Garching bei München, Germany, martins@mpe.mpg.de

Miguel Mas-Hesse Centro de Astrobiologia, LAEFF, Madrid, Spain, mm@laeff.inta.es

R.M. McDermid Sterrewacht Leiden, Universiteit Leiden, Postbus 9513, 2300 RA, Leiden, The Netherlands

Simona Mei University of Paris 7 Denis Diderot, 75205 Paris Cedex 13, France, simona.mei@obspm.fr; GEPI, Observatoire de Paris, Section de Meudon, 5 Place J. Janssen, 92195 Meudon Cedex, France

F. Meynadier LERMA, Paris Observatory, Paris, France, Frederic.Meynadier@obspm.fr

Roberto P. Mignani University College London, Mullard Space Science Laboratory, London, UK, rm2@mssl.ucl.ac.uk

W. Warren Miller Space Telescope Science Institute, 3700 San Martin Drive, Baltimore, MD 21218, USA

Jordi Miralda-Escudé Institut de Ciencies de l'Espai (CSIC-IEEC)/ICREA, Bellaterra, Spain

Bahram Mobasher Physics and Astronomy Department, University of California, 900 University Ave, Riverside, CA 92508, USA, mobasher@ucr.edu

V.M. Muñoz Marín Instituto de Astrofísica de Andalucía (CSIC), Granada, Spain

Claude Nicollier ESA/European Astronaut Centre, Cologne, Germany; Space Center EPFL, Ecole Polytechnique Fédérale de Lausanne, Lausanne, Switzerland, claude.nicollier@epfl.ch

Göran Östlin AlbaNova U. Centre, Stockholm Observatory, Stockholm, Sweden, ostlin@astro.su.se

Nino Panagia Space Telescope Science Institute, Baltimore, USA

R.F. Peletier Kapteyn Astronomical Institute, University of Groningen, 9700 AV Groningen, The Netherlands

E. Pérez Instituto de Astrofísica de Andalucía (CSIC), Granada, Spain

Artashes Petrosian Byurakan Astronomical Observatory, Erevan, Armenia, artptrs@yahoo.com

Nor Pirzkal Space Science Department of the European Space Agency, Space Telescope Science Institute, Baltimore, MD 21218, USA, npirzkal@stsci.edu

Alessandro Pizzella Università degli Studi di Padova, Dipartimento di Astronomia, Vicolo dell'Osservatorio, 3, 35122 Padova, Italy

Francesca Primas European Southern Observatory, Karl-Schwarzschildstr. 2, 85748 Garching, Germany, fprimas@eso.org

Sascha P. Quanz Max-Planck-Institut für Astronomie, Heidelberg, Germany

Thomas Rauch Institut für Astronomie und Astrophysik, Sand 1, 72076 Tübingen, Germany, rauch@astro.uni-tuebingen.de

Thomas P. Ray The Dublin Institute for Advanced Studies, 5 Merrion Square, Dublin 2, Ireland, tr@cp.dias.ie

Marina Rejkuba ESO, Karl-Schwarzschild-Strasse 2, 85748 Garching b. München, Germany, mrejkuba@eso.org

Alvio Renzini INAF, Osservatorio Astronomico di Padova, Padova, Italy, alvio.renzini@oapd.inaf.it

James Rhoads School of Earth and Space Exploration, Arizona State University, Tempe, USA

Adam G. Riess Space Telescope Science Institute, Baltimore, MD, USA, ariess@stsci.edu

Massimo Robberto Space Telescope Science Institute, Baltimore, USA

Martino Romaniello European Southern Observatory, Garching, Germany

Michael R. Rosa Space Telescope European Co-ordinating Facility, Karl-Schwarzschild-Str. 2, 85748 Garching, Germany, mrosa@stecf.org; ESA, Space Science Department, Noordwijk, The Netherlands

Jörn Rossa Department of Astronomy, University of Florida, Gainesville, FL, USA, jrossa@astro.ufl.edu

Renzo Sancisi INAF, Observatory of Bologna, Bologna, Italy, renzo.sancisi@oabo.inaf.it; Kapteyn Astronomical Institute, Groningen, Netherlands

A. Sandage Observatories of the Carnegie Institution of Washington, 813 Santa Barbara Street, Pasadena, CA 91101, USA

M. Sarzi Centre for Astrophysics Research, University of Hertfordshire, Hatfield, Herts AL10 9AB, UK

Daniel Schaerer Geneva Observatory, Geneva, Switzerland, Daniel.Schaerer@obs.unige.ch; Observatoire Midi-Pyrénées, Laboratoire d'Astrophysique, Toulouse, France

Markus Schmalzl Max-Planck-Institut für Astronomie, Heidelberg, Germany

H.R. Schmitt Remote Sensing Division, Naval Research Laboratory, Washington, DC, USA

Nick Scoville Astronomy Department, Caltech 105-24, Pasadena, CA 91125, USA, nzs@astro.caltech.edu

Richard A. Shaw National Optical Astronomical Observatory, Tucson, USA

Dominique Sluse Laboratoire d'Astrophysique, Ecole Polytechnique Fédérale de Lausanne (EPFL) Observatoire, 1290 Sauverny, Switzerland, dominique.sluse@epfl.ch

Letizia Stanghellini National Optical Astronomical Observatory, Tucson, USA

Jean Surdej Institut d'Astrophysique et de Géophysique, Université de Liège, Allée du 6 Août, 17, B5C, 4000 Liège, Belgium, surdej@astro.ulg.ac.be

G.A. Tammann Astronomisches Institut der Universität Basel, Venusstrasse 7, 4102 Binningen, Switzerland, G-A.Tammann@unibas.ch

Monica Tosi INAF, Osservatorio Astronomico di Bologna, Via Ranzani 1, Bologna, Italy, monica.tosi@oabo.inaf.it

G. Ven Institute for Advanced Study, Einstein Drive, Princeton, NJ 08540, USA

Eva Villaver Hubble Space Telescope Space Department of ESA, Space Telescope Science Institute, Baltimore, USA, villaver@stsci.edu

Klaus Werner Institut für Astronomie und Astrophysik, Sand 1, 72076 Tübingen, Germany

Bradley C. Whitmore Space Telescope Science Institute, 3700 San Martin Drive, Baltimore, MD 21218, USA

Tommy Wiklind ESA/STScI, 3700 San Martin Drive, Baltimore, MD 21218, USA, wiklind@stsci.edu

Chun Xu Shanghai Institute of Technical Physics, Shanghai, China

P.T. Zeeuw Sterrewacht Leiden, Universiteit Leiden, Postbus 9513, 2300 RA, Leiden, The Netherlands

Hans Zinnecker Potsdam Astrophysical Institute, Potsdam, Germany, hzinnecker@aip.de

G. Zotti INAF, Osservatorio Astronomico di Padova, Vicolo dell'Osservatorio 5, 35122 Padova, Italy, gianfranco.dezotti@oapd.inaf.it; Astrophysics Sector, SISSA/ISAS, Via Beirut 2-4, 34014 Trieste, Italy

Part I
Stars, Star Formation, Stellar Populations and Planets

Chapter 1
Hot Massive Stars: The Impact of HST

Paul A. Crowther

Abstract We review the contribution of Hubble Space Telescope to the study of hot, luminous stars. Optical and IR imaging have permitted spatially resolved observations of young, massive clusters within Local Group galaxies, such as R136, NGC 3603 and Arches, revealing unprecedented concentrations of very massive O stars. UV spectroscopy of field OB stars in the Magellanic Clouds have provided suitable templates for interpretation of metal-poor star-forming galaxies at high-redshift. Spectacular imaging provides the detailed structure of ejecta nebulae from individual stars, including the Homunculus associated with η Carinae and M1–67, associated with a Wolf–Rayet star. HST has permitted individual massive stars to be spatially resolved in giant HII regions located beyond the Local Group, such as NGC 604, plus individual clusters, dominated by the light of massive stars within starburst galaxies at larger distances, such as NGC 3125. UV spectroscopy of young, massive clusters in the extremely metal-poor HII galaxy I Zw 18 include signatures of large numbers of Wolf–Rayet stars.

1.1 Introduction

Massive stars distinguish themselves from their lower mass siblings by their exceptionally high main-sequence luminosities, such that their lifetimes are measured in Myr rather than Gyr. Individual stars may be readily studied in detail in external galaxies — it was not be accident that the first post-servicing mission WFPC2 image was obtained of the Wolf–Rayet star Melnick 34 in the 30 Doradus star-forming region of the LMC (News Release: STScI-1994-05). Here, selected HST results are discussed, together with their impact upon European astronomy.

1.2 Stellar Winds: Metallicity Dependent Winds

The first balloon and satellite missions provided the means of studying winds from hot stars, via ultraviolet P Cygni profiles, for which IUE provided a comprehensive sample of Milky Way OB stars (e.g. [17]). Comparable quality spectroscopy of

P.A. Crowther (✉)
Dept of Physics & Astronomy, University of Sheffield, Hounsfield Road, Sheffield, S3 7RH, UK
e-mail: Paul.Crowther@sheffield.ac.uk

F.D. Macchetto (ed.), *The Impact of HST on European Astronomy*,
Astrophysics and Space Science Proceedings,
DOI 10.1007/978-90-481-3400-7_1, © Springer Science+Business Media B.V. 2010

OB stars in the metal-poor Magellanic Clouds required the superior throughput and spatial resolution of HST [11, 33, 34].

Winds of early-type stars are predicted to be driven by radiation pressure through (CNO, Fe-peak) metal-lines (e.g. [32]), from which metal-poor stars are expected to possess lower density, slower winds. Indeed, wind velocities of early O stars in the SMC were established to be slower than those of comparable Galactic stars (e.g. [28]), from CIV 1550 observations, although it has taken large ground-based surveys, such as the VLT/FLAMES survey of massive stars in Milky Way, LMC and SMC clusters to quantify the metallicity-dependence of mass-loss rates, revealing $\mathrm{d}M/\mathrm{d}t \propto Z^{0.78\pm0.17}$ [21], in good agreement with theory $Z^{0.69\pm0.10}$ according to [32].

1.3 Ejecta Nebulae: Signatures of Mass Ejections

Amongst the many images obtained with HST, one of the most breathtaking has been the WFPC2 images of the Homunculus reflection nebula [22], produced by the 'eruption' of the prototype Luminous Blue Variable (LBV) η Carinae during the mid-19th Century, and now illuminated from within. The expanding emission lobes extend $8.5''$ (0.1 pc) from the central star, now known to be a 5.5 yr period binary system, for which WFPC2 achieved a spatial resolution of $\sim$115 AU, comparable with the size of our Solar System. A physical mechanism for the eruption, in which in excess of 10 $M_{\odot}$ were ejected over 20 years, remains unclear. STIS long-slit spectroscopy of the central star, reveals exceptional properties, with current mass-loss rates of $10^{-3}\ M_{\odot}\ \mathrm{yr}^{-1}$ for an adopted (infrared) luminosity of $5 \times 10^6\ L_{\odot}$ [16].

WFPC2 has also provided an unprecedented view of the young ejecta (or ring) nebula M1–67 associated with the Galactic Wolf–Rayet star WR124 [13]. The radial density distribution of this nebula, approximately $90''$ ($\sim$2 pc) in diameter with a $r^{-0.7}$ dependence, has enabled photo-ionization modelling of the central WR star [7] and may represent the immediate environment into which cosmological Gamma-Ray Bursts (GRBs) explode. M1–67 is compared to η Carinae on a common physical scale in Fig. 1.1.

1.4 Massive Binaries: Colliding Winds

High spatial resolution radio surveys of massive binaries reveal thermal (stellar wind) and non-thermal (colliding wind) components (e.g. [36]). Positions of stars within such systems have been established with WFPC2 imaging, enabling their relative wind strengths to be established [24].

Indeed, FGS has enabled searches for hitherto unknown massive binaries in regions of parameter space inaccessible to ground-based techniques. One such survey of 23 OB stars in Carina revealed five new binaries, including an apparent early O dwarf companion to HD 93129A, the prototype O2 supergiant, separated by only

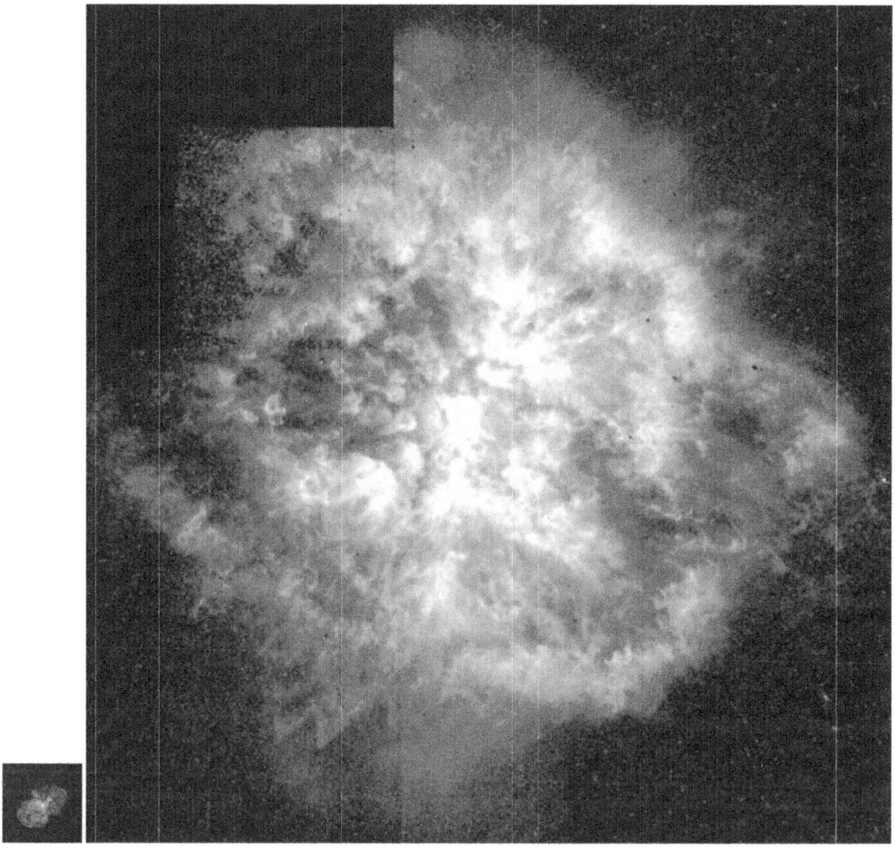

Fig. 1.1 (Color online) WFPC2 imaging, to scale, of the Homunculus nebula associated with η Carinae (*left*, 0.3 × 0.3 pc, STScI-1996-23, [22]) and M1–67, associated with a Wolf–Rayet star (*right*, 3 × 3 pc, STScI-1998-38, see [13])

55 mas (137 AU, [23]). Binarity was later confirmed by detection of a non-thermal component in the observed radio emission.

To date, the current heavyweight record holder is WR20a, a 3.7 day eclipsing binary system composed of two H-rich WN-type stars each of $\sim$82 $M_{\odot}$ [29].

1.5 Young Star Clusters: A Plethora of Hot Stars

Historically, R136 the central ionizing cluster of 30 Doradus in the LMC was considered as a potential supermassive star. 30 Doradus is the brightest giant HII region within the Local Group, responsible for the equivalent of 1,000 equivalent O7 V stars. The central source R136 was first resolved into multiple components by [35] using speckle imaging. Reference [8] provided confirmation with FOC, while [20] undertook FOS spectroscopy of individual sources within R136, revealing a multi-

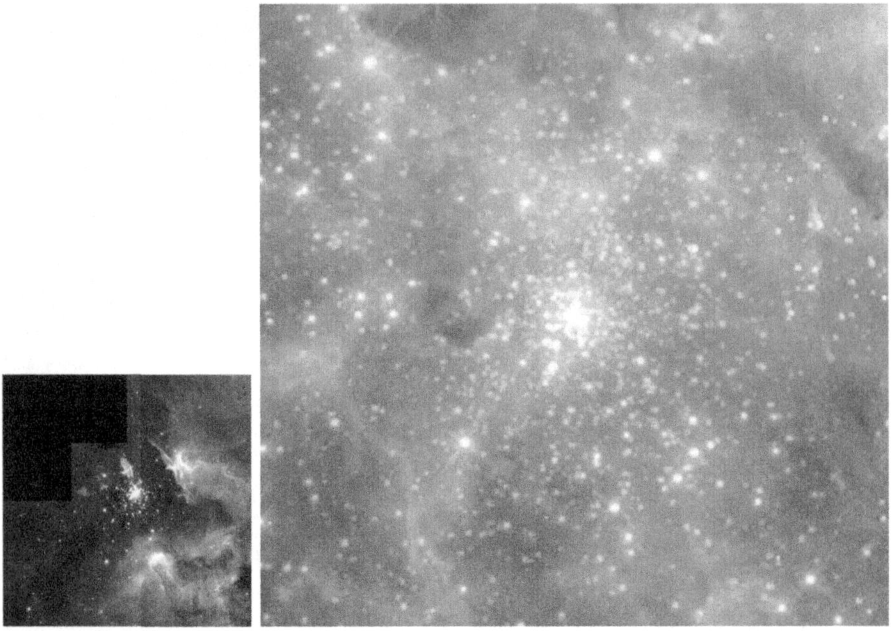

Fig. 1.2 (Color online) WFPC2 imaging, to scale, of the young star clusters NGC 3603 (4 × 4 pc, STScI-1999-20, [1]) and R136 (10 × 10 pc, ESA heic0416, [18])

tude of early O stars, indicating extreme youth (1–2 Myr) and apparently very high individual stellar masses, perhaps up to 120 $M_\odot$. The total stellar mass of R136 probably exceeds 5×10^4 $M_\odot$ [18]. NICMOS imaging of 30 Doradus has identified still younger, visibly obscured, young massive stars.

HST has also spatially resolved the Milky Way cluster NGC 3603 (see Fig. 1.2), again revealing many early O stars [10]. The central cluster is comparable to R136a within a radius of ~1 pc [5]. Elsewhere in the Milky Way, NICMOS has been used to spatially resolve the compact (<0.5 pc) Arches cluster close to the Galactic centre. The Arches cluster is apparently sufficiently young that the present-day mass function approximates the true IMF, from which an upper stellar mass limit of ~150 $M_\odot$ has been proposed [12].

At present, the most massive young Milky Way cluster known is Westerlund 1 (Wd 1), in which ~ 5×10^4 $M_\odot$ are contained within a radius of ~1.5 pc [4]. This cluster hosts a full menagerie of rare, massive stars, including Wolf–Rayet stars, red supergiants, yellow hypergiants, an LBV and a B[e] supergiant.

Such high mass, compact clusters are unusual for normal spiral galaxies. Large, scaled-up OB associations are more typical, for which WFPC2 imaging provided an excellent example of NGC 604 in M33 (see [37]), several hundred pc in diameter, containing the equivalent of several hundred O7 V stars. This is compared to ACS imaging of NGC 346 in the SMC [25] in Fig. 1.3, somewhat smaller and with the equivalent of 50 equivalent O7 V stars [19]. The high spatial resolution achieved with WFPC2 permitted [9] to identify individual Of and Wolf–Rayet emission line

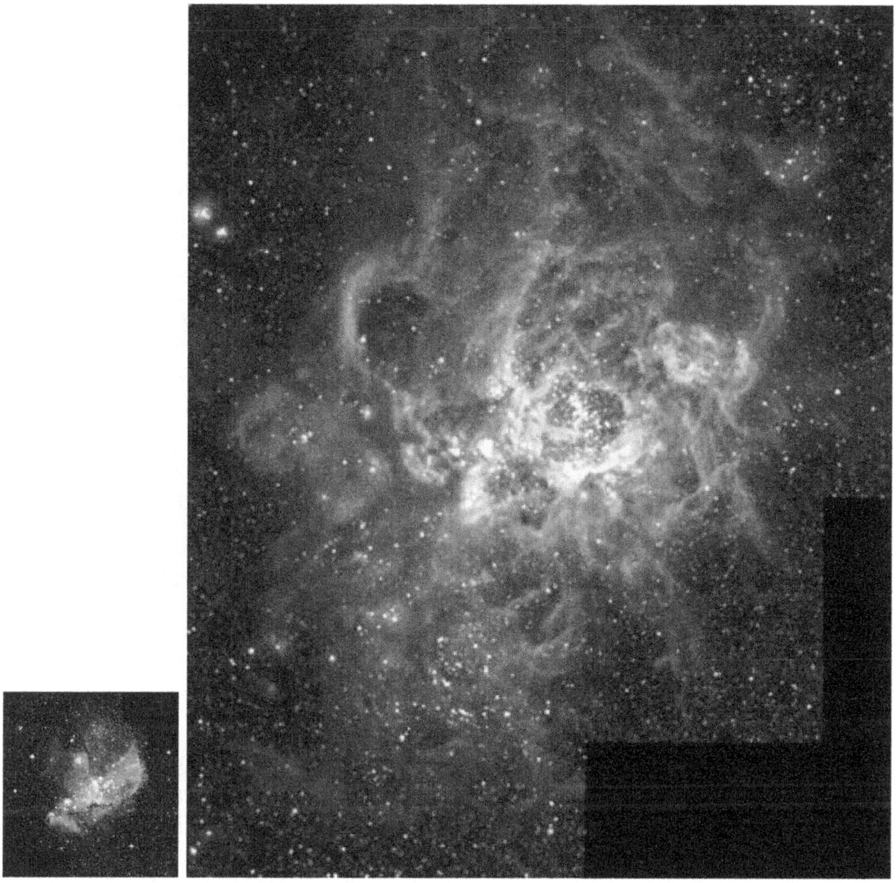

Fig. 1.3 (Color online) HST imaging, to scale, of the giant HII regions NGC 346 (SMC N11A) in the SMC (ACS: 90 × 90 pc, STScI-2005-04, [25]) and NGC 604 in M 33 (WFPC2: 360 × 450 pc, STScI-1996-27, [37])

stars in NGC 604 to be established. A similar ground-based study of Wolf–Rayet stars within a bright giant HII region in the southern barred spiral galaxy NGC 1313 by [15], comparable in size and ionizing output to NGC 604 demonstrates the invaluable role of ACS in disentangling the stellar content. Field Wolf–Rayet stars at the 4 Mpc distance of NGC 1313 can be detected from ground-based telescopes, but HST has proved decisive in detection core-collapse supernova progenitors from pre-SN imaging (e.g. [31]).

1.6 Starburst Knots: Templates for High-z Galaxies

Local starbursts, such as NGC 3125 (11 Mpc) host young massive clusters, significantly more massive than R136, as presented in Fig. 1.4. Starbursts from such

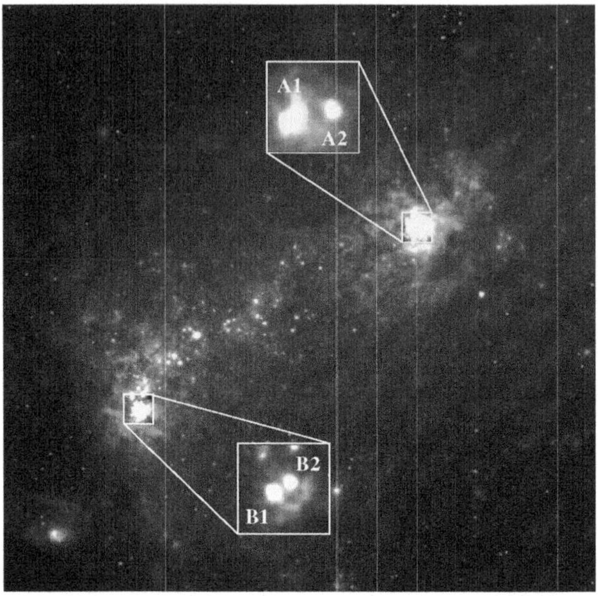

Fig. 1.4 (Color online) 20 × 20″ ACS imaging of the blue compact dwarf galaxy NGC 3125 (U, V, Hα, GO#10400, R. Chandar) in which young massive ($\sim 10^5$ $M_{\odot}$) clusters from knots A and B, ~500 pc apart [14] are indicated

galaxies provide an order of magnitude higher ionizing output than NGC 604 from within a much smaller region, e.g. the equivalent of 2,500 equivalent O7 V stars from knot A of NGC 3125. STIS ultraviolet spectroscopy of individual clusters enables robust age estimates of such clusters, based on comparisons between the OB P Cygni features and spectral synthesis predictions (e.g. [3]), as shown in Fig. 1.5 for NGC 3125-A1 from [14]. Host galaxies typically possess Magellanic Cloud metallicities, such that the use of template LMC/SMC OB stars (recall Sect. 1.2) has proved to be invaluable.

In extreme cases, far lower metallicities are sampled, such as the HII galaxy I Zw 18, possessing $1/30$ $Z_{\odot}$ based upon the recently reduced Solar oxygen abundance. Reference [2] undertook STIS ultraviolet spectroscopy across I Zw 18, revealing multiple clusters in which carbon-sequence Wolf–Rayet stars were observed. Such prominent WR signatures at low metallicity led [6] to conclude that I Zw 18 hosts substantial WR populations, wholly unexpected for single star evolutionary models at such low metallicity.

The composite rest-frame UV spectra of $z \sim 3$ Lyman break galaxies (LBG) [30] also includes spectral signatures of O and WR stars. Amongst the brightest (lensed) LBGs, MS1512-cB58 ($z \sim 2.7$) has been observed at rest-frame UV wavelengths with Keck, with which [26] analysed the interstellar lines, suggesting a Magellanic Cloud metallicity. Indeed, [27] could reproduce its stellar CIV 1550 feature using Magellanic Cloud template OB stars, where they had previously been unsuccessful using Milky Way template stars.

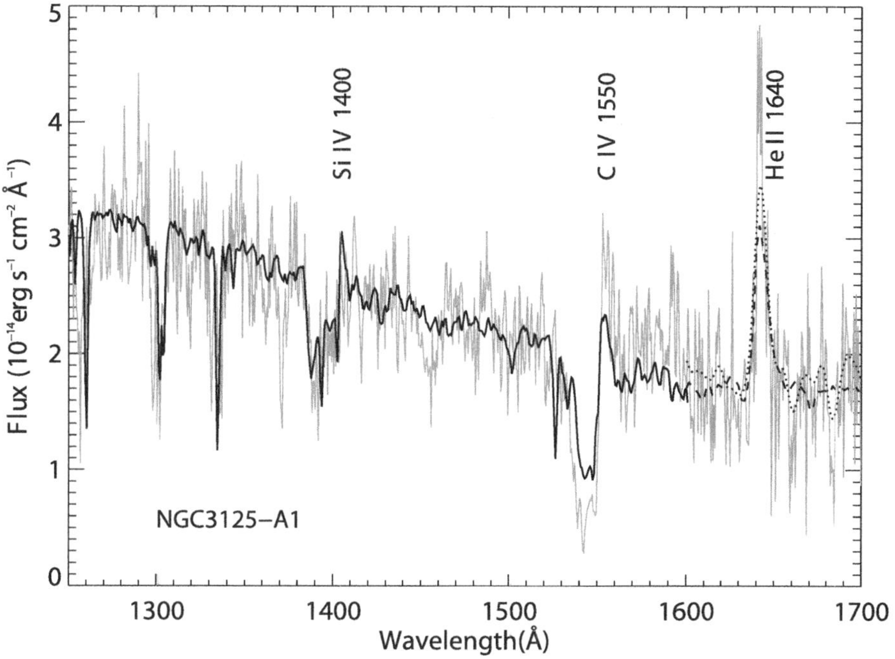

Fig. 1.5 Dereddened, slit-loss corrected STIS spectroscopy of NGC 3125–A1 (*thin solid line*), together with a 2×10^5 $M_\odot$ Starburst99 Magellanic Cloud 4 Myr instantaneous burst synthetic spectrum (*thick solid line*), plus LMC WN5–6 templates at He II λ1640 (*dashed line*) from [14]

1.7 Future Issues?

Perhaps the most significant unresolved aspect relating to high mass stars is their formation, for which future high-spatial resolution infrared and radio may prove decisive. Violent mass ejections, such as that experienced by η Carinae, producing the Homunculus nebula, also remains ill-explained. A further puzzle is how large numbers of Wolf–Rayet stars form in young, massive clusters within very metal-poor galaxies such as I Zw 18.

References

1. Brandner, W., Grebel, E.K., Chu, Y.-H., et al., 2000. Astron. J. 119, 292.
2. Brown, T.P., Heap, S.R., Hubeny, I., Lanz, T., Lindler, D., 2002. Astrophys. J. 579, L75.
3. Chandar, R., Leitherer, C., Tremonti, C.A., 2004. Astrophys. J. 604, 153.
4. Clark, J.S., Negueruela, I., Crowther, P.A., Goodwin, J.S., 2005. Astron. Astrophys. 434, 949.
5. Crowther, P.A., Dessart, L., 1998. Mon. Not. R. Astron. Soc. 296, 622.
6. Crowther, P.A., Hadfield, L.J., 2006. Astron. Astrophys. 449, 711.
7. Crowther, P.A., Pasquali, A., De Marco, O., et al., 1999. Astron. Astrophys. 350, 1007.
8. de Marchi, G., Nota, A., Leitherer, C., Ragazzoni, R., Barbieri, C., 1993. Astrophys. J. 419, 658.

9. Drissen, L., Moffat, A.F.J., Shara, M.M., 1993. Astron. J. 105, 1400.
10. Drissen, L., Moffat, A.F.J., Walborn, N.R., Shara, M.M., 1995. Astron. J. 110, 2235.
11. Evans, C.J., Lennon, D.J., Walborn, N.R., Trundle, C., Rix, S.A., 2004. Publ. Astron. Soc. Pac. 116, 909.
12. Figer, D.F., 2005. Nature 434, 192.
13. Grosdidier, Y., Moffat, A.F.J., Joncas, G., Acker, A., 1998. Astrophys. J. 506, L127.
14. Hadfield, L.J., Crowther, P.A., 2006. Mon. Not. R. Astron. Soc. 368, 1822.
15. Hadfield, L.J., Crowther, P.A., 2007. Mon. Not. R. Astron. Soc. 381, 418.
16. Hillier, D.J., Davidson, K., Ishibashi, K., Gull, T., 2001. Astrophys. J. 553, 837.
17. Howarth, I.D., Prinja, R.K., 1989. Astrophys. J. Suppl. Ser. 69, 527.
18. Hunter, D.A., O'Neil, E.J., Lynds, R., et al., 1996. Astrophys. J. 459, L27.
19. Kennicutt, R.C., 1984. Astrophys. J. 287, 116.
20. Massey, P., Hunter, D.A., 1998. Astrophys. J. 493, 180.
21. Mokiem, M.R., de Koter, A., Vink, J.S., et al., 2007. Astron. Astrophys. 465, 1003.
22. Morse, J.A., Davidson, K., Bally, J., et al., 1998. Astron. J. 116, 2443.
23. Nelan, E.P., Walborn, N.R., Wallace, D.J., et al., 2004. Astron. J. 128, 323.
24. Niemela, V.S., Shara, M.M., Wallace, D.J., Zurek, D.R., Moffat, A.F.J., 1998. Astron. J. 115, 2047.
25. Nota, A., Sirianni, M., Sabbi, E., Tosi, M., Clampin, M., et al., 2006. Astrophys. J. 640, L29.
26. Pettini, M., Rix, S.A., Steidel, C.C., et al., 2002. Astrophys. J. 569, 742.
27. Pettini, M., Rix, S.A., Steidel, C.C., Shapley, A.E., Adelberger, K.L., 2003. In: Proc. IAU Symp., vol. 212, p. 671.
28. Prinja, R.K., Crowther, P.A., 1998. Mon. Not. R. Astron. Soc. 300, 828.
29. Rauw, G., Crowther, P.A., De Becker, M., et al., 2005. Astron. Astrophys. 432, 985.
30. Shapley, A.E., Steidel, C.C., Pettini, M., Adelberger, K.L., 2003. Astrophys. J. 588, 65.
31. Smartt, S.J., Maund, J.R., Hendry, M.A., et al., 2004. Science 303, 499.
32. Vink, J.S., de Koter, A., Lamers, H.G.J.L.M., 2001. Astron. Astrophys. 369, 574.
33. Walborn, N.R., Lennon, D.J., Haser, S.M., Kudritzki, R.-P., Voels, S.A., 1995. Publ. Astron. Soc. Pac. 107, 104.
34. Walborn, N.R., Lennon, D.J., Heap, S.R., et al., 2000. Publ. Astron. Soc. Pac. 112, 1243.
35. Weigelt, G., Baier, G., 1985. Astron. Astrophys. 150, L18.
36. Williams, P.M., Dougherty, S.M., Davis, R.J., et al., 1997. Mon. Not. R. Astron. Soc. 289, 10.
37. Yang, H., Chu, Y.-H., Skillman, E.D., Terlevich, R., 1996. Astron. J. 112, 146.

Chapter 2
HST Spectroscopy of the Hottest White Dwarfs

Thomas Rauch and Klaus Werner

Abstract Spectral analysis needs the observation of lines of successive ionization stages in order to evaluate the ionization equilibrium (of a particular species) which is a sensitive indicator for the effective temperature ($T_{\rm eff}$). Since stars with $T_{\rm eff}$ as high as 100,000 K have their flux maximum in the extreme ultraviolet (EUV) wavelength range and due to the high degree of ionization, most of the metal lines are found in the ultraviolet (UV) range. Thus, high-S/N and high-resolution UV spectra are a pre-requisite for a precise analysis. Consequently, we employed the Faint Object Spectrograph (FOS), the Goddard High Resolution Spectrograph (GHRS), and the Space Telescope Imaging Spectrograph (STIS) aboard the Hubble Space Telescope (HST) in order to obtain suitable data. We present state-of-the-art analyses of the hottest (pre-) white dwarfs by means of NLTE model atmospheres which include the metal-line blanketing of all elements from hydrogen to nickel.

2.1 Introduction

In the early eighties of the last century, the evolution of "H-normal" post-AGB stars has been quite well understood, e.g. [12] and [2] have presented evolutionary calculations for these stars. At that time, neither standard evolutionary calculations nor model atmospheres could explain observations of H-deficient post-AGB stars.

In 1979 the discovery of PG 1159–035, the H-deficient prototype of the GW Vir variables, had shown the inadequacy of theory: the optical spectrum exhibits broad and shallow absorption lines of highly ionized species, e.g. He II and C IV, indicating $T_{\rm eff}$ to be much higher than 100,000 K. At this temperature regime, the assumption of local thermodynamical equilibrium (LTE) is not valid and thus, adequate fully metal line-blanketed NLTE model-atmospheres were required — but not available.

In Sect. 2.2 we describe briefly our NLTE model-atmosphere code *TMAP*, which has been developed over the last two decades and has been successfully used for the analysis of hot, compact stars. Such analyses have continuously provided constraints for evolutionary theory and, vice versa, predictions from evolutionary calculations have inspired us to search for lines of unidentified species in UV spectra (e.g. for Ne VII, F VI, Ar VII, and Ne VIII, respectively [16–18] and [19]) provided by the

T. Rauch (✉) · K. Werner
Institut für Astronomie und Astrophysik, Sand 1, 72076 Tübingen, Germany
e-mail: rauch@astro.uni-tuebingen.de

F.D. Macchetto (ed.), *The Impact of HST on European Astronomy*,
Astrophysics and Space Science Proceedings,
DOI 10.1007/978-90-481-3400-7_2, © Springer Science+Business Media B.V. 2010

HST and the Far Ultraviolet Spectroscopic Explorer (FUSE). The synergy effect of both satellites gave us the opportunity to precisely analyze strategic lines from the complete UV range (from the H I Lyman edge to the optical) and to determine photospheric properties with hitherto unprecedented accuracy. In Sects. 2.3 and 2.4, we give representative examples for our analyses of H-deficient and H-normal post-AGB stars.

2.2 NLTE Model Atmospheres

We use *TMAP*,[1] the Tübingen NLTE Model Atmosphere Package [10, 13, 15], for the calculation of plane-parallel, chemically homogeneous models in hydrostatic and radiative equilibrium. *TMAP* considers all elements from H to Ni [8, 9]. In the analysis of LS V $+46°21$ (Sect. 2.4), e.g., 686 levels are treated in NLTE, combined with 2,417 individual lines and about nine million iron-group lines.

2.3 Spectroscopy of PG 1159 Stars

PG 1159 stars are so-called "born-again post-AGB stars" [4], i.e. after their departure from the asymptotic giant branch (AGB) and at already declining luminosity, they experienced a (very) late thermal pulse (He-shell flash) and returned to the AGB. During the born-again phase, the entire H-rich envelope (10^{-4} $M_{\odot}$) was convectively mixed [1, 5] with the intershell material (10^{-2} $M_{\odot}$, located between He- and H-burning shells) and H is completely burned. The direct view on intershell matter (at the surface now) allows to conclude on details of nuclear and mixing processes in AGB stars. This is an important test for stellar evolutionary models [14].

Our analyses of PG 1159 stars revealed that their abundances of He, C, N, O, Ne, Mg, F, Si, and Ar are in line with predictions from evolutionary models. These models show also a Fe depletion due to n-captures within the s-process. In three observations of PG 1159 stars with FUSE, no iron lines are detectable which gives a surprisingly large Fe-deficiency of 1–2 dex [7]. An inspection of STIS observations of the same objects [6] shows that there is no increase of the Ni abundance and thus, it appears likely that the s-process has converted even Ni into trans iron-group elements. However, we do not have reliable atomic data to prove this. Other elements show deviations from theory, e.g. P appears roughly solar but the models predict a strong enhancement while S is expected to stay solar but shows large depletion (up to 2 dex). For a detailed review, see [14].

---

[1] http://astro.uni-tuebingen.de/~rauch/TMAP/TMAP.html.

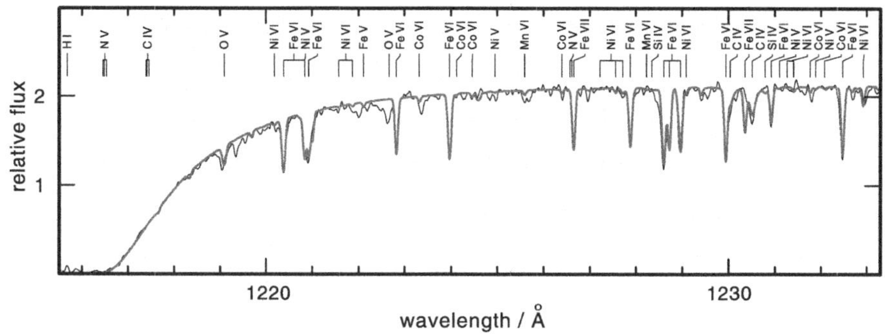

Fig. 2.1 (Color online) Section of the STIS observation of LS V + 46°21 compared to our final model. Identified lines are marked at the *top*

2.4 Spectroscopy of LS V + 46°21

LS V + 46°21 is the central star of the closest known ($d = 130$ pc, $\oslash = 1.6°$) planetary nebula Sh 2–216. We have observed LS V + 46°21 with STIS (5.5 ksec in 2,000). The STIS observation shows more than 1,000 absorption features (about 10% interstellar). 95% of these are identified. We have calculated the most detailed *TMAP* model-atmosphere ever [11] in order to reproduce the observed spectrum (an example is shown in Fig. 2.1). In the STIS observation, we identified Si V lines [6], Mg VI lines (for the first time in a post-AGB star), and Ar VI lines (for the first time in any star). Most of the determined abundances are in agreement with diffusion-model predictions [3].

2.5 TMAP in the Virtual Observatory

The HST with its UV spectroscopic capabilities has been crucial for these analyses and the development of *TMAP*. Hopefully, the Cosmic Origins Spectrograph (COS) will continue the work of its very successful precursors. The comparison of our synthetic spectra with the observations of hot, compact stars convinced us that theory works well and we have arrived at a high level of sophistication.

The spectral analysis, although to be done with sufficient care, has not to remain the field of specialists. Within the framework of *German Astrophysical Virtual Observatory* (GAVO, please note that the URLs given below will change to the GAVO portal[2] later) project, we provide grids of model-atmosphere fluxes (*TMAF*[3]) as well as a WWW interface (*TMAW*[4]) to calculate individual *TMAP* model atmospheres without detailed knowledge about theory etc.

---

[2]http://www.g-vo.org/portal/.

[3]http://astro.uni-tuebingen.de/~rauch/TMAF/TMAF.html.

[4]http://astro.uni-tuebingen.de/~rauch/TMAW/TMAW.html.

Since the reliability of synthetic spectra is strongly dependent on the accuracy of the atomic data which is used for their calculation, standard *TMAW* calculations use predefined model atoms which are provided within the Tübingen Model-Atom Database *TMAD*.[5]

While the use of the *TMAF* flux grids is the easiest way for a user of the Virtual Observatory, even individual analyses can easily be performed with appropriately adjusted model atoms.

Acknowledgements T.R. is supported by the *German Astrophysical Virtual Observatory* project of the German Federal Ministry of Education and Research (BMBF) under grant 05 AC6VTB.

References

1. Althaus, L.G., Miller Bertolami, M.M., Córsico, A.H., et al., 2005. Astron. Astrophys. 440, L1.
2. Blöcker, T., Schönberner, D., 1990. Astron. Astrophys. 240, L11.
3. Chayer, P., Vennes, S., Pradhan, A.K., et al., 1995. Astrophys. J. 454, 429.
4. Iben, I. Jr., Kaler, J.B., Truran, J.W., et al., 1983. Astrophys. J. 264, 605.
5. Herwig, F., Blöcker, T., Langer, N., et al., 1999. Astron. Astrophys. 349, L5.
6. Jahn, D., Rauch, T., Reiff, E., et al., 2007, Astron. Astrophys. 462, 281.
7. Miksa, S., Deetjen, J.L., Dreizler, S., et al., 2002. Astron. Astrophys. 389, 953.
8. Rauch, T., 1997. Astron. Astrophys. 320, 237.
9. Rauch, T., 2003. Astron. Astrophys. 403, 709.
10. Rauch, T., Deetjen, J.L., 2003. In: Hubeny, I., Mihalas, D., Werner, K. (Eds.), Stellar Atmosphere Modeling, The ASP Conference Series, vol. 288, p. 103.
11. Rauch, T., Ziegler, M., Werner, K., et al., 2007. Astron. Astrophys. 470, 317.
12. Schönberner, D., 1983. Astron. Astrophys. 272, 708.
13. Werner, K., 1986. Astron. Astrophys. 161, 177.
14. Werner, K., Herwig, F., 2006. Publ. Astron. Soc. Pac. 188, 183.
15. Werner, K., Dreizler, S., Deetjen, J.L., et al., 2003. In: Hubeny, I., Mihalas, D., Werner, K. (Eds.), Stellar Atmosphere Modeling, The ASP Conference Series, vol. 288, p. 31.
16. Werner, K., Rauch, T., Reiff, E., et al., 2004. Astron. Astrophys. 427, 685.
17. Werner, K., Rauch, T., Kruk, J.W., 2005. Astron. Astrophys. 433, 641.
18. Werner, K., Rauch, T., Kruk, J.W., 2007. Astron. Astrophys. 466, 317.
19. Werner, K., Rauch, T., Kruk, J.W., 2007. Astron. Astrophys. 474, 591.

[5] http://astro.uni-tuebingen.de/~rauch/TMAD/TMAD.html.

Chapter 3
Key Abundance Tracers in the UV: From the Lightest to the Heaviest

Francesca Primas

Abstract The uniqueness of the ultra-violet spectral region consists of giving access to atomic transitions of some of the very light and of the very heavy elements of the Periodic Table. The very light elements (here meant to be lithium, beryllium and boron) are not very abundant but they are a powerful diagnostic of the mixing mechanisms active in the outermost layers of stellar atmospheres, as well as a link to primordial epochs and primordial nucleosynthesis (lithium). The so-called "heavies", instead, are all the elements with atomic number $Z > 30$. They are also rare in number but they comprise the majority of the elements of the Periodic Table. This review aims at pointing out the most interesting and unique of these features and their importance to progress further in our understanding of galactic chemical evolution and stellar physics.

3.1 Introduction

Back in the early 1990s, the successful launch of the Hubble Space Telescope, equipped with a high resolution spectrograph, had a significant impact on high resolution abundance studies, giving access again (after the International Ultraviolet Explorer mission) to the very rich ultra-violet spectral region. This is a window clearly unaccessible from ground, and which contains a treasure of information for abundance studies of solar-type stars. The region between 250 and 300 nm is particularly rich of transitions of several different atomic and molecular species, some of which represent the only transition(s) of that specific element. In the following sections, I will briefly highlight some of the most interesting results obtained in this respect with HST and its spectrographs.

3.2 From the Very Light Elements ...

Knowledge of lithium, beryllium, and boron abundances in stars play a major role in our understanding of Big Bang nucleosynthesis, cosmic-ray physics, and stellar

---

F. Primas (✉)
European Southern Observatory, Karl-Schwarzschildstr. 2, 85748 Garching, Germany
e-mail: fprimas@eso.org

F.D. Macchetto (ed.), *The Impact of HST on European Astronomy*, Astrophysics and Space Science Proceedings,
DOI 10.1007/978-90-481-3400-7_3, © Springer Science+Business Media B.V. 2010

interiors. In the standard framework of the origin and evolution of the light elements, only D, ^{3}He and ^{7}Li are produced in significant amounts from Big Bang primordial nucleosynthesis. Spallation reactions of CNO nuclei during the propagation of cosmic rays (CR) in the Galaxy (with an extra α–α fusion channel for ^{6}Li and ^{7}Li) account for the other light-element isotopes, ^{6}Li, ^{9}Be, and ^{10}B [1], with the origin of ^{11}B being still under debate (because of the yet unknown contribution by ν-spallation in supernovae).

However, both Be and B observations have shown that a reverse spallation process (heavier nuclei impinging onto protons and α-particles) may be more important, especially at early epochs. Additionally, the canonical CR spallation underproduces B (and Be) at low metallicities, predicting $^{11}\mathrm{B}/^{10}\mathrm{B} \simeq 2.5$, which is instead observed to be around 4.

While lithium is within easy reach of ground-based telescopes (its main transition falls at 670.7 nm), Be and B have their main atomic transitions in the near-UV (313 nm) and UV (250 nm), thus requiring respectively instruments equipped with high quality UV sensitive detectors and space-based observing facilities. The recent availability of instruments like UVES at the ESO VLT and HIRES at Keck, and before that the launch of the Hubble Space Telescope equipped with medium-, high-res spectrographs (first the Goddard High Resolution Spectrograph, and now STIS) have had a strong impact on this field of research.

Apart from the strong interest in its own Galactic evolution (e.g. [2]), knowledge of B abundances in the same stars for which Li and Be are already available, offers a powerful diagnostic to probe and further constrain different stellar mixing mechanisms (e.g. [3]). This is because Li, Be, and B burn at low, but increasing temperatures (2.5, 3.5, and 5×10^6 K, respectively). Thus, in a standard stellar interior scenario, one would expect to burn lithium first (which is preserved in the outermost layer of the photosphere), then beryllium, and finally boron. Obviously, the interpretation of the observed data is more complicated than that, because of the likely simultaneous presence of more than just one mixing mechanism at a time. Moreover, depending on the nature of the mixing, one could indeed deplete only part of the lithium and affect already the layer where Be is preserved. In such a case, the boron abundance becomes a fundamental piece of information to discern between, e.g., depletion/mixing and production.

In our recent analysis of beryllium abundances [4] from high-res, high S/N, near-UV VLT spectra of a large sample of Galactic stars spanning a wide range of metallicities, we have identified a highly (Be) dispersed region at metallicities characteristic of the halo-disk transition ($-1.4 \leq$ [Fe/H] ≤ -0.7), which is in strong contrast with the well defined and narrow correlation found between Be and Fe at lower metallicities. Though reduced, the dispersion is still present also when these Be abundances are plotted versus oxygen. However, when the uncertainty on the O abundance determination is taken into account, the most dispersed objects (5) seem to share the same oxygen content. Hence, the large difference detected in their Be abundances cannot be ascribed to their O content (nor to stellar processes since the Li abundances are known to be un-depleted).

In order to disentangle the abundance pattern of these stars and to further constrain which mechanism(s) are the main responsible for the observed Be abundances, one clearly needs boron, which has been the main scientific goal of our Cycle12 HST-STIS observations. Galactic cosmic-ray spallation predicts a B/Be ratio close to 17 (e.g. [5]): if B is a pure CR spallation product, we expect to observe a B/Be ratio matching the one predicted by the theory. Otherwise, this new set of B abundances may hold important new clues on the contribution to the production of boron from ν-spallation or from some other mechanism. We are now in the process of verifying and thoroughly interpret the very preliminary results presented at this conference.

3.3 ... to the Very Heavy Ones

All heavy elements are made in neutron-bombardment reactions, which differ depending on the timescale of neutron capture compared to the beta-decay and thus they are classified as *slow* or *rapid*. The r-process is a signature of the element production happening at the end of the life of a high mass star, while the s-process can occur quietly during low- and intermediate-mass stellar evolution.

In the last decade, some very metal-poor stars were discovered and found to have strong enhancements of neutron-capture elements in their atmospheres. This has allowed to study for the first time the almost complete nucleosynthetic path of the chain of the heavy elements in the atmospheres of very old popII stars. Reference [6] have shown that at these early times, most of the neutron-capture element formation can be accounted for by the r-process, with the onset of the s-process occurring at higher metallicities. The synergy between ground- and space-based observations has allowed these authors to achieve a very detailed knowledge of the chemical composition and of the nucleosynthetic history of CS 22892-052, a halo giant with a metal content 1/1,000 less than solar (cf. Fig. 3.1).

What is unique to HST is the gained access to elements which have their main atomic transitions below 300 nm. Between 250 and 300 nm, there is a forest of transitions of many neutron-capture elements. Admittedly, those which can be studied solely from the ultra-violet are only a few (cf. Fig. 3.1). However, it is important to notice that some of the key neutron-capture elements (e.g. third-peak and actinides, i.e. elements with $Z > 75$) have a significantly higher number of transitions in the UV compared to the optical (e.g. platinum) which makes the derivation of their abundance significantly more robust. The relatively recent detections of radioactive elements such as thorium and uranium in a few r-process enhanced metal-poor stars [7] have clearly provided new opportunities to set lower limits on the age of the Galaxy and triggered a new wave of abundance analyses, which benefit the most from ground- and space-based combined observational efforts.

3.4 A Concluding Remark

Needless to say, the sudden halt of STIS operations has strongly impacted all programmes described in this short report. At time of writing, it is not clear yet which

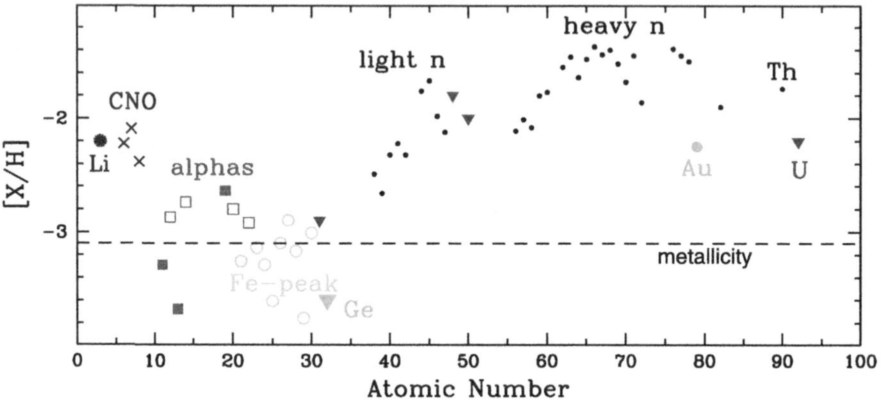

Fig. 3.1 (Color online) The comprehensive abundance analysis of CS 22892-052 [6]. The very few heavy elements with transitions only in the UV transitions are clearly marked to ease their identification (Ge and Au). Of the remaining *n*-capture elements, one should note that there are several that have *most* of their transitions in the UV

of the failed instruments currently on-board of HST will be served during the next repair mission (planned for 2008). Admittedly, every astronomer has her/his own preferences, depending on the science that she/he is most interested in. If repaired, STIS will continue playing a major role in the field of high resolution spectroscopy and abundance studies, and it will complement the excellent datasets now routinely obtained at 8–10 m class ground telescopes. Once HST stops functioning, the ultraviolet window will be unaccessible for a while, since none of the currently planned missions/projects foresees to have instruments specialized for this region.

Acknowledgements It has been a wonderful meeting, in honour of a top European scientist who has devoted his entire career to promoting European astronomy and who has fully supported ESA (HST, especially) facilities. Thank you, Duccio!

References

1. Meneguzzi, M., Audouze, J., Reeves, H., 1971. Astron. Astrophys. 15, 337.
2. Duncan, D.K., et al., 1997. Astrophys. J. 480, 784.
3. Primas, F., et al., 1999. Astron. Astrophys. 343, 545.
4. Primas, F., 2002. Astrophys. Space Sci. 281, 195.
5. Walker, T.P., Viola, V.E., Mathews, G.J., 1985. Astrophys. J. 299, 745.
6. Sneden, C., et al., 2003. Astrophys. J. 591, 936.
7. Hill, V., et al., 2002. Astron. Astrophys. 387, 560.

Chapter 4
UV Spectroscopy of Metal-Poor Massive Stars in the Small Magellanic Cloud

Daniel J. Lennon

Abstract The Hubble Space Telescope has provided the first clear evidence for weaker winds of metal-poor massive stars in the Small Magellanic Cloud, confirming theoretical predictions of the metallicity dependence of mass-loss rates and wind terminal velocities. For lower luminosity O-type stars however, derived mass-loss rates are orders of magnitude lower than predicted, and are at present unexplained.

4.1 Introduction

Massive stars may loose mass throughout their lifetimes either as steady winds or in episodic outbursts during a short-lived Luminous Blue Variable (LBV) phase. To first order, the product of a massive star's mass-loss rate during its core-hydrogen or core-helium burning phase times the duration of these evolutionary phases can amount to a significant fraction of the star's initial mass. Similar considerations apply to the shorter-lived LBV phases. Mass-loss therefore plays a crucial role in the evolution of stars more massive than about 20 solar masses, uncertainties of even a factor of two being significant for the star's evolution. There is general agreement that the steady winds of massive stars are driven by radiation pressure on millions of metal absorption lines, these winds therefore depend on metallicity (Z). Calculations [9] indicate that in the range $0.1\ Z_{\odot} < Z < 3\ Z_{\odot}$ the mass-loss rate $\dot{M} \sim Z^{x}$ where $x \sim 0.7$, while the ratio of wind terminal velocity to escape velocity ($v_{\infty}/v_{\rm esc}$) is rather insensitive to Z, though dependent on effective temperature. The Small Magellanic Cloud with $Z \sim 0.2\ Z_{\odot}$ is therefore the ideal laboratory in which to study the wind properties and evolution of massive stars at low metallicity.

4.2 Weak Winds in the Small Magellanic Cloud

There were several attempts to confirm theoretical expectations using the IUE satellite to observe massive stars in the Magellanic Clouds, however with mixed re-

D.J. Lennon (✉)
Isaac Newton Group of Telescopes, La Laguna, Tenerife, Spain
e-mail: djl@ing.iac.es

D.J. Lennon
Instituto de Astrofisica de Canarias, La Laguna, Tenerife, Spain

F.D. Macchetto (ed.), *The Impact of HST on European Astronomy*, Astrophysics and Space Science Proceedings,
DOI 10.1007/978-90-481-3400-7_4, © Springer Science+Business Media B.V. 2010

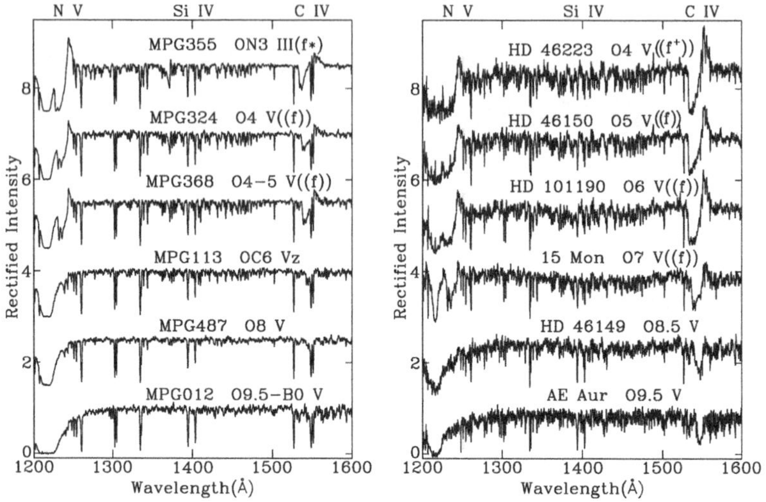

Fig. 4.1 Spectra of O-type main-sequence stars. *Left panel*: SMC sources observed with HST/STIS, all of which lie in the cluster NGC 346. *Right panel*: Galactic sources observed with IUE at high resolution [12]. The features identified at the *top* are N V λλ1239, 1243; Si IV λλ1394, 1403; and C IV λλ1548, 1551. A characteristic of O-type dwarfs in the SMC is the absence of saturated C IV P-Cygni profiles, in sharp contrast to the Galactic stars. In addition wind features are extremely weak or absent for dwarfs later than O5. The strong N V feature in MPG 355 reflects the large nitrogen enrichment in this object

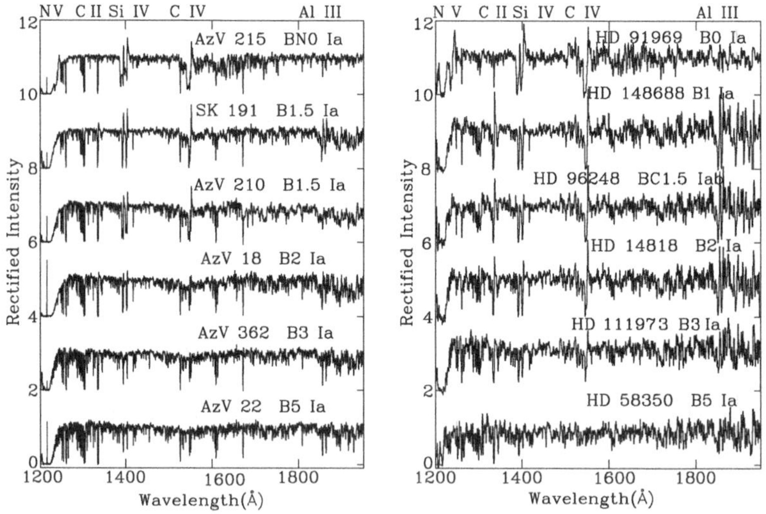

Fig. 4.2 Spectra of B-type supergiant stars. *Left panel*: SMC sources observed with HST/STIS. *Right panel*: Galactic sources observed with IUE at high resolution [13]. The features identified at the *top* are N V λλ1239, 1243; C II λλ1334, 1336; Si IV λλ1394, 1403; C IV λλ1548, 1551; and Al III λλ1855, 1863. Note the extreme weakness of these features in the SMC stars compared to their Galactic counterparts

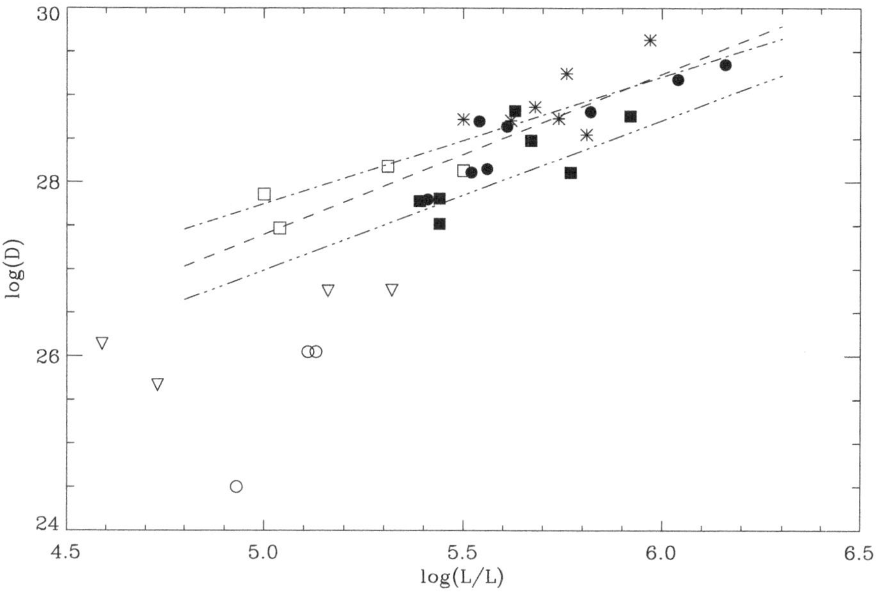

Fig. 4.3 Modified wind-momentum ($D = \dot{M} v_{\infty} \sqrt{R}$) versus luminosity for SMC OB stars observed with HST [1, 5, 10, 11]; *filled symbols* represent stars with measured terminal velocities, *open symbols* are stars with assumed terminal velocities, *circles* and *squares* represent O- and B-type stars respectively. Additional OB supergiants from [4] are marked with *asterisks*. In all cases we refer to un-clumped mass-loss rates. The *lines* are the empirical fits derived for O-type stars [7] (*dashed line*), early B-type supergiants [10] (*dash-dot*) and mid B-type supergiants [10] (*dash-dot-dot-dot*). The three O-type stars (*open circles*) well below the expected relationship, plus the upper limits (*inverted triangles*) from [6] represent the anomalous weak wind class of O-type stars

sults. The Hubble Space Telescope (HST) revolutionized our UV view of massive stars in the Magellanic Clouds, with the first high s/n UV spectroscopic campaign (Kudritzki, GOs:2233/4110) yielding startling spectra from the Faint Object Spectrograph (FOS), whose morphology confirmed the presence of weak winds in the O-type main-sequence stars in the SMC [14]. Subsequent quantitative analysis confirmed that mass-loss rates were indeed lower in the SMC sample than in LMC and Galactic counterparts [8], and in reasonable agreement with theory. This prompted a more systematic approach to the SMC, enlarging the previous sample of nine stars with an additional 30, the use of high resolution with STIS, the inclusion of B-type supergiants, and an extension to lower luminosity late-O dwarfs (Lennon, GOs:7437/9116; [2, 15]). The UV spectra for O-type dwarfs and B-type supergiants are displayed and discussed in Figs. 4.1 and 4.2 respectively.

Many of these stars have been analysed to derive mass-loss rates confirming the expected theoretical trend of mass-loss with metallicity [1, 4, 5, 10, 11] while theoretical expectations for terminal velocities have also been confirmed [3]. However, a small number of stars were found to have mass-loss rates which are an order of magnitude or more less than theoretical predictions [1], see Fig. 4.3. Additional members of this group have been found in the SMC [6], in all cases these stars are low luminosity O-type dwarfs having anomalous weak winds. The nature of their winds is not as yet understood, but possibly indicate an incomplete understanding of mass-loss in the weak wind limit, or perhaps arise as a result of the lack of accurate diagnostics.

References

1. Bouret, J.-C., Lanz, T., Hillier, D.J., Heap, S.R., Hubeny, I., Lennon, D.J., Smith, L.J., Evans, C.J., 2003. Astrophys. J. 595, 1182.
2. Evans, C.J., Lennon, D.J., Walborn, N.R., Trundle, C., Rix, S.A., 2004. Publ. Astron. Soc. Pac. 116, 909.
3. Evans, C.J., Lennon, D.J., Trundle, C., Heap, S.R., Lindler, D.J., 2004. Astrophys. J. 607, 451.
4. Evans, C.J., Crowther, P.A., Fullerton, A.W., Hillier, D.J., 2004. Astrophys. J. 610, 1021.
5. Hillier, D.J., Lanz, T., Heap, S.R., Hubeny, I., Smith, L.J., Evans, C.J., Lennon, D.J., Bouret, J.-C., 2003. Astrophys. J. 588, 1039.
6. Martins, F., Schaerer, D., Hillier, D.J., Heydari-Malayeri, M., 2004. Astron. Astrophys. 420, 1087.
7. Mokiem, M.R., de Koter, A., Vink, J.S., Puls, J., Evans, C.J., Smartt, S.J., Crowther, P.A., Herrero, A., Langer, N., Lennon, D.J., Najarro, F., Villamariz, M.R., 2007. Astron. Astrophys. 473, 603.
8. Puls, J., Kudritzki, R.-P., Herrero, A., Pauldrach, A.W.A., Haser, S.M., Lennon, D.J., Gabler, R., Voels, S.A., Vilchez, J.M., Wachter, S., Feldmeier, A., 1996. Astron. Astrophys. 305, 171.
9. Puls, J., Springmann, U., Lennon, M.A., 2000. Astron. Astrophys. Suppl. Ser. 141, 23.
10. Trundle, C., Lennon, D.J., 2005. Astron. Astrophys. 434, 677.
11. Trundle, C., Lennon, D.J., Puls, J., Dufton, P.L., 2004. Astron. Astrophys. 417, 217.
12. Walborn, N.R., Nichols-Bohlin, J., Panek, R.J., 1985. International ultraviolet explorer atlas of O-type spectra. NASA RP1155, NASA, Washington.
13. Walborn, N.R., Parker, J.W., Nichols, J., 1995. International ultraviolet explorer atlas of B-type spectra. NASA RP1363, NASA, Washington.
14. Walborn, N.R., Lennon, D.J., Haser, S.M., Kudritzki, R.-P., Voels, S.A., 1995. Publ. Astron. Soc. Pac. 107, 104.
15. Walborn, N.R., Lennon, D.J., Heap, S.R., Lindler, D.J., Smith, L., Evans, C.J., Parker, J.W., 2000. Publ. Astron. Soc. Pac. 112, 1243.

Chapter 5
Star Formation Histories of Resolved Galaxies

Monica Tosi

Abstract The impact of HST photometry and European astronomy in studies concerning the star formation histories of resolved galaxies is described. Our current knowledge of the star formation history of systems within 10–20 Mpc, as derived from the colour–magnitude diagrams of their resolved stellar populations, is reviewed, as well as the impact of these results on our understanding of galaxy evolution.

5.1 Introduction

One of the astronomy research fields which have been more impressively impacted by the advent of HST is the study of the star formation histories (SFH) of resolved stellar populations. The superior photometric performances of HST have allowed the resolution of individual stars to unprecedented (and still unequalled) levels of faintness, distance and crowding. Such an outstanding resolving power has triggered a large number of HST programs aimed at the derivation of the SFH of galaxies in the local Universe, involving many people on both sides of the Atlantic Ocean. These studies have significantly improved our understanding of galaxy formation and evolution. The impacts of European Astronomy and of HST in the field are briefly summarized in the following.

5.2 The Impact of European Astronomy

The whole business of deriving the SFH of nearby galaxies from the colour–magnitude diagrams (CMDs) of their resolved stars started in Europe, when the synthetic CMD method was developed several years ago [7, 20, 33, 36]. The method is based on the well known circumstance that the location of any star in the CMD is uniquely related to its mass, age and chemical composition. This makes the CMD the best tool to infer the age of stellar systems. In the case of simple stellar populations, i.e. coeval stars with the same chemical composition, isochrone fitting is the most frequently used method to infer the system age. In the case of galaxies, with

M. Tosi (✉)
INAF, Osservatorio Astronomico di Bologna, Via Ranzani 1, Bologna, Italy
e-mail: monica.tosi@oabo.inaf.it

F.D. Macchetto (ed.), *The Impact of HST on European Astronomy*, Astrophysics and Space Science Proceedings,
DOI 10.1007/978-90-481-3400-7_5, © Springer Science+Business Media B.V. 2010

rather complicated mixtures of different stellar generations, the age determination is less straightforward, but their CMDs remain the best tool to derive the SFH.

Figure 5.1 shows the effect of different SFHs on the CMD of a hypothetical galactic region with the intrinsic and photometric properties typical of a region in the SMC observed with HST/WFPC2. If the star formation rate (SFR) has been constant for all the galaxy lifetime, the CMD of the region is expected to have the morphology of the top-left panel, with a prominent blue plume mostly populated by main-sequence (MS) stars and an equally prominent red plume resulting from the overposition of old subgiants, red giant branch (RGB), horizontal branch (HB) and asymptotic giant branch (AGB) stars, with younger stars in the AGB, blue loops and red supergiant phases. If we change the SFH, not dramatically, but simply introducing quiescent and/or more intense phases, or assuming a shorter e-folding time for the SFR, the morphology of the CMD changes significantly, as visible in the other panels of Fig. 5.1. The tight dependence of the CMD morphology on the SFH is the cornerstone of the synthetic CMD method, which consists in comparing the observational CMD of a galactic region with synthetic CMDs created via Monte Carlo extractions on stellar evolution tracks or isochrones for a variety of SFHs, IMFs and age–metallicity relations (see e.g. [5, 21, 36] for a detailed description of different procedures).

The method does not provide unique solutions, but significantly reduces the possible SFH scenarios. Since its first applications to photometric data from ground-based, moderate size telescopes, the synthetic CMD method demonstrated its power, showing that even in tiny galaxies such as Local Group dwarf irregulars (dIrrs) the SFH varies from one region to the other and that their star formation regime is rather continuous, with long episodes of moderate activity, separated by short quiescent intervals (the so-called gasping regime, [19, 20, 24, 33, 36]).

5.3 The Impact of HST

When the first non-aberrated images were acquired with HST, the impressive improvement in the achievable photometric resolution and depth, and the corresponding quantum leap in the quality of the CMDs, triggered a worldwide burst of interest in the derivation of the SFHs of nearby galaxies and in the synthetic CMD method. Many people developed their own procedures and a few years later already $\sim$10 different groups participated to an experiment organized in Coimbra (Portugal) to compare with each other the results from different procedures (see [31] and references therein). The experiment showed that most of the procedures provided consistent results and brought further interest to the field. Since then, many studies have been performed to infer the SFH of almost all the dwarf galaxies in the Local Group (e.g. [10, 15, 22, 25, 32]).

The high spatial resolution of the HST cameras also allows us to spatially resolve the SFH in the closest galaxies. For instance [13] and [14] have resolved and measured the SF activity over the last 0.5 Gyr in the various sub-regions of the dIrrs Gr8 and Sextans A, close to the borders of the Local Group. The resulting space/time

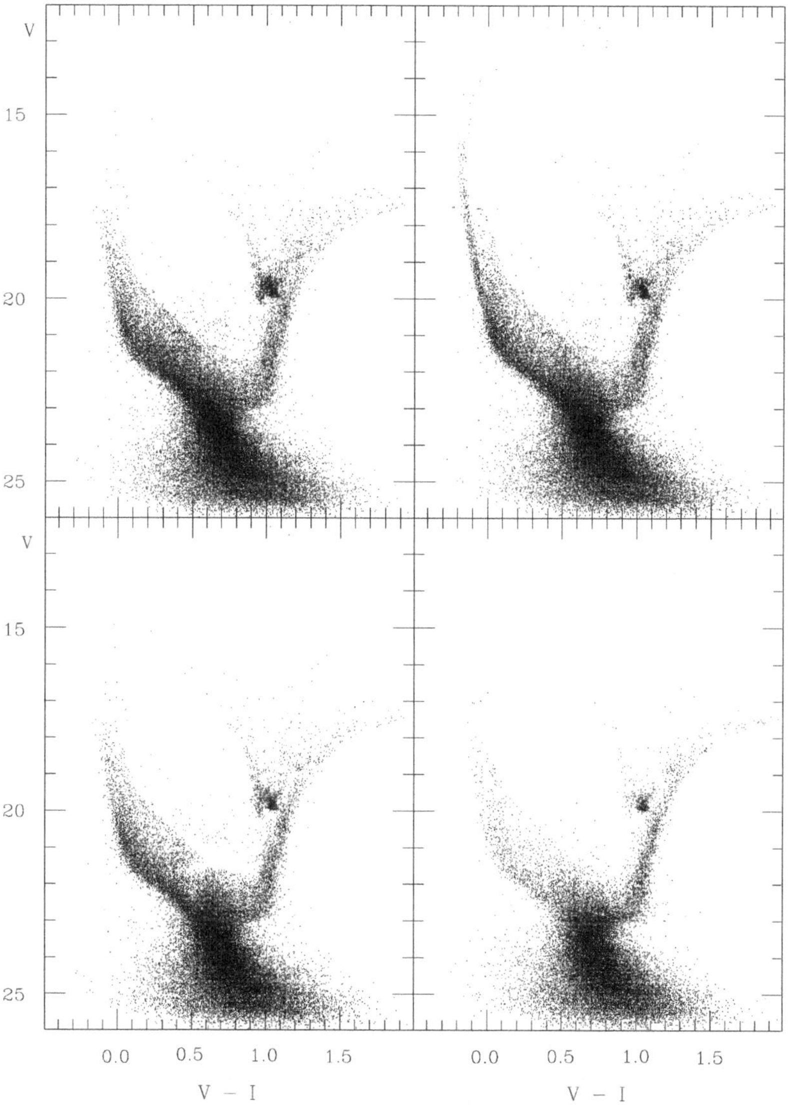

Fig. 5.1 The effect of the SFH on the CMD of a hypothetical galactic region with $(m - M)_0 = 19$, $E(B - V) = 0.08$, and with the photometric errors and incompleteness typical of HST/WFPC2 photometry. All the shown synthetic CMDs contain 50,000 stars and are based on the $Z = 0.004$ Padova models [18]. *Top-left panel*: the case of a SFR constant from 13 Gyr ago to the present epoch. *Top-right panel*: the effect of concentrating in the last 20 Myr the formation of all the stars which were born in the last 100 Myr in the previous case. The CMD has a much brighter and thicker blue plume. *Bottom-left panel*: same constant SFR as in the first case, but with a quiescent interval between 3 and 2 Gyr ago; a gap appears in the CMD region delimited by the 2 and 3 Gyr isochrones. *Bottom-right panel*: exponentially decreasing SFR over the whole Hubble time, with e-folding time 5 Gyr; the blue plume contains much fewer stars than in the constant SFR case, and the late evolutionary phases (RGB, red clump, blue loops) are also differently populated, with fewer stars in the younger sequences

distribution of the SF, with lightening and fading of adjacent cells, is intriguingly reminiscent of the predictions of the stochastic self-propagating SF theory proposed by [30] almost 30 years ago.

The HST/ACS provides spectacularly deep and spatially resolved images, such as those of the star forming region NGC346 in the SMC, where [26] and [27] have been able to measure 85,000 stars, from very old to very young ones, including, for the first time in the SMC, pre-main-sequence objects of mass from 3 to 0.6 $M_{\odot}$. The famous CMD of an Andromeda region by [8] showed that detailed SFHs can be finally derived also for external spirals.

In galaxies beyond the Local Group, distance makes crowding more severe, and even HST cannot resolve stars as faint as the MS turn-off of old populations. The higher the distance, the worse the crowding conditions, and the shorter the lookback time reachable even with the deepest, highest resolution photometry (see Fig. 2 in [35]). Depending on distance and intrinsic crowding, the reachable lookback time in galaxies more than 1 Mpc away ranges from several Gyr (in the best cases, when the RGB or even the HB are clearly identified), to several hundreds Myr (when AGB stars are recognized), to a few tens Myr (when only the brightest supergiants are resolved).

Since the Local Group doesn't host all types of galaxies, with the notable and unfortunate absence of the most (ellipticals) and the least (Blue Compact Dwarfs, BCDs) evolved ones, a few groups have embarqued in the challenging task of deriving the SFH of more distant galaxies. In spite of the larger uncertainties and the shorter lookback time, these studies have led to quite interesting results, which wouldn't have been possible without HST.

The case of NGC1705 is particularly instructive. The HST/WFPC2 photometry was deep and good enough to let us resolve individual stars from the most central regions to the extreme outskirts [37]. We have been able to divide the galaxy in roughly concentric regions, all sufficiently populated by resolved stars, and derive the SF history of each region. Young massive stars are concentrated at the center and their percentage rapidly decreases outwards, while faint red stars are increasingly visible towards the outer regions. The latter circumstance doesn't necessarily imply that old stars are absent at the center; it simply means that crowding is too severe there to let us resolve them. In the outer regions, where crowding is definitely not a problem, the CMDs present a well defined upper portion of the RGB, whose tip is also very well defined and allowed us to accurately derive the galaxy distance [37].

By applying to each region the synthetic CMD method, [4] have inferred their SFHs, summarized in Fig. 5.2, where the SFR per unit area is plotted as a function of age. It can be seen that, except for the innermost region, where crowding does not allow us to reach old lookback times, all the regions have been forming stars since at least 5 Gyr. On average, the SF appears to have been rather continuous: there are evidences for interruptions in the SF activity, but always shorter than a few Myr or tens of Myr, at least in the age range where we do have this time resolution (i.e. in the last 1 Gyr or so). Quiescent phases of 100 Myr or longer would have appeared as gaps in the empirical CMDs of stars younger than 1 Gyr, and such gaps are absent. The SF history of NGC1705 shows three striking features: one is the burst

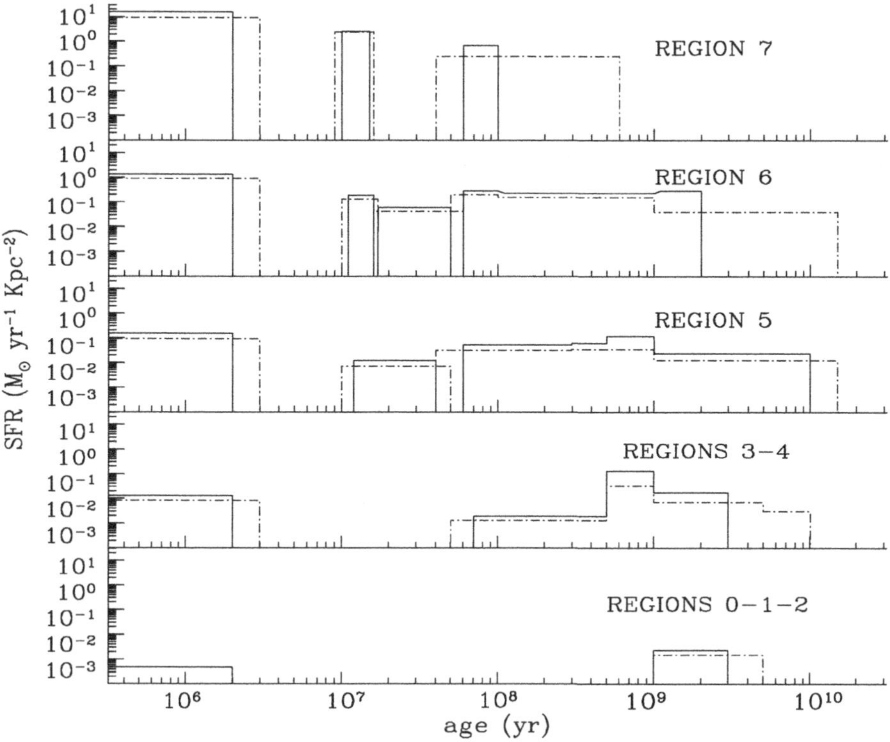

Fig. 5.2 SFR per unit area vs age for NGC1705's concentric regions (decreasing numbers from the center outwards). Notice that both *axes* are on logarithmic scale. The different *linetypes* show the uncertainty on the inferred SFH [4]

occurred in the central regions 10–15 Myr ago, when the central Super Star Cluster also formed and when the observed galactic wind is supposed to have originated; the second is the quiescent phase with no SF anywhere in the galaxy right after such burst, probably due to the gas sweeping by shocks and winds triggered by the explosions of the burst supernovae; and the third is the new, even stronger, SF activity occurring everywhere in NGC1705 (but much higher in the inner regions) in the last 2 Myr. The latter event puts interesting constraints on the cooling timescales of the gas heated by the supernovae generated in the previous burst and on the modeling of SF processes.

Figure 5.3 sketches the global SFHs derived by various authors for some of the late-type dwarfs studied so far. The lookback time is indicated and in all cases stars with that age were detected.

The latter is one of the most interesting results of all the studies of BCDs whose individual stars have been resolved by HST and the SFH has been derived with the synthetic CMD method: all of them have turned out to be already active at the lookback time reached by the photometry (see [1, 2, 4, 11, 12, 16, 17, 23, 28, 29]). None appears to be experiencing now its first star formation activity, including the

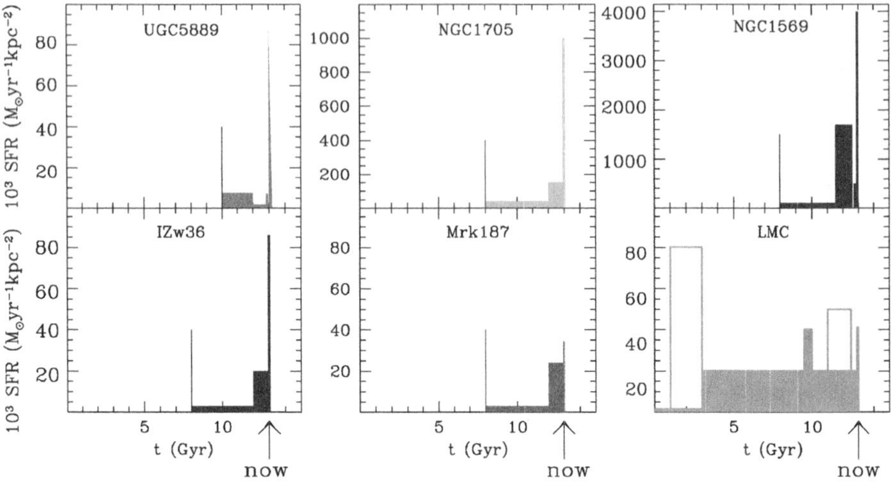

Fig. 5.3 (Color online) SFH (SF rate per unit area vs time) from the CMDs of a few late-type galaxies resolved by HST: three BCDs (NGC 1705 [4], IZw36 [29], Mrk187 [28]), one starburst dIrr (NGC 1569 [3, 21]), one LSB (UGC5889 [40]), and the LMC bar used for the Coimbra experiment [38]. The lookback time reached by the photometry is indicated in each case with a *black vertical line*. Notice that these are global SFHs (except for the LMC bar), while Fig. 5.2 shows the SFH of individual regions

most metal poor, gas rich ones, such as SBS1415 and IZw18 (see [2, 25, 39] and Chap. 22 by Aloisi in this book).

Equally interesting is that the galaxy with stronger SF activity is the dIrr NGC1569, the only one with a rate comparable to the 1 $M_\odot$ yr^{-1} required by [6] to let a late-type galaxy contribute to the blue galaxy excess observed in counts at intermediate redshift. All the other dwarfs, independently of being classified as BCDs or irregulars or low surface brightness (LSB), present much lower SFRs.

We also notice that all the galaxies of Fig. 5.3, but the LMC, have similar global SFHs, with a recent burst overimposed on a quiter, more continuous regime. If we consider also the other late-type dwarfs in the Local Group studied so far, we can speculate that the only apparent difference in the SFH of BCDs and dIrrs is that the former always show a very recent SF burst, the latter not always. This is presumably due to selection effects that made it possible in the past to discover distant dwarfs only if featured with strong HII region emissions, whilst nearby dwarfs were discoverable also if the peak of their SF activity occurred long ago.

5.4 The Impact on Galaxy Evolution

The main results from all the studies performed on the SF histories of dwarf galaxies are: (1) no evidence of long interruptions in the SF activity has been found, except in early-type galaxies; (2) no galaxy currently experiencing its first SF activity has been found yet; (3) strong bursts don't seem to be frequent; (4) the SF regime seems to be a gasping rather than a bursting one in all late-type dwarfs, both in the Local Group and outside it; (5) no significant difference has been found in the SFH and in the stellar populations of BCDs and dIrrs, except that the former always have a recent SF burst.

These results have profound effects on our understanding of the evolution of dwarf galaxies. In addition, the combination of HST photometry with high-performance spectroscopy is providing in the last years a first in-depth view of their chemical evolution. The detailed chemical abundances measured in many of their stars have solved the age–metallicity degeneracy affecting the RGB colours and have led to the derivation of age–metallicity relations not only in the LMC (e.g. [9]) but also in other nearby dwarf spheroidals (e.g. [34]). As a consequence, for the first time detailed chemical evolution models can be computed for individual dwarf galaxies, an application performable until recently only for spirals, where the wealth and quality of the observational data were sufficient to properly constrain the models. The new generation models can now adopt SF laws and initial mass functions which are not free parameters, but are provided by the SFH studies, and their predictions can be compared with the chemical abundances not only of very young objects, but also of stars of various ages. Such a decrease in the number of free parameters opens the route to a much more reliable use of theoretical models and, ultimately, to a much better understanding of galaxy evolution.

Acknowledgements I'm grateful to all the persons I have studied the SFHs with, in particular to A. Aloisi, L. Angeretti, F. Annibali, L. Greggio, A. Nota and E. Sabbi. This work has been partially supported by the PRIN-INAF-2005.

References

1. Aloisi, A., Tosi, M., Greggio, L., 1999. Astron. J. 118, 302.
2. Aloisi, A., van der Marel, R.P., Mack, J., Leitherer, C., Sirianni, M., Tosi, M., 2005. Astron. J. 631, L45.
3. Angeretti, L., Tosi, M., Greggio, L., Sabbi, E., Aloisi, A., Leitherer, C., 2005. Astron. J. 129, 2203.
4. Annibali, F., Greggio, L., Tosi, M., Aloisi, A., Leitherer, C., 2003. Astron. J. 126, 2752.
5. Aparicio, A., Gallart, C., 2004. Astron. J. 128, 1465.
6. Babul, A., Ferguson, H.C., 1996. Astrophys. J. 458, 100.
7. Bertelli, G., Mateo, M., Chiosi, C., Bressan, A., 1992. Astrophys. J. 388, 400.
8. Brown, T.M., Ferguson, H.C., Smith, E., Kimble, R.A., Sweigart, A.V., Renzini, A., Rich, R.M., VandenBerg, D.A., 2002. Astrophys. J. 592, L17.
9. Cole, A.A., Tolstoy, E., Gallagher, J.S., Smecker-Hane, T.A., 2005. Astron. J. 129, 1465.
10. Cole, A.A., et al., 2007. Astrophys. J. 659, L17.
11. Crone, M.M., Schulte-Ladbeck, R.E., Hopp, U., Greggio, L., 2000. Astrophys. J. 545, L31.
12. Crone, M.M., Schulte-Ladbeck, R.E., Greggio, L., Hopp, U., 2002. Astrophys. J. 567, 258.
13. Dohm-Palmer, R.C., Skillman, E.D., Gallagher, J.S., Tolstoy, E., Mateo, M., Dufour, R.J., Saha, A., Hoessel, J., Chiosi, C., 1998. Astron. J. 116, 1227.
14. Dohm-Palmer, R.C., Skillman, E.D., Mateo, M., Saha, A., Dolphin, A., Tolstoy, E., Gallagher, J.S., Cole, A.A., 2002. Astron. J. 123, 813.

15. Dolphin, A.E., et al., 2003. Astron. J. 126, 187.
16. Drozdovsky, I.O., Schulte-Ladbeck, R.E., Hopp, U., Crone, M.M., Greggio, L., 2001. Astrophys. J. 551, L135.
17. Drozdovsky, I.O., Schulte-Ladbeck, R.E., Hopp, U., Greggio, L., Crone, M.M., 2002. Astron. J. 124, 811.
18. Fagotto, F., Bressan, A., Bertelli, G., Chiosi, C., 1994. Astron. Astrophys. Suppl. Ser. 105, 39.
19. Ferraro, F.R., Fusi Pecci, F., Tosi, M., Buonanno, R., 1989. Mon. Not. R. Astron. Soc. 241, 433.
20. Gallart, C., Aparicio, A., Bertelli, G., Chiosi, C., 1996. Astron. J. 112, 1950.
21. Greggio, L., Tosi, M., Clampin, M., De Marchi, G., Leitherer, C., Nota, A., Sirianni, M., 1998. Astrophys. J. 504, 725.
22. Harris, J., Zaritsky, D., 2004. Astron. J. 127, 153.
23. Lynds, R., Tolstoy, E., O'Neil, E.J. Jr., Hunter, D.A., 1998. Astron. J. 116, 146.
24. Marconi, G., Tosi, M., Greggio, L., Focardi, P., 1995. Astron. J. 109, 173.
25. Momany, Y., et al., 2005. Astron. Astrophys. 439, 111.
26. Nota, A., et al., 2006. Astrophys. J. 640, L29.
27. Sabbi, E., et al., 2007. Astron. J. 133, 44.
28. Schulte-Ladbeck, R.E., Hopp, U., Greggio, L., Crone, M.M., 2000. Astron. J. 120, 1713.
29. Schulte-Ladbeck, R.E., Hopp, U., Greggio, L., Crone, M.M., Drozdovsky, I.O., 2001. Astron. J. 121, 3007.
30. Seiden, P.E., Schulman, L.S., Gerola, H., 1979. Astrophys. J. 232, 709.
31. Skillman, E.D., Gallart, C., 2002. ASP Conf. Ser. 274, 535.
32. Skillman, E.D., Tolstoy, E., Cole, A.A., Dolphin, A.E., Saha, A., Gallagher, J.J., Dohm-Palmer, R.C., Mateo, M., 2003. Astrophys. J. 596, 253.
33. Tolstoy, E., 1996. Astrophys. J. 462, 684.
34. Tolstoy, E., Venn, K.A., Shetrone, M., Primas, F., Hill, V., Kaufer, A., Szeifert, T., 2003. Astron. J. 125, 707.
35. Tosi, M., 2007. In: Vallenari, A., Tantalo, R., Portinari, L., Moretti, A. (Eds.), From Stars to Galaxies, ASP Conf. Ser., vol. 374, p. 221.
36. Tosi, M., Greggio, L., Marconi, G., Focardi, P., 1991. Astron. J. 102, 951.
37. Tosi, M., Sabbi, E., Bellazzini, M., Aloisi, A., Greggio, L., Leitherer, C., Montegriffo, P., 2001. Astron. J. 122, 127.
38. Tosi, M., Greggio, L., Annibali, F., 2002. ASP Conf. Ser. 274, 529.
39. Tosi, M., Aloisi, A., Mack, J., Maio, M., 2006. In: IAU Symp., vol. 235, p. 65.
40. Vallenari, A., Schmidtobreick, L., Bomans, D.J., 2005. Astron. Astrophys. 435, 821.

Chapter 6
HST's View of the Youngest Massive Stars in the Magellanic Clouds

M. Heydari-Malayeri, M.R. Rosa, V. Charmandaris, L. Deharveng, F. Martins, F. Meynadier, D. Schaerer, and H. Zinnecker

Abstract Accurate physical parameters of newborn massive stars are essential ingredients to shed light on their formation, which is still an unsolved problem. The rare class of compact H II regions in the Magellanic Clouds (MCs), termed "high-excitation blobs" (HEBs), presents a unique opportunity to acquire this information. These objects ($4''$ to $10''$, $\sim$1 to 3 pc, in diameter) harbor the youngest massive stars

M. Heydari-Malayeri (✉) · V. Charmandaris · F. Meynadier
LERMA, Paris Observatory, Paris, France
e-mail: m.heydari@obspm.fr

F. Meynadier
e-mail: Frederic.Meynadier@obspm.fr

M.R. Rosa
ST-ECF, ESO, Garching, Germany
e-mail: mrosa@stecf.org

V. Charmandaris
University of Crete, Heraklion, Greece
e-mail: vassilis@physics.uoc.gr

L. Deharveng
Marseille Observatory, Marseille, France
e-mail: Lise.Deharveng@oamp.fr

F. Martins
MPE, Garching bei München, Germany
e-mail: martins@mpe.mpg.de

D. Schaerer
Geneva Observatory, Geneva, Switzerland
e-mail: Daniel.Schaerer@obs.unige.ch

D. Schaerer
Observatoire Midi-Pyrénées, Laboratoire d'Astrophysique, Toulouse, France

H. Zinnecker
Potsdam Astrophysical Institute, Potsdam, Germany
e-mail: hzinnecker@aip.de

F.D. Macchetto (ed.), *The Impact of HST on European Astronomy*, Astrophysics and Space Science Proceedings,
DOI 10.1007/978-90-481-3400-7_6, © Springer Science+Business Media B.V. 2010

of the OB association/molecular cloud complexes in the MCs accessible through high-resolution near-IR and optical techniques. We present a brief overview of the results obtained with *HST* mainly on two HEBs, one in the LMC (N159-5) and the other in the SMC (N81).

6.1 Introduction

Massive stars (>8 $M_{\odot}$) play a key role in several fields of astrophysics. However their formation process is still an unsolved problem in spite of progress, both in theory and observation, in recent years.

In order to better understand their formation one needs photospheric lines to access accurate effective temperatures ($T_{\rm eff}$), luminosities (L), masses (M), and gravities (g). $T_{\rm eff}$ and L are crucial for placing the stars on the HR diagram and to establish whether they are ZAMS or evolved stars. This is very important since massive stars evolve fast, and the most massive ones can become supergiants while still enshrouded in clumps of gas and dust. Spectral signatures are also needed to understand how and when the stellar wind develops. Moreover, mass loss may affect the accretion rate and play an important role in massive star formation process. Other important issues are multiplicity and the IMF of the cluster members. High-resolution radio continuum and IR observations yield indispensable information about UCHII regions but do not have direct access to newborn massive stars.

Only observations in the near-IR and optical (mainly spectroscopy) of nascent stars can accurately provide these physical parameters. This can be achieved in the visible by using the traditional methods of spectral classification [28], or by using the more recent method of line fitting, based on atmosphere models [12, 20]. In the near-IR similar methodology makes use of 2 μm spectroscopy [1], or profile fitting [20, 21].

The MCs offer valuable opportunities for this study since they are seen face-on and have well-determined distances. This facilitates obtaining accurate absolute magnitudes and fluxes. Moreover, their overall extinction is low. Being metal-poor, they provide important templates for studying star formation in distant metal-poor galaxies which cannot be observed with comparable spatial resolution.

6.2 High-Excitation Blobs

The youngest massive stars in the MCs accessible to IR and optical observations are found in High-Excitation Blobs (HEBs), see below for details. The reason for this terminology is that no spatial features could be distinguished with ground-based telescopes. This is a rare class of H II regions in the MCs; so far only six members have been detected in the LMC and three in the SMC.

For massive stars the accretion time-scale is larger than the Kelvin–Helmholtz time-scale. This means that massive stars reach the main sequence while accretion is still going on. Moreover, they evolve very fast. Therefore, obtaining the physical parameters of massive stars "at birth" may be an unattainable task! Consequently, HEBs offer a compromise between stars inside ultra-compact H II regions and the exciting stars of evolved H II regions.

In contrast to the typical H II regions of the MCs, which are extended structures (sizes of several arc minutes corresponding to more than 50 pc, powered by a large number of exciting stars), HEBs are very dense and small regions ($\sim 4''$ to $10''$ in diameter corresponding to $\sim$1–3 pc). They have a higher degree of excitation ([O III]/Hβ]) with respect to the typical H II regions, and are, in general, heavily affected by local dust. In comparison with Galactic regions, some of HEBs are similar to classical H II regions and some look like compact H II regions [18]. However, HEBs should be considered in the context of massive star formation in the MCs. Compared with other compact H II regions of the same sizes in the MCs, they constitute a distinctly detached group with high excitations and luminosities [22].

6.3 LMC N159-5 (Papillon)

This prototype of HEBs is situated in the H II complex N159 some 500 pc south of 30 Dor [2]. N159 is associated with one of the most important concentrations of molecular gas in the LMC [17] and contains several signposts of ongoing star formation (cocoon stars, IR sources, masers). Several physical aspects of this object were studied using ground-based observations in the visible, IR, and radio [3, 13, 15, 16, 25, 26].

Our high-resolution *HST* observations show N159 to be a turbulent environment typical of massive star formation regions with stellar winds, cavities sculpted in the ionized gas as well as shocks, absorption features, filaments, and arcs [5]. The compact H II region is resolved for the first time into a butterfly-shaped structure with the wings separated by $\sim 2''.3$ (0.6 pc). Two subarcsec features show up in the wings, a kind of "smoke ring" and a "globule" or sort of cometary structure. These features suggest a dynamic environment, the physical details of which are still unclear. The butterfly shape may be due to a bipolar phenomenon, but with the present data we cannot verify. Regarding the exciting star(s), from the Hβ flux we expect a star of type O8 at least. The exciting star(s) is(are) not detected since the extinction is larger than $A_V = 5$ mag. High-resolution near-IR observations are necessary to penetrate into the dust concentration. However, our *JHK* VLT/ISAAC observations were not able to resolve the blob [23]. Nonetheless, we could study the stellar populations associated with the object and we find that there are two main stellar populations: a massive, young population about 3 Myr old, and an older population of low mass stars with an age between 1 and 10 Gyr. The position of the blob on the color–magnitude diagram can be explained by the presence of a $\sim 50\,M_\odot$ star, affected by an extinction of $A_V \sim 7$ mag.

6.4 SMC N81

This object is situated in the wing of the SMC, south-east of the main bar [4, 15, 16, 25, 26]. SMC N81 [6]. Six stars are grouped in an area 2″ wide. The two brightest stars, the main exciting sources, are only 0″.27 apart (0.08 pc). Two prominent dark lanes divide the nebula into three lobes. A magnificent curved plume of some 15″ (4.5 pc) in length is apparently linked to an absorption hole lying towards the center of the H II region. It is uncertain whether the plume is due to an outflow or an infall of dust. The absorption hole has a size of 0″.25 (0.07 pc). The H II region is in contact with a dense medium toward the west where two ridges, representing ionization/shock fronts, are present. This idea is corroborated by high-resolution ATCA interferometer observations in the radio continuum at 3 cm, which show compressed radio emission in that direction [14], most probably due to the presence of a molecular cloud.

We also used *HST*/STIS to obtain UV spectra of 8 stars associated with N81 [10]. Analysis of those spectra shows that all of them are massive stars of types O6–O8. However, they appear to be sub-luminous, by ~2 mag in comparison with typical O dwarfs in the MCs. An important property of these stars is their weakness of the stellar winds as evidenced by the line profiles. Mass loss rates as low as $10^{-9}\,M_{\odot}\,yr^{-1}$, i.e. ~2 orders of mag weaker compared with O stars of the same type [19]. These stars may belong to the class of Vz stars: O dwarfs lying on or close to the ZAMS [27]. The weakness of the winds cannot be attributed to metallicity effects. It may be due to the youth of the stars (younger stars have weaker winds?) and if confirmed warrants further investigation.

6.5 Other Objects

Our *HST* observations allowed us also to resolve other HEBs for the first time. For example SMC N88A, which in comparison with SMC N81, is more compact (size ~3″.5, ~1 pc), denser, and of higher excitation, shows two components in interaction [7]. The extinction rises to more than 3.5 mag in the visible towards its tight core. Such a high extinction is unprecedented for an H II region in the metal-poor SMC. The exciting star(s) of N88A, which are certainly more massive/hotter than those of N81, are not detected in the optical due to the heavy extinction.

For LMC N11A, our *HST* data revealed its previously unknown nebular and stellar features, notably the presence of an embedded cluster of stars. Five of the stars are packed in an area less than 2″ (0.5 pc), with the most luminous one being a mid O type star [8, 24].

LMC N83B displays an impressive cavity, with an estimated age of only ~30,000 yr, sculpted in the ionized gas by the powerful winds of a massive star. The observations bring also to light two compact H II blobs of only 2″.8 (0.7 pc) and 1″ (0.3 pc) in size. The former harbors the presumably hottest star of the burst and is also strongly affected by dust with a strong extinction. These features plus an outstanding ridge are formed in the border zone between the molecular cloud and the ionized gas possibly in a sequential process triggered by the ionization front of an older H II region [9].

In addition, the two compact regions A1 and A2 lying toward the giant H II region N160 were for the first time resolved and their stellar content and morphology revealed. A1, being of higher excitation, is powered by a single massive star whose strong wind has created a surrounding bubble. A2 harbors several stars enshrouded by large quantities of dust [11].

Apart from the papers cited above, our *HST* projects have resulted in six *HST* NASA/ESA Press Releases (July 23, 1998; June 10, 1999; October 5, 2000; March 28, 2001; December 19, 2001; September 12, 2002).

6.6 Conclusions

Physical parameters of massive stars are needed to better understand massive star formation. The exciting stars of HEBs, being the youngest massive stars reachable through IR and optical techniques, offer a good compromise between massive stars "at birth" and those in evolved H II regions. HEBs are also particularly attractive in the context of massive star formation in the MCs. An unexpected and intriguing result is the weak winds of O type stars in SMC N81. Are weak winds a general property of the massive stars embedded in HEBs? Future high-resolution near-IR observations (spectroscopy & imaging) are needed to make progress. In the near future, near-IR observations with JWST and E-ELT will yield the photospheric signatures of newborn massive stars enshrouded in $A_V = 100$–200 mag [29] blobs at even earlier stages of evolution.

References

1. Hanson, M.M., et al., 2005. Astrophys. J. Suppl. Ser. 161, 154.
2. Heydari-Malayeri, M., Testor, G., 1982. Astron. Astrophys. 111, L11.
3. Heydari-Malayeri, M., Testor, G., 1985. Astron. Astrophys. 144, 98.
4. Heydari-Malayeri, M., Le Bertre, Th., Magain, P., 1988. Astron. Astrophys. 195, 230.
5. Heydari-Malayeri, M., et al., 1999. Astron. Astrophys. 352, 665.
6. Heydari-Malayeri, M., et al., 1999. Astron. Astrophys. 344, 848.
7. Heydari-Malayeri, M., et al., 1999. Astron. Astrophys. 347, 841.
8. Heydari-Malayeri, M., et al., 2001. Astron. Astrophys. 372, 527.
9. Heydari-Malayeri, M., et al., 2001. Astron. Astrophys. 372, 495.
10. Heydari-Malayeri, M., et al., 2002. Astron. Astrophys. 381, 951.
11. Heydari-Malayeri, M., et al., 2002. Astron. Astrophys. 381, 941.
12. Hillier, D.J., Miller, D.L., 1998. Astrophys. J. 496, 407.
13. Hunt, M.R., Whiteoak, J.B., 1994. Proc. — Astron. Soc. Aust. 11, 68.
14. Indebetouw, R., Johnson, K.E., Conti, P., 2004. Astron. J. 128, 2206.
15. Israel, F.P., Koornneef, J., 1988. Astron. Astrophys. 190, 21.
16. Israel, F.P., Koornneef, J., 1991. Astron. Astrophys. 248, 404.
17. Johansson, L.E.B., Greve, A., Booth, R.S., et al., 1998. Astron. Astrophys. 331, 857.
18. Martin-Hernandez, N.L., Vermeij, R., van der Hulst, J.M., 2005. Astron. Astrophys. 433, 205.

19. Martins, F., et al., 2004. Astron. Astrophys. 420, 1087.
20. Martins, F., Schaerer, D., Hillier, D.J., 2005. Astron. Astrophys. 436, 1049.
21. Martins, F., Genzel, R., Hillier, D.J., et al., 2007. Astron. Astrophys. 468, 233.
22. Meynadier, F., Heydari-Malayeri, M., 2007. Astron. Astrophys. 461, 565.
23. Meynadier, F., Heydari-Malayeri, M., Deharveng, L., et al., 2004. Astron. Astrophys. 422, 129.
24. Parker, J.Wm., Garmany, C.D., Massey, P., Walborn, N.R., 1992. Astron. J. 103, 1205.
25. Vermeij, R., et al., 2002. Astron. Astrophys. 390, 649.
26. Vermeij, R., van der Hulst, J.M., 2002. Astron. Astrophys. 391, 1081.
27. Walborn, N.R., Parker, J.W., 1992. Astrophys. J. 399, L87.
28. Walborn, N.R., 2006. In: Proc. Joint Discussion, IAU. astro-ph/0610837.
29. Zinnecker, H., 2006. In: Whitelock, P., et al. (Eds.), The Scientific Requirements for Extremely Large Telescopes, Proc. of the IAU Symp., vol. 232, p. 324.

Chapter 7
Planetary Nebulae and Their Central Stars in the Magellanic Clouds

Eva Villaver, Letizia Stanghellini, and Richard A. Shaw

Abstract During the last decade the *Hubble Space Telescope* (HST) has allowed us to extend late stellar evolution studies to nearby galaxies where the effect of the environment can be quantified. Using HST we have observed over a hundred Planetary Nebulae (PNe) in the Magellanic Clouds, where its known distance has allowed us to determine accurate masses for their central stars. We find an average central star mass of $\langle M_{CS}, LMC \rangle = 0.65 \pm 0.07\ M_{\odot}$ in the Large Magellanic Cloud, higher than the one reported in the literature for both white dwarfs and the central stars of PNe in the Galaxy. Higher central star masses are expected in a lower metallicity environment as a consequence of the reduced mass-loss rates during the Asymptotic Giant Branch. We present the first observational evidence from PNe progenitors of this effect.

7.1 Introduction

Low- and intermediate-mass stars after experiencing heavy winds at the end of the Asymptotic Giant Branch (AGB) phase, leave behind a core that is below the Chandrasekhar mass limit. This core eventually illuminates the stellar remnant giving birth to a planetary nebula (PN). The final mass reached by the central star of a PN depends mainly on the stellar mass during the main-sequence phase and on the mass-loss during the AGB phase. Mass-loss during the AGB phase has a strong dependency on metallicity, as it is thought to be driven mainly by dust.

The final mass of the central star alone determines the energy budget and the timescale of evolution of the nebular shell [1, 2]. An accurate measurement of the central star properties (mass, effective temperature and luminosity) is then crucial to understanding how the formation and evolution of PNe relates to the initial mass of the progenitor star. Moreover, central stars of PNe are the immediate stellar precursors of white dwarfs and in principle the determination of their masses should provide further constraints to fundamental empirical quantities such as the initial–final

E. Villaver (✉)
Hubble Space Telescope Space Department of ESA, Space Telescope Science Institute, Baltimore, USA
e-mail: villaver@stsci.edu

L. Stanghellini · R.A. Shaw
National Optical Astronomical Observatory, Tucson, USA

F.D. Macchetto (ed.), *The Impact of HST on European Astronomy*, Astrophysics and Space Science Proceedings,
DOI 10.1007/978-90-481-3400-7_7, © Springer Science+Business Media B.V. 2010

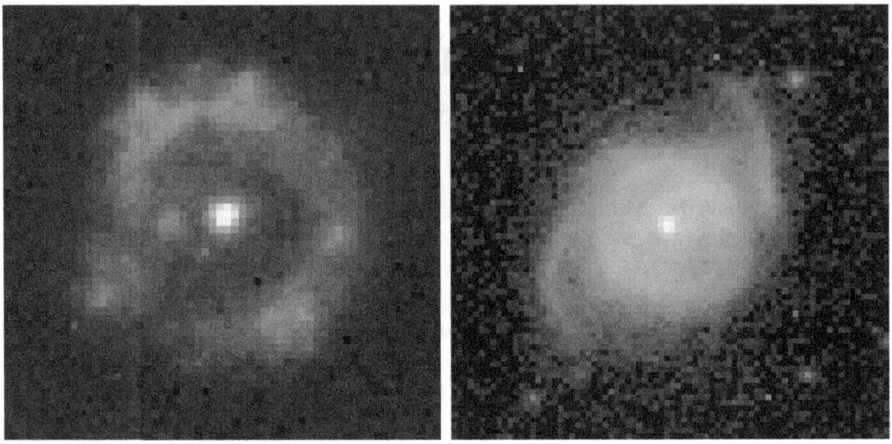

Fig. 7.1 (Color online) The figure shows the PNe SMC MG 13 (*right*) and LMC SMP 10 (*left*). The central star is clearly visible in both images. The image was obtained using the clear (50CCD) bandpass with HST/STIS with a log intensity stretch. The field of view is 3″ on each side

mass relation which rely primarily on the measurements of white dwarfs masses. Although the central stars of PNe can be up to 10^4 times more luminous than white dwarfs, their mass determination in our Galaxy is extremely uncertain primarily due to the poor knowledge of their individual distances. The central star evolutionary tracks in the HR diagram depend primarily on mass and show very little variation with the central star luminosity. In the best case scenario, the horizontal part of the HR diagram, the dependency of the central star mass (M_{CS}) with its luminosity (L_{CS}) goes as, $[M_{CS}/M_{\odot}] = 0.5 + 1.8 \times 10^{-5}$ $[L_{CS}/L_{\odot}]$ [3], which means that in order to accurately determine the central star mass the distance has to be known to better than 10%.

7.2 Our HST Programs

In order to overcome the problem with the distance uncertainty we have observed PNe in the Magellanic Clouds. Only the Hubble Space Telescope (*HST*) allows one to spatially resolve PNe at the distance of the Magellanic Clouds and separate the nebula from the central star.

Our group has acquired and analyzed the first *HST* data samples of PNe in the Magellanic Clouds, aimed at detecting the central star through broad-band imaging [4, 5] (see Fig. 7.1). Our analysis of the central stars includes a sample of 89 PNe in the Large Magellanic Cloud (LMC) [6, 7] and 27 PNe in the Small Magellanic Cloud [8]. Often the central star is not detected above the nebular continuum, either because it has evolved to luminosities below our detection limits or because the nebular continuum is brighter than the stellar one. As a consequence the properties of 14 (out of 27 PNe observed in the SMC) and 37 (out of 89 PNe in the LMC) central stars were determined.

We have used standard aperture photometry techniques to determine the central star magnitude. The stellar extinction correction has been estimated from the nebular Balmer decrement and the effective temperature of the star by using the method originally developed by Zanstra.

7.2.1 The Average Mass of the Central Stars in the LMC

The mean and the median of the mass distribution of the 37 central stars of PNe in the LMC are 0.65 ± 0.07 and 0.64 ± 0.06 $M_{\odot}$ respectively. The average mass obtained from a sample that contained ~200 central stars of PNe in the Galaxy was found to be $\langle M_{CS}, GAL \rangle = 0.60 \pm 0.13$ $M_{\odot}$ [9]. Recently, from a large sample of 350 white dwarfs from the Palomar Green Survey an average white dwarf mass of $\langle M_{WD}, GAL \rangle = 0.603 \pm 0.134$ $M_{\odot}$ was found [10], in good agreement with the results of [9]. The average mass of the total sample of LMC central stars in our LMC sample (37 objects) is slightly higher than the average mass of both white dwarfs and central stars of PNe in the Galaxy reported in the literature.

As mentioned in the introduction the central star mass depends mainly on the stellar mass during the Main-Sequence phase (hereafter, the initial mass), and on the mass-loss during the AGB phase. Mass-loss during the AGB phase has a strong dependency on metallicity as it is thought to be driven mainly by dust. The dust formation process depends on the chemical composition of the gas: the lower the metallicity the smaller the amount of dust formed, and the lower the efficiency of the momentum transfer to the gas. Thus, low metallicity stars with dust-driven winds are expected to loose smaller amounts of matter and therefore are expected to end-up with higher central star masses.

The metallicity of the LMC is on average half that of the solar mix. Therefore, it is well expected that the efficiency of mass-loss during the AGB phase will be reduced in the LMC. As a result, a higher central star mass should be the outcome of the evolution for a given initial progenitor.

If we believe the higher average central star mass of PNe we find in the LMC, we might have the first observational evidence from PNe progenitors for reduced mass-loss rates in a lower metallicity environment. The consequences are very important in terms of Galactic chemical enrichment, as a higher fraction of main sequence stars should reach the Chandrasekhar mass limit in the LMC than in the Galaxy.

7.2.2 The Central Star Mass–Nebular Morphology Relation

It has been suggested from chemical enrichment studies that the initial mass of the progenitor star determines the morphology of the PN. In particular, the N and O chemical enrichment found in the Galactic bipolar and extremely asymmetric morphological classes, together with their lower average distance from the Galactic plane, suggests that the bipolar class might evolve from more massive progenitors. From the initial–final mass relation we should expect that massive progenitors will end up as massive central stars. In our Galaxy, the correlation between the central star mass and the PN morphology has been explored by several authors who have found slightly different mass distributions for the central stars of symmetric and axisymmetric PNe.

We have revisited this issue using the samples of PNe observed in the LMC and we have not found any correlation between the mass of the central star and the morphology of the nebula [6, 7]. Although the number of objects with bipolar/quadrupolar morphology and central star masses measured is small, it still represents 24% of the sample.

7.3 Conclusions

The known distance to the Magellanic Clouds combined with the *HST*'s spatial resolution has allowed us to separate the central star from the nebula and obtain reliable physical parameters for the central stars of a sizable sample of PNe in the Large and Small Magellanic Clouds. We do not find any significant relation between the morphology and the central star mass in the LMC sample.

We derive an average mass $\langle M_{CS}, LMC \rangle = 0.65 \pm 0.07\ M_{\odot}$ for the total sample of central stars analyzed in the LMC (37 objects when we combine the 21 central star masses from [7] with the 16 obtained in [6]). This average mass suggests that the average mass of the central stars of PNe in the LMC is higher than that in the Galaxy. Although the uncertainties in the determination of Galactic central stars masses hampers our conclusions, if the initial mass distribution in the galaxies in the 1–5 $M_{\odot}$ range were the same, we would expect to find exactly this effect. That is, higher final masses in the LMC compared to the Galaxy, which is a consequence of the reduced mass-loss rate expected in a lower metallicity environment.

References

1. Villaver, E., García-Segura, G., Manchado, A., 2002. Astrophys. J. 571, 880.
2. Villaver, E., Manchado, A., García-Segura, G., 2002. Astrophys. J. 581, 1204.
3. Vassiliadis, E., Wood, P.R., 1994. Astrophys. J. Suppl. Ser. 92, 125.
4. Shaw, R.A., Stanghellini, L., Mutchler, M., Balick, B., Blades, J.C., 2001. Astrophys. J. 548, 727.
5. Stanghellini, L., Shaw, R.A., Mutchler, M., Palen, S., Balick, B., Blades, J.C., 2002. Astrophys. J. 575, 178.
6. Villaver, E., Stanghellini, L., Shaw, R.A., 2003. Astrophys. J. 597, 298.
7. Villaver, E., Stanghellini, L., Shaw, R.A., 2007. Astrophys. J. 656, 831.
8. Villaver, E., Stanghellini, L., Shaw, R.A., 2004. Astrophys. J. 614, 716.
9. Stanghellini, L., Villaver, E., Manchado, A., Guerrero, M.A., 2002. Astrophys. J. 576, 285.
10. Liebert, J., Bergeron, P., Holberg, J.B., 2005. Astrophys. J. Suppl. Ser. 156, 47.

Chapter 8
A Look at Neutron Stars with HST: From Positions to Physics

Patrizia Caraveo

Abstract Easy to spot at radio wavelengths, neutron stars are harder to detect in X rays, and even harder at optical wavelengths. Notwithstanding the detection challenge, the study of neutron stars' optical emission is an important tool to secure their identification and to understand the emission mechanisms at work.

8.1 Introduction

Isolated Neutron Stars (INSs) are challenging targets for optical telescopes. Their ability to accelerate particles makes them radio emitters as well as bright sources of X and gamma-ray photons, but, apart for the two youngest objects, is of no use to produce optical radiation. What matters at optical wavelengths is their thermal emission, but the coupling between high surface temperatures with the stars' tiny dimension conspire to make them faint optical emitters. Thus, the sample of Isolated Neutron Stars (INSs) detected by HST contain young (and powerful) objects as well as unconspicuous ones that happen to be within few hundred parsecs from us.

Chapter 9 by Mignani in this book summarizes the INSs results obtained by HST over its lifetime spanning from multicolour photometry to timing to spectral analysis. However, in view of the targets faintness, multicolour photometry has been the most widely used approach yielding images with all the instruments available on board over the years. Figure 8.1 provides a compilation of all the images collected by HST on classical rotation powered neutron stars. All, but Geminga [1], belong to the pulsar family. Apart from providing the source magnitude, and eventually information on the variability of Pulsar Wind Nubulae surrounding Crab [7] and PSR0540-69, such images have yielded accurate source positioning making it possible to measure NSs' proper motions and (when at all possible) parallactic distances.

8.2 The Power of Astrometry on Pulsars' Astronomy

Since neutron stars are fast moving objects, HST angular resolution has been instrumental to nail down the displacements of their optical counterparts. Measuring

P. Caraveo (✉)
INAF–IASF, Via Bassini 15, 20133 Milan, Italy
e-mail: pat@iasf-milano.inaf.it

F.D. Macchetto (ed.), *The Impact of HST on European Astronomy*, Astrophysics and Space Science Proceedings,
DOI 10.1007/978-90-481-3400-7_8, © Springer Science+Business Media B.V. 2010

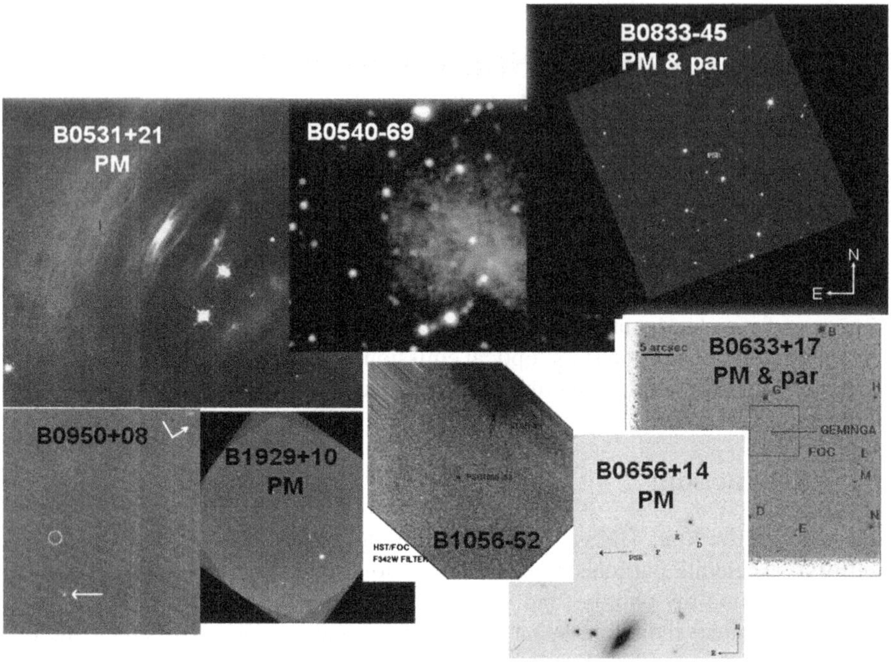

Fig. 8.1 (Color online) Compilation of the images collected by HST on classical rotation powered neutron stars. Object have been arranged *clockwise* as a function of their kinematic age, starting at *top left* with the Crab and ending at *bottom left* with the oldest one. Note that only the two youngest pulsars, namely Crab and PSR0540-69 do have a nebulosity. All the other objects are just point sources. *PM* means that proper motion is available while *Par* refers to parallax. References *clockwise* [3–6, 15–17, 19]

proper motions provide an effective way to confirm a tentative INS identification, based only on positional coincidence, also when the faintness of the counterpart makes it difficult to measure the source pulsation. Indeed, a number of optical identification have been confirmed through proper motion detection. It has been the case for Geminga [3], PSR 0656+14 [16], and PSR 1929+10 [17].

Apart from straight identification, proper motion has also been exploited in conjunction with high resolution radio and X-ray images to study the alignment between jet-like features, characterizing the pulsar wind nebulae surrounding the youngest pulsars, and the source direction of motion. Tantalizing examples have been found for the Crab [2], with the proper motion direction (later slightly revised using data collected over a much longer period by [18]), and Vela [5, 8] pulsars. Such alignments point towards the existence of a physical link between the pulsar rotation axis (somehow responsible for the jets) and the proper motion direction, due to a directional kick during the supernova explosion or immediately after it.

Moreover, in few cases, it has been possible to measure the source parallactic displacements allowing a direct assessment of its distance. Although this has been possible only for the nearest objects, it has allowed to measure the distance of neutron stars beyond the capabilities of radio observations either owing to their radio silence (as in the case of Geminga [3]) or to their position not easily covered by radio VLBI measurements (as in the case of Vela [5]). A later effort to measure the distance to the Vela pulsar using radio techniques [9] yielded a value of 287^{+19}_{-17} pc fully consistent with the previous HST one of 294^{+76}_{-50} pc. Since the distance to Geminga can only be gauged at optical wavelengths, the parallax measurement was recently repeated using the ACS instrument. Its larger field of view contains many more field stars, a very important commodity when frames have to be very accurately aligned. However, in spite of the undeniable improvements of the new sets of images, the resulting distance of 250^{+120}_{-62} pc [10] is affected by an error much bigger than that attached to the previous value (159^{+60}_{-34} pc) obtained with the WFPC2 by [3]. While the two values are fully compatible, it remains to be understood why a more sensitive instrument, providing many more reference stars, could not provide a more accurate result.

8.3 ... and on Radio Quiet INSs Astronomy

As for the case of Geminga, the role played by the study of the optical counterparts can be fully appreciated when dealing with neutron stars with no radio emission. Seven such objects have been discovered in the ROSAT data and the study of their optical counterparts has been instrumental to understand their nature and to try to extract some clue on their physics. Originally thought to be nonpulsating, the magnificent 7 have now been shown to be slow rotator with periods of several seconds. At variance with the classical pulsars, which are characterized by composite spectra containing both thermal and nonthermal components, they exhibit pure thermal spectra. Such a peculiarity prompted the idea to use them to constrain the different families of neutron star equation of state EOS. Since the luminosity of a purely thermal emitter depends just on its radius, the precise knowledge of the star distance would make it possible to use the X-ray flux (and temperature) to compute the neutron star radius, thus gaining important leverage on its EOS [21]. A fascinating idea which resulted in a very focused effort to search, and later study, the INSs optical counterparts. Figure 8.2 summarizes the HST contribution to this subject. At first, attention went to RXJ 1856.5-37545, the brightest of the X-ray sources. A fast moving counterpart was found at a parallactic distance of 61 pc [20], making it the nearest neutron star, and pointing to a radius smaller than 8 km, an uncomfortably small value. The parallax distance was soon revised to 140 pc [11] or 117 ± 12 pc [22], bringing the NS radius value to about 12 km, a value fully consistent with a number of equations of state, rendering vane all hopes to constraining the families of neutron stars' EOS. The same happens for RXJ0720.5-3754 which has been found to be at 360^{+170}_{-90} pc [14], a value too uncertain to be of any use to exclude EOS families.

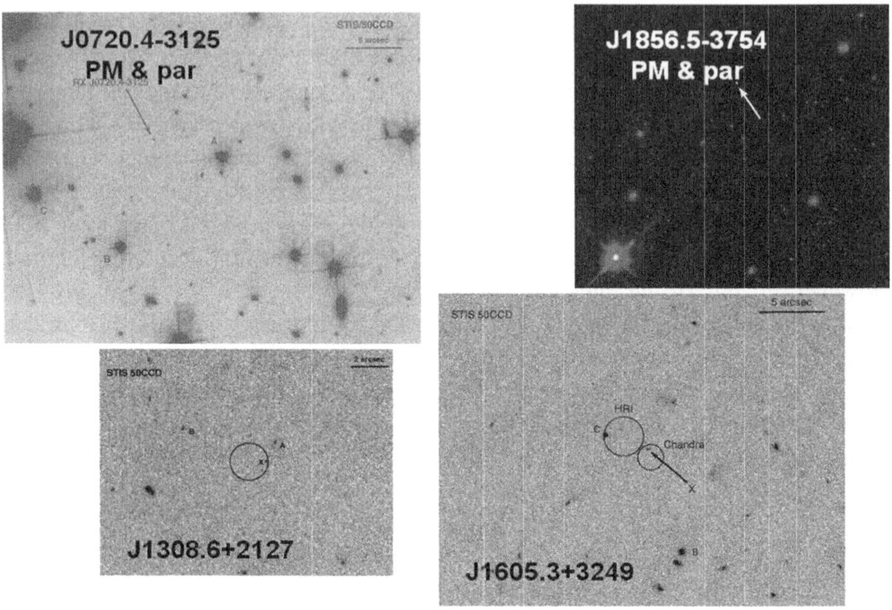

Fig. 8.2 (Color online) HST images used to spot counterparts of radio silent INSs. For both RXJ 1856.5-3754 and J0720.4-3125 proper motion as well as parallax have been measured [11, 14, 22]. Possible counterparts have been proposed for RXJ1308+2127 [12] and RXJ 1605.3+3249 [13]

8.4 Conclusions

In spite of the important results obtained for classical pulsars, HST has proved to be most effective to unravel the cases of radio quiet isolated neutron stars, be they Geminga-like or magnificent 7-like. With gamma-ray telescopes finally in orbit (as in the case of the AGILE Italian mission) or ready to be launched (as in the case of NASA GLAST mission) the number of Geminga-like neutron stars is expected to grow dramatically, providing many more challenging sources to be investigated by the instruments on board HST.

References

1. Bignami, G.F., Caraveo, P.A., 1996. Annu. Rev. Astron. Astrophys. 34, 331.
2. Caraveo, P.A., Mignani, R., 2001. Astron. Astrophys. 344, 367.
3. Caraveo, P.A., et al., 1996. Astrophys. J. 461, L91.
4. Caraveo, P.A., et al., 2000. arXiv:astro-ph/0009035v1.
5. Caraveo, P.A., et al., 2001. Astrophys. J. 561, 930.
6. Hester, J.J., et al., 1995. Astrophys. J. 448, 240.
7. Hester, J.J., et al., 2002. Astrophys. J. 577, L49.
8. De Luca, A.J., et al., 2000. Astron. Astrophys. 354, 1011.
9. Dodson, R., et al., 2003. Astrophys. J. 596, 1137.

10. Faherty, J., Walter, F., 2007. Astrophys. Space Sci. 308, 225.
11. Kaplan, D.L., et al., 2002. Astrophys. J. 571, 477.
12. Kaplan, D.L., et al., 2002. Astrophys. J. 579, L29.
13. Kaplan, D.L., et al., 2003. Astrophys. J. 588, L33.
14. Kaplan, D.L., et al., 2007. Astrophys. J. 660, 1428.
15. Mignani, R.P., et al., 1997. Astrophys. J. 474, L51.
16. Mignani, R.P., et al., 2000. Astrophys. J. 543, 318.
17. Mignani, R.P., et al., 2002. Astrophys. J. 580, L147.
18. Ng, C.Y., Romani, R.W., 2006. Astrophys. J. 644, 445.
19. Pavlov, G.G., et al., 1996. Astrophys. J. 467, 370.
20. Walter, F., 2001. Astrophys. J. 549, 433.
21. Walter, F., Mattheus, L.D., 1997. Nature 389, 358.
22. Walter, F., et al., 2002. Astrophys. J. 576, L145.

Chapter 9
The HST Contribution to Neutron Star Astronomy

Roberto P. Mignani

Abstract While isolated neutron stars (INSs) are among the brightest γ-ray sources, they are among the faintest ones in the optical, and their study is a challenging task which require the most powerful telescopes. *HST* has lead neutron star optical astronomy yielding nearly all the identifications achieved since the early 1990s. Here, the major *HST* contributions in the optical studies of INSs and their relevance for neutron stars' astronomy are reviewed.

9.1 Introduction

Before the launch of *HST*, optical studies of INSs were the exception. In the first 20 years since the pulsars discovery, only the Crab and Vela pulsars were identified [5, 13], while optical pulsations were detected from an unidentified source at the center of SNR B0540-69 in the LMC [14], and only a candidate counterpart was found for the mysterious γ-ray source Geminga [1]. This score was expected to be considerably improved by *HST*, thanks to its much larger sensitivity with respect to ground based telescopes, and to the sharp spatial resolution of the *WFPC* as well as to the near-UV view of the ESA's *FOC*. Unfortunately, the spherical aberration of the *HST* optics affected the execution of most approved proposals, except for those aimed at the brightest targets. So, in the early 1990s the leadership in the INSs optical astronomy was still in the hands of ground-based observatories, mainly in those of the ESO *NTT* which secured the identification of Geminga through the proper motion of its counterpart, a technique soon become the standard one, and the likely identifications of the optical pulsar in SNR B0540-69 and of PSR B0656+14 [17]. However, the refurbishment of *HST* in SM-1 (Dec. 2003) brought its performance back to the original expectations and gave it a leading role in INSs' optical astronomy, maintained even after the advent of the 10-m class telescopes. Since *HST* has provided eight new INSs identifications, against the two of the *VLT* and the *KECK* [18], boosting the identification rate by a factor 4. This could have been higher if not for the *FOC* removal in SM-3B (March 2002) and for the *STIS* failure (Aug. 2004), which alone have yielded nearly all the *HST* INSs identifications, depriving the telescope of its near-UV view. Thus, *HST* observations have opened wide a new,

R.P. Mignani (✉)
University College London, Mullard Space Science Laboratory, London, UK
e-mail: rm2@mssl.ucl.ac.uk

F.D. Macchetto (ed.), *The Impact of HST on European Astronomy*, Astrophysics and Space Science Proceedings,
DOI 10.1007/978-90-481-3400-7_9, © Springer Science+Business Media B.V. 2010

important observing window on INSs and triggered the interest of a larger and larger fraction of the neutron star community.

9.2 *HST* Observations of Neutron Stars

Given their faintness ($m > 25$), most INSs observations have been performed in image mode with the *FOC*, the *WFPC2* and the *STIS*, while *NICMOS* has been rarely used, also for being idle in 1999–2002, and *ACS* has not been used much before its Jan. 2007 failure. Due to their higher near-UV QE, the *FOC* and the *STIS* have been the best instruments for pathfinding identification programs, while both the *WFPC2* and *ACS*, with their wider field of view, have been preferred for astrometry (see Chap. 8 by Caraveo in this book). *FOS* spectroscopy has been performed only for the brightest objects, while medium-resolution *STIS*-MAMA spectra have been obtained in most cases. Timing observations have been hampered by the *HSP* removal in SM-1, leaving *HST* without timing facilities until the *STIS* installation in SM-2 (Feb. 1997), unfortunately idle since Aug. 2004. Strangely enough, *HST* polarimetry observations have been rarely performed.

Most observations have been aimed at classical rotation-powered pulsars. Multiband photometry allowed to study their Spectral Energy Distribution (SED) in detail. For the young Vela pulsar [15] and PSR B0540-69 [25] *WFPC2* observations unveiled a power-law continuum ($F_\nu \propto \nu^{-\alpha}$) confirming, like for the Crab, the magnetospheric origin of the optical emission. In the near-UV, *STIS* observations provided the first spectrum of the Crab [7] since the *IUE* one and of the Vela pulsar [24]. Observations of the middle-aged Geminga and PSR B0656+14 with the *FOC* [2, 16, 20] and the *STIS* [10, 11, 26] allowed to identify for the first time a Rayleigh–Jeans spectral component ($T \sim 10^5$ K), likely originated by the cooling neutron star surface. The detection of near-UV thermal emission is critical for neutron star physics. Coupled with the distance, the joint fit to the near-UV–to–X-rays thermal spectrum yields the neutron star thermal map, which is crucial to study the neutron star's conductivity, hence its chemical composition and physical conditions as well as the magnetic field topology, and to constrain the neutron star equation of state. This also yields a better measure of the surface temperature which, coupled with the neutron star spin-down age, allows to test cooling models. Last, studying the near-UV thermal emission is crucial to explore temperatures too low for the X-rays and to constrain neutron star cooling curves above 1 Myr, where different slopes are predicted, or to pinpoint evidence for re-heating of the neutron star core as, e.g. for PSR J0437-4715 whose derived temperature ($\approx 10^5$ K) largely exceeds any expectation for a 5 Gyr old INS [11]. In the near-IR, *NICMOS* has observed PSR B0656+14 and Geminga [12], the first INSs detected at these wavelengths after the Crab. In both cases the near-IR SED is consistent with a power-law ($\alpha > 0$), suggesting that the emission at longer wavelength is magnetospheric. Interestingly, for PSR B0656+14 the power-law steepness might also suggest that the near-IR emission is due to a debris disk made of fallback material from the supernova explosion (e.g. [23]). Unfortunately, *Spitzer* observations could not resolve the pulsar emission from that of the crowdy background and thus constrain its mid-IR spectrum. So far, polarimetry observations have been performed only for the Crab pulsar. However, phase-resolved *HSP* polarimetry [6] has shown, for the first time, that the pulsar polarization properties are wavelength independent. Phase-averaged optical polarization observations of the Crab have been also performed with the *WFPC2* and *ACS* and are now in progress for PSR B0540-69 and the Vela pulsar. As shown by [19], polarization observations are a powerful diagnostic to test neutron star magnetosphere models, to constrain the pulsar rotation and magnetic axis, and to investigate pulsar/ISM magneto-dynamical interactions. Timing observations have been performed for nearly all the brightest INSs. *HSP* discovered the wavelength dependence of the Crab pulsar light curve profile [22] and provided a very precise braking index measurement for PSR B0540-69 [3], the only one obtained in the optical. The *STIS*-MAMAs detected for the first time near-UV pulsations from Geminga [11], PSR B0656+14 [26] and the Vela pulsar [24], showing the lightcurve dependence on the underlying spectrum.

HST observations have been fundamental to understand the nature of radio-silent INSs whose high-energy emission is not rotation-powered. This is the case, e.g. of the INSs with purely thermal X-ray emission (*XTINS*), discovered by *ROSAT* (e.g. [8]) and originally thought to be extinted radio pulsars, re-heated by ISM accretion. However, astrometry of the *XTINS* optical counterparts discovered by *HST* yielded space velocities too high for ISM accretion, thus favouring younger ages and a natural cooling scenario. *HST* observations have also allowed to characterize their thermal optical SED and to build, thanks to the measured parallactic distances, the surface thermal map, with a cooler and larger region and a warmer and smaller one, emitting the optical and X-ray radiation respectively. Together with the detection of X-ray pulsations, this helped to explain the claimed, surprising inconsistencies with the neutron star model cooling curves which apparently predict too low temperatures for the measured *XTINS* ages.

HST observations have also allowed to resolve for the first time the structure of the synchrotron nebulae powered by the neutron star relativistic winds around the Crab pulsar [9] and PSR B0540-69 [4], with morphologies similar to the X-ray ones. For the former, a continuous monitoring with the *WFPC2* has also shown evidence for an expanding equatorial wind, later confirmed in X-rays by *Chandra*. Very recently, with the *WFPC2* we have found evidence of variability also in the B0540-69 nebula, which could be attributed to an expanding jet from the pulsar. So far, evidence for an expanding pulsar jet was found for the Vela pulsar only, through *Chandra* X-ray observations [21].

9.3 Future Perspectives

Currently, only the *WFPC2* and *NICMOS* are available as imaging instruments, both older than ten years. *WFC3* will be installed in SM-4, providing a complete near-UV–to–near-IR spectral coverage over an equally large field of view. However, while the *WFC3* QE is higher in the near-IR with respect to *NICMOS*, in the near-UV and in the optical is lower with respect to the *STIS*-MAMAs and *ACS*. This encourages repairing both *STIS* and *ACS* in SM-4. Furthermore, *STIS* is the only instrument suited for INSs near-UV timing, while *ACS* allows to carry out polarimetry observations, not possible with the *WFC3*. Thus, an upgraded and fully-refurbished *HST* is critical to maintain its established world-leading role in neutron star astronomy.

Acknowledgements The author thanks C. Nicollier and the SM-1 crew, whose work changed his life.

References

1. Bignami, G.F., et al., 1987. Astrophys. J. 319, 358.
2. Bignami, G.F., et al., 1996. Astrophys. J. 456, L111.
3. Boyd, P.T., et al., 1995. Astrophys. J. 448, 365.
4. Caraveo, P.A., et al., 2001. In: Proc. of "A Decade of HST Science", STScI, vol. 105, p. 9.
5. Cocke, W.J., Disney, M.J., Taylor, D.J., 1969. Nature 221, 525.
6. Graham-Smith, F., et al., 1996. Mon. Not. R. Astron. Soc. 282, 1354.
7. Gull, T.R., et al., 1998. Astrophys. J. 495, 51.
8. Haberl, F., 2007, Astrophys. Space Sci. 308, 181.
9. Hester, J.J., et al., 1995. Astrophys. J. 448, 240.
10. Kargaltsev, O., Pavlov, G.G., 2007. Astrophys. Space Sci. 308, 287.
11. Kargaltsev, O.Y., et al., 2005. Astrophys. J. 625, 307.
12. Koptsevich, A.B., et al., 2001. Astron. Astrophys. 370, 1004.
13. Lasker, B., 1976. Astrophys. J. 203, 193.
14. Middleditch, J., Pennypacker, C., 1985. Nature 313, 659.
15. Mignani, R.P., Caraveo, P.A., 2001. Astron. Astrophys. 376, 213.
16. Mignani, R.P., Caraveo, P.A., Bignami, G.F., 1998. Astron. Astrophys. 332, L37.
17. Mignani, R.P., Caraveo, P.A., Bignami, G.F., 2000. Messenger 99, 22.
18. Mignani, R.P., De Luca, A., Caraveo, P.A., 2004. In: Proc. IAU Symp., vol. 218, p. 391.
19. Mignani, R.P., et al., 2007. Astron. Astrophys. 467, 1157.
20. Pavlov, G.G., Welty, A.D., Cordova, F.A., 1997. Astrophys. J. 489, 75.
21. Pavlov, G.G., et al., 2001. Astrophys. J. 554, L189.
22. Percival, J.W., et al., 1993. Astrophys. J. 407, 276.
23. Perna, R., Hernquist, L., Narayan, R., 2000. Astrophys. J. 541, 344.
24. Romani, R.W., Kargaltsev, O., Pavlov, G.G., 2005. Astrophys. J. 627, 383.
25. Serafimovich, N.I., et al., 2004. Astron. Astrophys. 425, 1041.
26. Shibanov, Yu.A., et al., 2005. Astron. Astrophys. 440, 693.

Chapter 10
Exotic Populations in Galactic Globular Clusters

Francesco R. Ferraro

Abstract Hubble Space Telescope high-resolution observations of the central region of Galactic globular clusters have shown the presence of a large variety of exotic stellar objects whose formation and evolution may be strongly affected by dynamical interactions. In this paper I briefly describe the main properties of two classes of exotic objects: the so-called Blue Stragglers and the optical companions to Millisecond pulsar. Both these classes of objects are invaluable tools to investigate the evolution of binary systems in very dense environments and are powerful tracers of the dynamical history of the parent cluster.

10.1 Introduction

Galactic Globular Clusters (GGCs) are very efficient "kilns" for forming exotic objects, such as low-mass X-ray binaries, cataclysmic variables, millisecond pulsars (MSP), blue straggler stars (BSS), etc. Most of these stars cannot be interpreted within the standard contest of single mass stellar evolution, and are thought to be produced by the evolution of primordial binaries, and/or by the effect of some dynamical processes. The advent of the Hubble Space Telescope (HST) thanks to its spatial distribution and its superior imaging capabilities in the UV has given a crucial contribution to a deeper understanding of binary evolution, cluster dynamics, and the complex interplay between stellar and dynamical evolution in dense stellar systems.

In this paper I briefly comment the main properties of two anomalous sequences observed in the color–magnitude diagram (CMD) of GGCs which highly benefited by HST observations: (i) the so-called BSS sequence; (ii) the sequence defined by MSP companions.

10.1.1 Blue Stragglers Stars

Commonly defined as those stars brighter and bluer (hotter) than the Main Sequence (MS) turnoff stars, BSS lie along an extrapolation of the MS, thus mimicking a rejuvenated stellar population. First discovered by [14] in M3, their nature has been

F.R. Ferraro (✉)
Dipartimento di Astronomia, Via Ranzani 1, 40127 Bologna, Italy
e-mail: francesco.ferraro3@unibo.it

F.D. Macchetto (ed.), *The Impact of HST on European Astronomy*,
Astrophysics and Space Science Proceedings,
DOI 10.1007/978-90-481-3400-7_10, © Springer Science+Business Media B.V. 2010

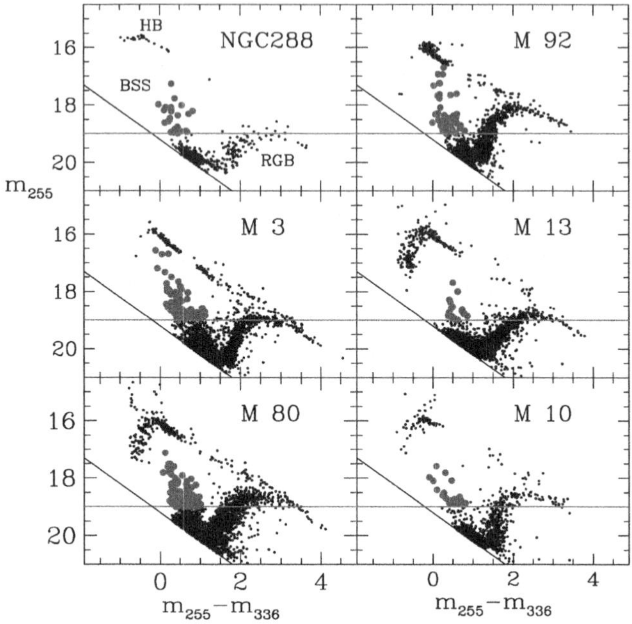

Fig. 10.1 (Color online) $(m_{255}, m_{255} - U)$ CMDs for NGC288, M3, M80, M92, M13, and M10. Horizontal and vertical shifts have been applied to the other clusters to match M3. BSS are plotted as *big filled circles*. Some of our results are puzzling: the largest BSS specific frequencies are found in our most and least dense clusters (M80 and NGC288, respectively); almost identical clusters (M3 and M13) have radically different BSS populations. (From [8])

a puzzle for many years, and their formation mechanism is still not completely understood. BSS are more massive than the normal MS stars [15], thus indicating that some process which increases the initial mass of single stars must be at work. Such processes could be related either to mass transfer between binary companions, the coalescence of a binary system, or the merger of two single or binary stars driven by stellar collisions.

According to [11] (see also [5]), BSS in different environments could have different origins. In particular, BSS in loose GGCs might be produced by mass-transfer/coalescence of primordial binaries (hereafter MT-BSS), whereas in high density GGCs (depending on survival-destruction rates for primordial binaries) BSS might arise mostly from stellar collisions (COL-BSS), particularly those that involve binaries. As shown by [8] (see also Fig. 10.1) the two formation channels can have comparable efficiency in producing BSS in their respective typical environment (see the case of M80 [6] and NGC288 [2]). Moreover, these formation mechanisms could also act simultaneously within the same cluster, with efficiencies that depend on the radial regions, corresponding to widely different stellar densities. This is suggested by the bimodality of the BSS radial distribution observed in a few clusters (M3, 47 Tuc, NGC6752, and M5), where the BSS specific frequency has been found to

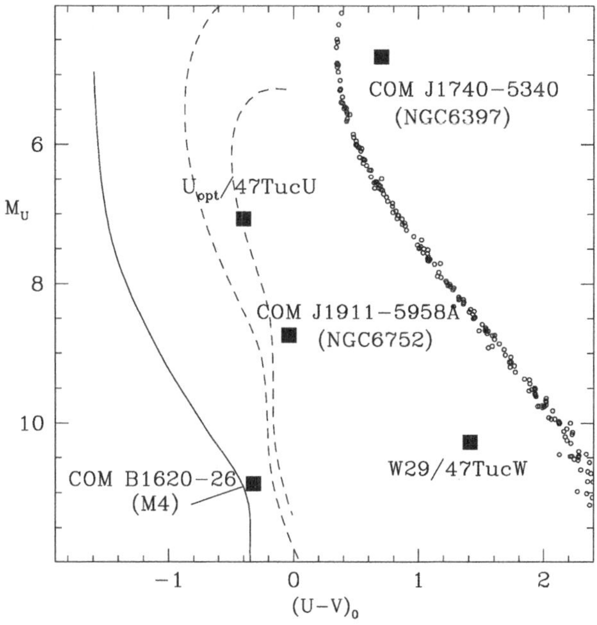

Fig. 10.2 All the optical counterparts to MSP companion detected so far in GGCs are plotted (as *large filled squares*) in the $(M_U, U-V)$ plane. The cooling sequences for He-WD (from [16]) and for the CO-WD are also replotted. MS stars of NGC6752 are plotted for reference. (From [9])

be highly peaked in the cluster center, rapidly decreasing at intermediate radii and rising again outward (see [13]).

Despite all the most recent observational efforts, one can hardly say that the BSS problem is solved; indeed, in some ways it is more puzzling than ever. Extending the UV searches for BSS to a larger sample of GGCs, characterized by different structural and dynamical properties and extensive survey of BSS surface abundance patterns (see the results by [10] in 47 Tuc) is therefore crucial to finally unveil the nature and the formation mechanisms of these peculiar objects.

10.1.2 Binary Millisecond Pulsar in GGCs

MSPs are formed in binary systems containing a neutron star (NS), which is eventually spun up through mass accretion from the evolving companion (e.g. [1]). Even though the disk of the Galaxy has a total mass 100 times larger than the GGC system, more than 50% of the entire MSP population has been found in the latter. This is not surprising because in the galactic field the only viable formation channel for MSPs is the evolution of primordial binaries. At variance, in the ultra-dense stellar environment of a GGC core, dynamical interactions can promote the formation of various kinds of binaries suitable for recycling the neutron stars into MSPs (e.g. [4]). The search for the optical counterparts to MSP companions in GGCs has just begun: only five optical companions to MSPs have been identified up to now in GGCs (see Fig. 10.2 and [9]). Among them, at least three objects seem to be helium

White Dwarf (WD), roughly located on the same-mass cooling sequence. If further supported by additional cases, this evidence could confirm that in GGCs, like in the field, the favored by-product of the recycling process of MSPs is a low mass ($M = 0.15$–$0.2\ M_{\odot}$) helium WD orbiting a MSP. However, a noticeable exception has been discovered in the center of NGC 6397: it is a MSP binary system containing a tidally deformed star with mass $M = 0.1$–$0.4\ M_{\odot}$ [7]. This is the first case of a MSP in a binary system with such a companion, and no similar objects have been detected in the galactic field to date. It could represent a newly born MSP, or the result of an exchange interaction. The latter possibility is in agreement with the predictions of [12], who suggested that most of the current companions to eclipsing pulsars in GGCs would be the swollen descendants of MS turnoff stars, which have replaced the original WD companion of the pulsar in an exchange interaction in the cluster core. In particular, [3] suggested that the anomalous position of the companion star in the CMD is consistent with an evolved Sub Giant Branch star orbiting the NS and loosing mass.

References

1. Bhattacharya, D., van den Heuvel, E.P.J., 1991. Phys. Rep. 203, 1.
2. Bellazzini, M., et al., 2002. Astron. J. 123, 1509.
3. Burderi, L., D'Antona, F., Burgay, M., 2002. Astrophys. J. 574, 325.
4. Davies, M.B., Hansen, B.M.S., 1998. Mon. Not. R. Astron. Soc. 301, 15.
5. Davies, M.B., Piotto, G., De Angeli, F., 2004. Mon. Not. R. Astron. Soc. 349, 129.
6. Ferraro, F.R., Paltrinieri, B., Rood, R.T., Dorman, B., 1999. Astrophys. J. 522, 983.
7. Ferraro, F.R., Possenti, A., D'Amico, N., Sabbi, E., 2001. Astrophys. J. 561, L93.
8. Ferraro, F.R., et al., 2003. Astrophys. J. 588, 464.
9. Ferraro, F.R., Possenti, A., Sabbi, E., D'Amico, N., 2003. Astrophys. J. 596, L211.
10. Ferraro, F.R., et al., 2006. Astrophys. J. 647, L53.
11. Fusi Pecci, F., 1992. Astron. J. 104, 1831.
12. King, A.R., Davies, M.B., Beer, M.E., 2003. Mon. Not. R. Astron. Soc. 345, 678.
13. Lanzoni, B., et al., 2007. Astrophys. J. 663, 267.
14. Sandage, A.R., 1953. Astron. J. 58, 61.
15. Shara, M.M., Saffer, R.A., Livio, M., 1997. Astrophys. J. 489, L59.
16. Serenelli, A.M., 2002. Mon. Not. R. Astron. Soc. 337, 1091.

Chapter 11
The Stellar Mass Function in Globular Clusters

Guido De Marchi

Abstract Globular clusters in our own Milky Way formed at redshift $z \simeq 5$ or more, when the physical conditions of the environment, such as pressure, density, temperature and chemical composition, were very different from those found in current star forming regions. The end product of this massive star formation is a stellar initial mass function (IMF) that holds the secret to the making of stars in the primeval universe. Over time, under the effects of stellar evolution and dynamical interactions, the stellar IMF of globular clusters has evolved to become what we can now accurately measure with the HST down to very small masses near the Hydrogen burning limit. But how does this present mass function compare with the original IMF and what can we learn from it about star formation at high redshift? I report here on the discovery of a surprising correlation between the shape of the current mass function of globular clusters and their central concentration, which suggests that our understanding of their dynamical evolution might not yet be complete.

The advancements made possible by the HST for our understanding of the structure and evolution of globular clusters (GCs) are dramatic, as discussed by Alvio Renzini in this book (Chap. 16). To realise why this was possible, one just needs to take a brief look at Fig. 11.1, showing a region near the half-light radius of NGC 6397 as observed by us with the HST shortly after the first servicing mission [19]. Although NGC 6397 is a very dense cluster, background galaxies can be easily seen! With data of this type it became finally possible to study the luminosity and mass function (MF) of stars in GCs down to very small masses, close to the H-burning limit, without the troubles that had vexed previous attempts from the ground, particularly as regards the very uncertain correction for photometric completeness (see e.g. [20]).

After NGC 6397, many more GCs were studied in detail with the HST, by us and others, and there are today of order 20 objects with a well characterised stellar MF down to very small masses ($<0.3\ M_{\odot}$). More should soon become available when the complete results of the ACS GCs survey are released [22]. Yet the fundamental result of our original work [19], namely that the stellar MF does not continue to rise at low masses, but rather flattens out (and probably drops off) well before reaching the H-burning limit, has been consistently upheld by later works (see e.g. [15, 21]),

G. De Marchi (✉)
ESA, Space Science Department, Keplerlaan 1, 2200 AG Noordwijk, Netherlands
e-mail: gdemarchi@rssd.esa.int

F.D. Macchetto (ed.), *The Impact of HST on European Astronomy*, Astrophysics and Space Science Proceedings,
DOI 10.1007/978-90-481-3400-7_11, © Springer Science+Business Media B.V. 2010

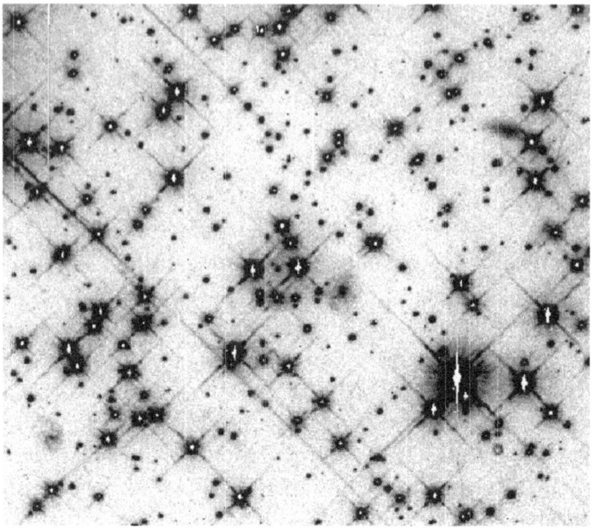

Fig. 11.1 First WFPC2 image of the cluster NGC 6397 after the first HST servicing mission, in the *I* band. The region, located near the cluster's half-light radius, spans 50 arcsec on a side. North is inclined 45° left of the vertical. Although the cluster is relatively dense, three background galaxies are clearly visible along the diagonal

in spite of the initial skepticism. The picture that has emerged over the past 15 years is one in which very low-mass stars ($\sim$0.1 $M_\odot$ or less) contribute insignificantly to the total mass budget of GCs, as the typical global mass function (GMF) of cluster stars is well described by a low-normal [2, 17] or tapered power-law [7] distribution, with a characteristic mass near $\sim$0.35 $M_\odot$.

Such a GMF is remarkably different from that found in young Galactic clusters and star forming regions, which suggest a much smaller value of the characteristic mass, close to $\sim$0.15 $M_\odot$ [2]. If the present GMF of GCs is in any way representative of their IMF, this would imply that the physical conditions of the environment, which were undoubtedly different at redshift $z \simeq 5$ when GCs formed, must play a major role in the star formation process, possibly suggesting that the IMF is not universal. Thus, studying the GMF of GCs becomes a very powerful tool to probe the properties of star formation in the primeval universe.

Unfortunately, GCs are not really fossil records of star formation at high redshift, but rather living records that have evolved, however slowly, over a Hubble time. Stars within a cluster interact gravitationally with each other, thereby altering over time the structure of the cluster and the distribution of matter in it. Furthermore, the interaction with the tidal field of the Galaxy will lead to mass loss via tidal stripping in a way that depends on the cluster's orbit. The ultimate goal of our research, then, is to understand and model all these effects, so that we can extract information on the IMF and the properties of the star formation process from the measured stellar GMF.

The dynamical evolution of GCs is governed by the two-body relaxation process, whereby stars exchange energy via repeated distant encounters (see [12, 23] for a review). Two-body encounters lead to the expansion of the outer regions of the cluster, while driving the stellar density in the central regions to increase dramatically towards an infinite value during the so-called core-collapse. This should result in a central cusp in the surface brightness profile [10] and an increase in the central concentration parameter $c = \log(r_t/r_c)$, i.e. the logarithmic ratio of tidal and core radii. Therefore, c has traditionally been seen as a gauge of the dynamical state of a cluster, with values of c in excess of ~ 2 indicating a post-core-collapse phase [11, 24].

Equipartition of energy through two-body relaxation also leads to mass segregation. This implies that the local stellar MF will change with time and place. Even if the IMF was the same everywhere when the cluster formed, after a few relaxation times there will be proportionately more low-mass stars in the cluster periphery and proportionately less in the core. The first tentative evidence of mass segregation in 47 Tuc [3] has been fully confirmed by early HST observations in this and other clusters (see e.g. [4, 18]) and is now acknowledged in all observed GCs.

Finally, since two-body relaxation drives the velocity distribution towards a Maxwellian, an increasing number of stars will acquire enough energy to exceed the escape velocity and evaporate from the cluster. This process, greatly enhanced by the presence of the tidal field of the Galaxy, is the leading cause of mass loss for most GCs (see e.g. [13]) and results is a selective depletion of low-mass stars, since these have typically higher velocities. This effect, integrated over the orbit and time, implies a progressive departure of the GMF from the stellar IMF [25], namely a flattening at low masses that is now well established observationally [1, 5, 6, 8, 16].

If this scenario is correct, as GCs evolve dynamically the increase in their central concentration should correspond to the flattening of their GMF, since both effects result from the same two-body relaxation process. In other words, one would expect that clusters with dense cores should have lost more low-mass stars than loose clusters. Although perfectly reasonable, this prediction had so far never been verified observationally.

To test this hypothesis, we have built a sample of 20 GCs for which reliable estimates exist of both c and the shape of the GMF (see [9] for details). The former comes from accurate surface photometry [14], while the latter has been determined by us using high-quality HST and VLT photometry. In this work we have limited our analysis of the GMF to the mass range 0.3–0.8 $M_\odot$, in which a power-law distribution of the type $dN/dm \propto m^\alpha$ appears to adequately reproduce the observations. Below $\sim 0.3\ M_\odot$ the number of clusters with reliable photometry at those masses is still limited.

In Fig. 11.2 we show the run of the GMF index α as a function of c for the clusters in our sample, as indicated by the labels. It is immediately obvious that the data do not follow the expected correlation or trend between flattening GMF and increasing central concentration mentioned above, as no high-concentration cluster with a shallow GMF is present. The median value of c (1.4) splits the cluster population in two groups. The mean GMF index of clusters with $c < 1.4$ is $\alpha = -0.3 \pm 0.7$, while the others have a much tighter distribution with $\alpha = -1.4 \pm 0.3$ (see large thick crosses in Fig. 11.2). The relationship $\alpha + 2.5 = 2.3/c$, shown as a dashed line in Fig. 11.2, is a simple yet satisfactory eye-ball fit to the distribution.

Figure 11.2 suggests that a relatively low concentration is a necessary condition (although probably not also a sufficient one) for a depleted GMF. It appears,

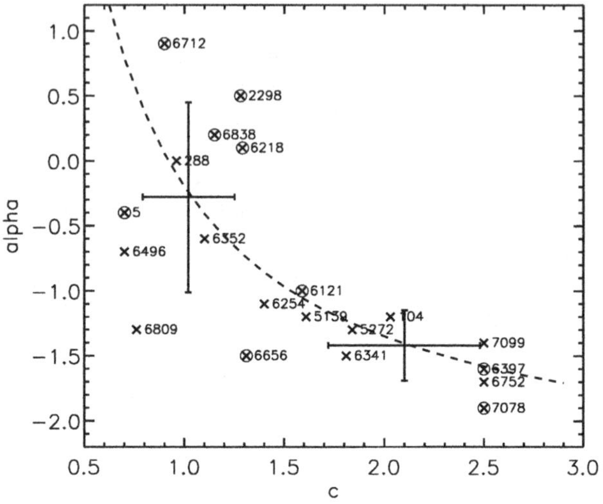

Fig. 11.2 Observed trend between the MF index α and the central concentration parameter c. Clusters are indicated by their NGC (or Pal) index number. Objects for which a GMF index is available are marked with a *circled cross*. For all others, the value of α is that of the MF measured near the half-light radius

therefore, that mass loss, even severe, via evaporation and tidal truncation has not triggered core-collapse for low-concentration clusters. This finding is unexpected and counter-intuitive. Presently, no satisfactory explanation exists for the observed behaviour, but several preliminary hypotheses can be put forth, ranging from IMF variations to the role of binaries in cluster cores to halt or delay their collapse (see [9] for more details).

The apparently simple dependence of α from c in Fig. 11.2 might also suggest that the observed distribution in practice represents an evolutionary sequence. In this scenario, the value of c at the time of cluster formation determines its evolution along two opposite directions of increasing and decreasing concentration. Clusters born with sufficiently high concentration ($c > 1.5$) evolve towards core-collapse. Mass loss can be important via stellar evolution in the first ~1 Gyr, and to a lesser extent via evaporation or tidal stripping throughout the life of the cluster, but the GMF at any time does not depart significantly from the IMF. Clusters with $c < 1.5$ at birth also evolve towards core-collapse, but mass loss via stellar evolution and, most importantly, via relaxation and tidal stripping proceeds faster, particularly if their orbit has a short perigalactic distance or frequent disc crossings. Therefore, as the tidal boundary shrinks and the cluster loses preferentially low-mass stars, the GMF progressively flattens. This speeds up energy equipartition, but c still decreases, since the tidal radius shrinks more quickly than the luminous core radius (although the central density, particularly that of heavy remnants, is increasing). These clusters could eventually undergo core-collapse, but this might only affect a few stars in the core and thus be very hard to detect.

In summary, while no conclusive explanation still exists for the unexpected observed trend between central concentration and shape of the GMF, Fig. 11.2 should serve as a warning that the surface brightness profile and the central concentration parameter of GCs are not reliable indicators of their dynamical state. A more solid assessment of a cluster's dynamical state will require the study of the complete radial variation of its stellar MF and of the properties of its stellar population, particularly in the core.

References

1. Andreuzzi, G., De Marchi, G., Ferraro, F., et al., 2001. Astron. Astrophys. 372, 851.
2. Chabrier, J., 2003. Publ. Astron. Soc. Pac. 115, 763.
3. Da Costa, G., 1982. Astron. J. 87, 990.
4. De Marchi, G., Paresce, F., 1995. Astron. Astrophys. 304, 202.
5. De Marchi, G., Pulone, L., 2007. Astron. Astrophys. 467, 107.
6. De Marchi, G., Leibundgut, B., Paresce, F., Pulone, L., 1999. Astron. Astrophys. 343, 9L.
7. De Marchi, G., Paresce, F., Portegies Zwart, S., 2005. In: Corbelli, E., Palla, F., Zinnecker, H. (Eds.), ASSL, vol. 327, p. 77. Springer, Dordrecht.
8. De Marchi, G., Pulone, L., Paresce, F., 2006. Astron. Astrophys. 449, 161.
9. De Marchi, G., Paresce, F., Pulone, L., 2007. Astrophys. J. 656, L65.
10. Djorgovski, S., King, R., 1986. Astrophys. J. 305, L61.
11. Djorgovski, S., Meylan, G., 1993. In: Djorgovski, S. Meylan, G. (Eds.), Structure and Dynamics of Globular Clusters, ASP Conf. Ser., vol. 50, p. 325. ASP, San Francisco.
12. Elson, R., Hut, P., Ingaki, S., 1987. Annu. Rev. Astron. Astrophys. 25, 565.
13. Gnedin, O., Ostriker, J., 1997. Astrophys. J. 474, 223.
14. Harris, W., 1996. Astron. J. 112, 1487.
15. King, I., Anderson, J., Cool, A., et al., 1998. Astrophys. J. 492, L37.
16. Koch, A., Grebel, E., Odenkirchen, M., et al., 2004. Astron. J. 128, 2274.
17. Paresce, F., De Marchi, G., 2000. Astrophys. J. 534, 870.
18. Paresce, F., De Marchi, G., Jedrzejewski, R., 1995. Astrophys. J. Lett. 442, L57.
19. Paresce, F., De Marchi, G., Romaniello, M., 1995. Astrophys. J. 440, 216.
20. Richer, H., Fahlman, G., Buonanno, R., et al., 1991. Astrophys. J. 381, 147.
21. Richer, H., et al., 2008. Astron. J. 135, 214.
22. Sarajedini, A., Bedin, L., Chaboyer, B, et al., 2002. Astron. J. 133, 1658.
23. Spitzer, L., 1987. Dynamical Evolution of Globular Clusters. PUP, Princeton.
24. Trager, S., Djorgovski, S., King, I., 1995. Astron. J. 109, 218.
25. Vesperini, E., Heggie, D., 1997. Mon. Not. R. Astron. Soc. 289, 898.

Chapter 12
Early Phases of Protoplanetary Disk Evolution

Inga Kamp

Abstract It is widely accepted that planetary systems form from protoplanetary disks, and observations of the dust reveal significant grain growth over timescales of a few million years. However, we know little about the gas processing in the first 10–20 Myr of disk evolution. This is the phase where protoplanetesimals form and accrete into planetary cores. One outstanding question is whether gas dispersal is coeval with the formation of planetesimals. If the gas dispersal would precede the dust evolution, the formation of gas giant planets through the standard core-accretion scenario would be impeded. HST's contribution to our understanding of disk dispersal processes will be summarized and the [O I] 6,300 Å line as tracer of disk structure in early phases of disk evolution discussed.

12.1 Introduction

Protoplanetary disks are generally thought to be the birthplaces of planetary systems. Before HST was launched, radio telescopes have revealed the existence of elongated rotating structures around young stars through dust continuum (e.g. [5, 12]) and molecular line observations (e.g. [8]).

The launch of the Hubble Space Telescope in 1990 opened a new window for research on protoplanetary disks. The high spatial resolution from space enabled the detailed imaging of the Orion nebula, a star forming region at a distance of 450 pc [9]. These images were taken with the Wide Field Camera in several optical narrow band filters such as Hα, [O III], [O I] and [S II]. This data shows not only that such protoplanetary disks are ubiquitous around newly formed stars (50% of stars show disk detection), but reveals the impact of disk irradiation and erosion by nearby hot O and B stars. The observed disks have typical sizes of 50 to 200 AU similar to what was previously observed from the ground. Reference [4] obtained even deeper narrow band images with the WFPC2 camera, showing dark silhouette disks and the ionized and irradiated disk surfaces. They also detected numerous outflows that accompany these early stages of star formation.

I. Kamp
Space Telescope Science Division of ESA, STScI, Baltimore, MD 21218, USA

I. Kamp (✉)
Kapteyn Astronomical Institute, PO Box 800, 9700 AV Groningen, The Netherlands
e-mail: kamp@astro.rug.nl

F.D. Macchetto (ed.), *The Impact of HST on European Astronomy*,
Astrophysics and Space Science Proceedings,
DOI 10.1007/978-90-481-3400-7_12, © Springer Science+Business Media B.V. 2010

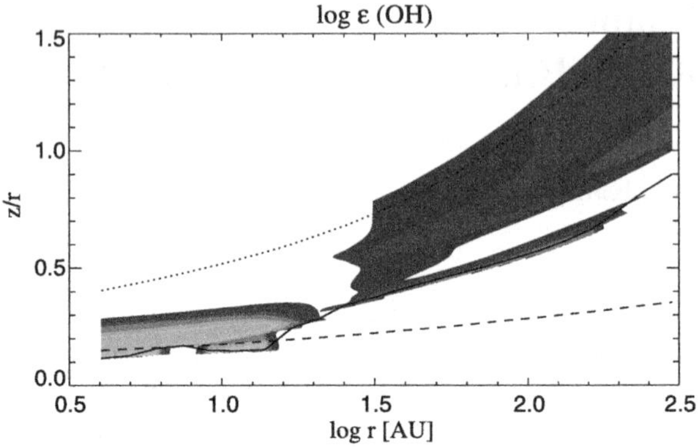

Fig. 12.1 (Color online) Logarithm of OH abundance in the Herbig Ae disk. Abundances are high in a narrow surface region just above the $\log \tau = 1$ layer out to distances of 20 AU

The images of the irradiated, photoevaporating disks in Orion triggered a whole series of models by [6] and [10, 11] which study the structure of these photoevaporating disks and aim at characterizing the disk erosion through an external star. Recent modeling by [2, 3] study the viscous disk evolution together with photoevaporation by the central star itself.

12.2 Disk Surfaces as Diagnostics

Narrow band [O I] 6,300 Å observations in Orion show that the emission arises from the surface of the protoplanetary disk [4]. Reference [13] demonstrate that the emission can arise from OH photodissociation, that leaves the neutral Oxygen atom in the $^1D_2$ excited state, which is the upper level of the forbidden line at 6,300 Å. They use a one dimensional PDR code to estimate the gas temperature at the disk surface and find that OH photodissociation is more important than thermal excitation for the origin of the [O I] line.

12.2.1 Modeling of the 6,300 Å Line

Thermochemical disk models by [7] predict the existence of a thin hot surface OH layer for irradiated protoplanetary disks around T Tauri stars. These models have been extended to Herbig Ae stars and Fig. 12.1 illustrates the distribution of OH in the disk. The OH gas is generally very warm, a few 100 to 1,000 K and located just above the dust $\log \tau = 1$ layer (Fig. 12.2).

Reference [1] observed the [O I] 6,300 Å line in the disk of several Herbig stars. Some of the line profiles, e.g. those of the stars HD100546 and HD97048, show a double-peaked profile and point to a disk origin. The authors find that the emission arises from OH in the flaring disk surface and the observed line fluxes can be explained by an overall OH abundance of 5×10^{-7}.

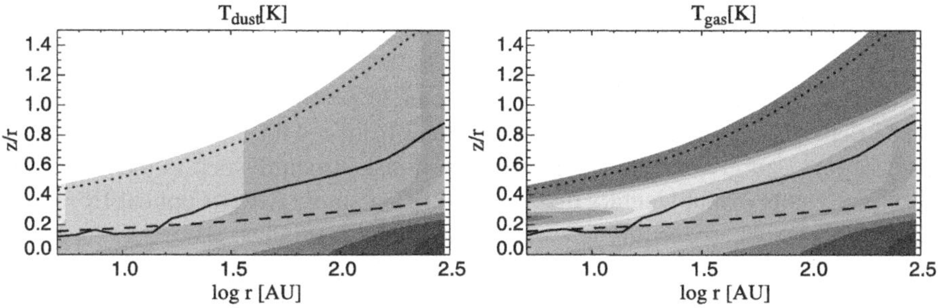

Fig. 12.2 (Color online) Dust and gas temperature in the Herbig Ae disk. The *solid line* indicates the location, below which gas and dust temperatures agree to within 10%. The *dotted line* marks the transition between the surrounding cloud and the disk. The *color levels* correspond to the following temperatures: 10, 20, 30, 40, 50, 60, 70, 80, 90, 100, 200, 500, 1,000, 2,000, and 5,000 K

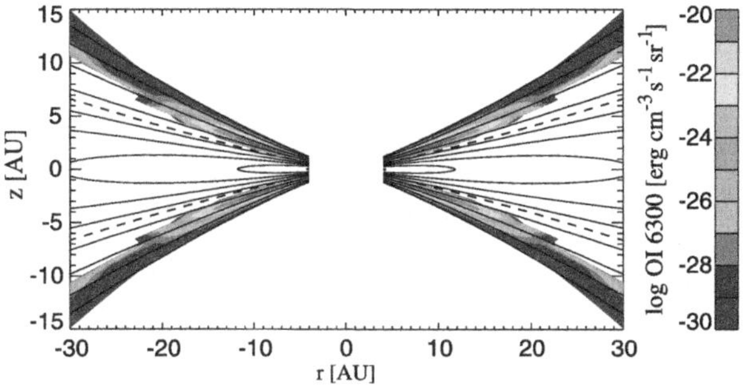

Fig. 12.3 (Color online) OI 6,300 Å emission from the protoplanetary disk around a Herbig Ae star ($T_{\mathrm{eff}} = 9{,}500$ K, $\log g = 4.0$, $L = 42\ \mathrm{L}_\odot$. The *color contours* show the logarithm of emitted [O I] line intensity. The *solid lines* are logarithmic density contours starting with $n = 10^4$ in steps of factor 10. The *dashed line* denotes the position of the continuum $\log \tau = 1$ surface

Using a thermo-chemical model of a disk around a Herbig Ae star with stellar properties similar to HD97048, the emission of the 6,300 Å line is computed directly from the OH photodissociation rate

$$I(\mathrm{OI}\ 6{,}300) = A_{6{,}300\,\text{Å}}\,\Gamma_{\mathrm{OH}} \frac{0.55}{8.5 \times 10^{-3}} \frac{hc}{4\pi 6.3 \times 10^{-5}} \tag{12.1}$$

Here, 0.55 is the fraction of neutral O atoms left in the upper state of the 6,300 Å transition. The denominator 8.5×10^{-3} is the sum of all downward transition probabilities from this level. The Einstein A coefficient of the 6,300 Å line is $6.3 \times 10^{-3}\ \mathrm{s}^{-1}$. The total intensity emitted from the disk is shown in Fig. 12.3. This is several orders of magnitude smaller than the observed and predicted emission by [1]. The reason for this is (a) that the OH abundance computed in our model is generally smaller than 5×10^{-7} and (b) the OH photodissociation rate as computed directly from the stellar spectrum is several orders of magnitude smaller than the one assumed in [1]. Since the gas temperature in the self-consistent thermo-chemical models is very high, thermal emission may actually significantly contribute to the total [O I] 6,300 Å line emission.

12.2.2 Conclusion

The forbidden [O I] 6,300 Å line presents an important probe of the irradiation of the disk surface. Detailed modeling in combination with high spatial resolution observations can give feedback on the vertical disk structure such as the amount of flaring in young protoplanetary disks. The WFC3 camera with its optical/IR narrow band filters opens an exciting new tool for this kind of studies.

References

1. Acke, B., van den Ancker, M.E., Dullemond, C.P., 2005. Astron. Astrophys. 436, 209.
2. Alexander, R.D., Clarke, C.J., Pringle, J.E., 2006. Mon. Not. R. Astron. Soc. 369, 216.
3. Alexander, R.D., Clarke, C.J., Pringle, J.E., 2006. Mon. Not. R. Astron. Soc. 369, 229.
4. Bally, J., O'Dell, C.R., McCaughrean, M.J., 2000. Astron. J. 119, 2919.
5. Beckwith, S., Sargent, A.I., Scoville, N.Z., Masson, C.R., Zuckerman, B., Phillips, T.G., 1986. Astrophys. J. 309, 755.
6. Hollenbach, D.J., Yorke, H.W., Johnstone, D., 2000. In: Mannings, V., Boss, A., Russell, S. (Eds.), Protostars and Planets IV, p. 401. University of Arizona Press, Tucson.
7. Kamp, I., Dullemond, C., 2004. Astrophys. J. 615, 991.
8. Koerner, D.W., Sarget, A.I., Beckwith, S.V.W., 1993. Icarus 106, 2.
9. O'Dell, C.R., Wen, Z., Hu, X., 1993. Astrophys. J. 410, 696.
10. Richling, S., Yorke, H.W., 2000. Astrophys. J. 539, 258.
11. Richling, S., Yorke, H.W., 2003. Commun. Konkoly Obs., 103.
12. Sargent, A.I., Beckwith, S., 1987. Astrophys. J. 323, 294.
13. Störzer, H., Hollenbach, D., 1998. Astrophys. J. 495, 853.

Chapter 13
Unveiling the Role of Jets in Star Formation

Deirdre Coffey, Francesca Bacciotti, Thomas P. Ray, and Jochen Eislöffel

Abstract We describe the role of HST in providing the first and long-awaited observational backing for the role of bipolar jets in the star formation process.

13.1 Introduction

The striking phenomena of jets/outflows from protostars and pre-main sequence stars was wholly unanticipated by theorists, who are still striving to understand the basic mechanisms involved. Progress is hindered by long-standing observational difficulties due to small spatial scales, and the fact that many such objects are heavily embedded.

For example, proper motion studies of outflows, at the resolution of standard non-AO ground-based telescopes, demand relatively long time intervals between observations ($\sim$5 years for the nearest objects). Meanwhile, the outflow from XZ Tau observed with *HST*/WFPC2 at one-year intervals exhibited dramatic changes on such timescales. These observations allowed measurements of outflow velocity *and deceleration*, and revealed an extraordinary brightening of the star itself in rarely observed EXor-type fashion [1]. Furthermore, given that jet widths are typically about $1''$, ground-based non-AO observations cannot attempt to differentiate between models currently proposed for jet generation. However, *HST*/STIS spectroscopic observations, of more evolved (and hence less embedded) optically visible pre-main sequence stars, allow detailed analysis of the jet internal structure and kinematics and, as a result, have contributed significantly to our understanding of the role of jets in the star formation process, Sects. 13.2 and 13.3.

D. Coffey (✉) · F. Bacciotti
INAF, Osservatorio Astrofisico di Arcetri, Largo E. Fermi 5, 50125 Firenze, Italy
e-mail: dac@arcetri.astro.it

F. Bacciotti
e-mail: fran@arcetri.astro.it

T.P. Ray
The Dublin Institute for Advanced Studies, 5 Merrion Square, Dublin 2, Ireland
e-mail: tr@cp.dias.ie

J. Eislöffel
Thüringer Landessternwarte Tautenburg, Sternwarte 5, 07778 Tautenburg, Germany
e-mail: jochen@tls-tautenburg.de

F.D. Macchetto (ed.), *The Impact of HST on European Astronomy*, Astrophysics and Space Science Proceedings,
DOI 10.1007/978-90-481-3400-7_13, © Springer Science+Business Media B.V. 2010

13.2 Determining Jet Mass Flux

Given that supersonic jets and outflows are observed to transport significant amounts of energy and momentum away from the central source, the most important parameter in any outflow model is mass flux. This requires a knowledge of the total gas density, which is not directly measurable. However, the electron density can be easily obtained from the [S II]$\lambda\lambda$6718,6731 doublet, and the ionisation fraction can be found using a technique developed in recent years to determine jet physical conditions by comparing optical forbidden emission line ratios [2]. Hence the total density can be obtained. This technique was applied to jet observations using a *HST*/STIS multiple-slit configuration. The resulting 3D datacube allowed extraction of 2D images for given velocity intervals, Fig. 13.1. These revealed an onion-like kinematic structure, in which the gas is more collimated at higher velocities and excitations [3, 4]. From these maps, emission line ratios then give information about the gas physics [5]. For example, the electron density is higher closer to the star, closer to the flow axis, and at higher velocities. At the jet base, the temperature is $8 \times 10^3 - 2 \times 10^4$ K, and the ionisation fraction is 0.01–0.4 yielding a total gas density of $\sim 10^6$ cm^{-3}. From these values, the initial mass flux in the jet, $\dot{M}_{\rm jet}$, is found to be of the order of 10^{-7} M$_\odot$ yr^{-1}, with the colder and slower external layers contributing most. According to magneto-centrifugal models, $\dot{M}_{\rm jet}$ should be about 5–10% of the mass accretion rate through the disk onto the central object. These observations confirm this prediction in all the cases studied.

13.3 Indications of Jet Rotation

Collimated jets are believed to extract a substantial amount of the excess angular momentum from the star/disk system, thus allowing disk accretion. Several models propose apparently feasible mechanisms of jet launching [6, 7] but, since this occurs within a few AU of the star, the observational evidence favouring one model over another has remained elusive.

In recent years, a new parameter for constraining theoretical models has emerged facilitated only by the high angular resolution of *HST*, namely indications of *rotation* of jets about their symmetry axis. This interpretation followed from the detection of systematic radial velocity differences of typically 10–25 km s^{-1} across the jet in *HST*/STIS optical and near-ultraviolet spectra, Fig. 13.2, [8–11]. These studies constitute a survey of eight jets, and reveal that transverse gradients in Doppler shift are common among these objects. Furthermore, in the two cases where both lobes of the bipolar jet were observed, there is agreement in the sense of Doppler gradient, as would be expected from a rotating bipolar flow. There is also agreement in measurements made at optical and near-ultraviolet wavelength ranges, and there is agreement in results for the two cases observed with both parallel and perpendicular slit configurations. Finally, there is agreement in the sense of Doppler shift between the *jet and disk* in two of four cases observed (i.e. DG Tau [8, 12], and CW Tau [9, 13]), a further one case remains inconclusive (i.e. HH 30 [9, 14]), and

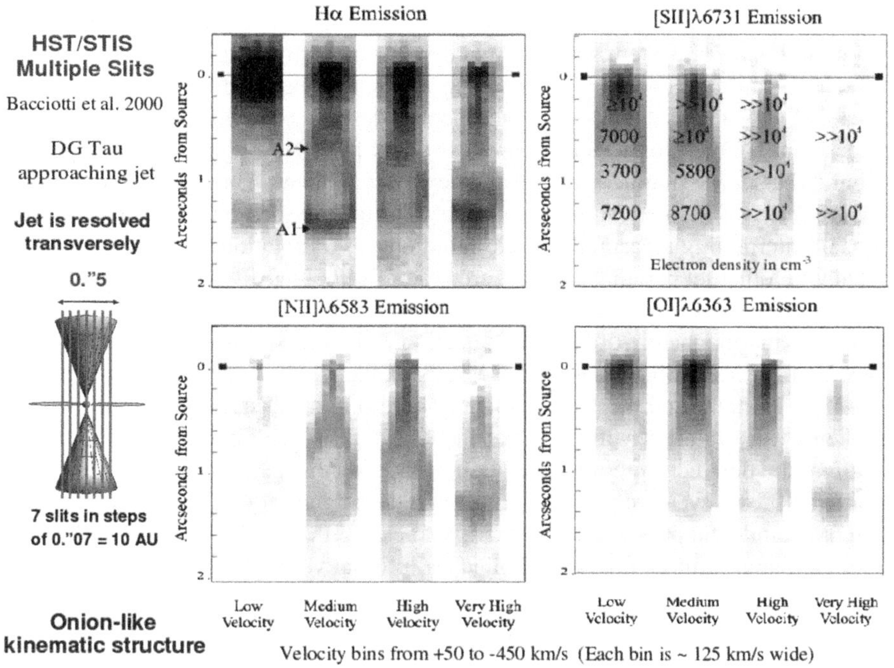

Fig. 13.1 (Color online) HST/STIS observations allowing the jet to be resolved transversely and providing a 3D datacube from which 2D channel maps in different velocity bins could be obtained

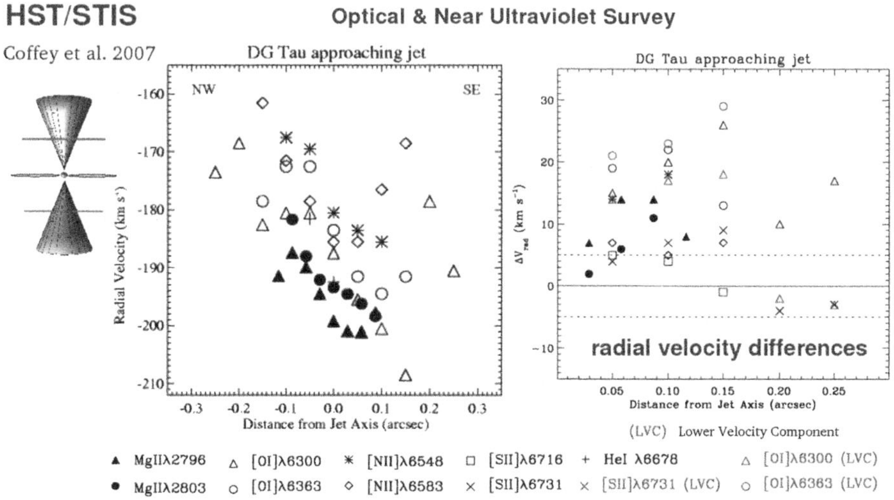

Fig. 13.2 (Color online) HST/STIS observations of gradients in Doppler shifted radial velocities across the jet base, indicative of jet rotation close to the launching point

a fourth case shows a disagreement (i.e. RW Aur [9, 15]), although this latter case is a complex triple system showing evidence of tidal-stripping and so unidentified influences could be coming into play. An alternative explanation for a gradient in radial velocity across the jet base could be the presence of asymmetric shocks or jet precession, but these explanations are less likely [11]. Nevertheless, it is important to build the survey sample, especially in terms of agreement between jet and disk.

A number of theoretical analyses have subsequently demonstrated the implications of interpreting measured Doppler gradients as jet rotation. That is, when combined with other jet parameters, measured radial velocities differences allow the extraction of important information such as the jet launching radius, under the assumption of steady magnetohydrodynamic ejection [10, 16–18]. The results consistently show that, at the resolution of the observations, the jet is launched from a region of 0.1–4 AU from the star on the disk plane. Therefore, if indeed the detected Doppler gradients trace rotation within the jet then, under the assumption of steady MHD ejection, the derived footpoint radii support the existence of magnetized warm disk winds [18]. However, since we do not resolved the innermost layers of the flow, we cannot exclude the possibility that there also exists a stellar wind [6] or an X-wind [7] component. Importantly, the implied jet toroidal velocities cannot be explained by an X-wind model alone, nor by a cold disk-wind [18]. Although we cannot probe the inner axial part of the flow at the current resolution, this may be possible in the near future via interferometry.

Combining measures of jet toroidal velocities with jet mass flux allows an estimate of the total amount of angular momentum flux of the jet for the observed region. Calculations show that the jet can extract 60–80% of the excess angular momentum the disk needs to loose in order to accrete onto the star [10] and hence determining observationally for the first time that the jet's role in the star formation process is a considerable one.

References

1. Coffey, D., Downes, T.P., Ray, T.P., 2004. Astron. Astrophys. 419, 593.
2. Bacciotti, F., Eislöffel, J., 1999. Astron. Astrophys. 342, 717.
3. Bacciotti, F., Mundt, R., Ray, T., Eislöffel, J., Solf, J., Camezind, M., 2000. Astrophys. J. 537, L49.
4. Woitas, J., Ray, T., Bacciotti, F., Davis, C.J., Eislöffel, J., 2002. Astrophys. J. 580, 336.
5. Bacciotti, F., 2002. Rev. Mex. Astron. Astrofis. 13, 8.
6. Pudritz, R.E., Ouyed, R., Fendt, C., Brandenburg, A., 2007. In: Reipurth, B., Jewitt, D., Keil, K. (Eds.), Protostars & Planets V, p. 277. Univ. Arizona Press, Tucson.
7. Shang, H., Li, Z.-Y., Hirano, N., 2007. In: Reipurth, B., Jewitt, D., Keil, K. (Eds.), Protostars & Planets V, p. 261. Univ. Arizona Press, Tucson.
8. Bacciotti, F., Ray, T., Mundt, R., Eislöffel, J., Solf, J., 2002. Astrophys. J. 576, 222.
9. Coffey, D., Bacciotti, F., Woitas, J., Ray, T., Eislöffel, J., 2004. Astrophys. J. 604, 758.
10. Woitas, J., Bacciotti, F., Ray, T., Marconi, A., Coffey, D., Eislöffel, J., 2005. Astron. Astrophys. 432, 149.
11. Coffey, D., Bacciotti, F., Ray, T., Eislöffel, J., Woitas, J., 2007. Astrophys. J. 663, 350.
12. Testi, L., Bacciotti, F., Sargent, A.I., Ray, T., Eislöffel, J., 2002. Astron. Astrophys. 394, 31.

13. Cabrit, S., 2006. Private communication.
14. Pety, J., Gueth, F., Guilloteau, S., Dutrey, A., 2006. Astron. Astrophys. 458, 841.
15. Cabrit, S., Pety, J., Pesenti, N., Dougados, C., 2006. Astron. Astrophys. 452, 897.
16. Anderson, J., Li, Z., Krasnopolsky, R., Blandford, R., 2003. Astrophys. J. 590, L107.
17. Pesenti, N., Dougados, C., Cabrit, S., et al., 2004. Astron. Astrophys. 416, L9.
18. Ferreira, J., Dougados, C., Cabrit, S., 2006. Astron. Astrophys. 453, 785.

Chapter 14
A *Hubble* View of Star Forming Regions in the Magellanic Clouds

Dimitrios A. Gouliermis, Thomas Henning, Wolfgang Brandner, Michael R. Rosa, Andrew E. Dolphin, Markus Schmalzl, Eva Hennekemper, Hans Zinnecker, Nino Panagia, You-Hua Chu, Bernhard Brandl, Sascha P. Quanz, Massimo Robberto, Guido De Marchi, Robert A. Gruendl, and Martino Romaniello

Abstract The Magellanic Clouds (MCs) offer an outstanding variety of young stellar associations, in which large samples of low-mass stars (with $M \lesssim 1\ M_{\odot}$ currently in the act of formation can be resolved and explored in detail only with the *Hubble* Space Telescope. These pre-main sequence (PMS) stars provide a unique snapshot of the star formation process, as it is being recorded for the last 20 Myr,

D.A. Gouliermis (✉) · T. Henning · W. Brandner · M. Schmalzl · E. Hennekemper · S.P. Quanz
Max-Planck-Institut für Astronomie, Heidelberg, Germany
e-mail: dgoulier@mpia-hd.mpg.de

M.R. Rosa
Space Telescope European Coordinating Facility, ESO, Garching, Germany

G. De Marchi
ESA, Space Science Department, Noordwijk, The Netherlands

A.E. Dolphin
Raytheon Corporation, Waltham, USA

H. Zinnecker
Astrophysikalisches Institut Potsdam, Potsdam, Germany

N. Panagia · M. Robberto
Space Telescope Science Institute, Baltimore, USA

Y.-H. Chu · R.A. Gruendl
Department of Astronomy, University of Illinois, Urbana, USA

B. Brandl
Sterrewacht Leiden, Leiden University, Leiden, The Netherlands

M. Romaniello
European Southern Observatory, Garching, Germany

F.D. Macchetto (ed.), *The Impact of HST on European Astronomy*, Astrophysics and Space Science Proceedings,
DOI 10.1007/978-90-481-3400-7_14, © Springer Science+Business Media B.V. 2010

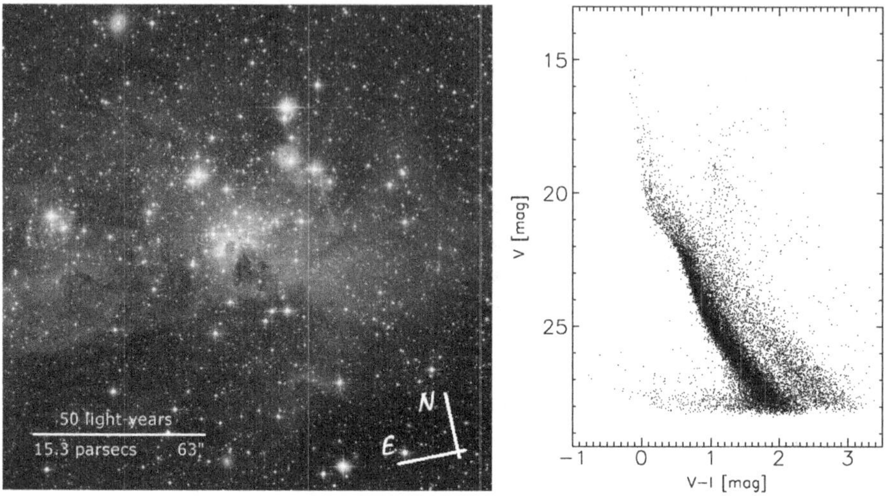

Fig. 14.1 (Color online) Color-composite image from ACS/WFC observations in the filters $F555W$ and $F814W$ of the LMC star-forming region LH 95/N 64 (*left*), and the corresponding $V - I$, V CMD of the detected stars (*right*). These observations allow us to explore the scientific gain that can be achieved for MCs studies using high spatial resolution photometry from *Hubble*. This CMD includes more than 2,500 PMS stars, easily distinguished as a secondary red sequence, almost parallel to the faint part of the main sequence. Image credit: NASA, ESA and D.A. Gouliermis (MPIA). Acknowledgments: Davide de Martin (ESA/Hubble)

and they give important information on the low-mass Initial Mass Function (IMF) of their host environments. We present the latest results from observations with the *Advanced Camera for Surveys* (ACS) of such star-forming regions in the MCs, and discuss the importance of *Hubble* for a comprehensive understanding of the most recent low-mass star formation and the low-mass IMF in the MCs.

14.1 A New View of MCs Associations

Stellar associations contain the richest sample of young bright stars in a galaxy. Consequently our knowledge on the young massive stars of both the Large and Small Magellanic Cloud (LMC, SMC) has been collected from photometric and spectroscopic studies of young stellar associations. However, the picture of these stellar systems changed when *Hubble* observations revealed that MCs associations are not mere aggregates of young bright stars alone, but they also host large numbers of faint PMS stars. An investigation on the main-sequence IMF of the LMC association LH 52 with HST/WFPC2 observations by [2] revealed $\sim$500 low-mass candidate PMS stars easily distinguishable in the $V - I$, V Color–Magnitude Diagram (CMD) [3]. More recently, deeper observations with the Wide-Field Channel (WFC) of ACS of another LMC association (the star-forming region LH 95/N 64) revealed the coexistence of PMS stars and early-type stars in such stellar systems (Fig. 14.1, *left*). These one-of-a-kind observations dramatically changed the picture

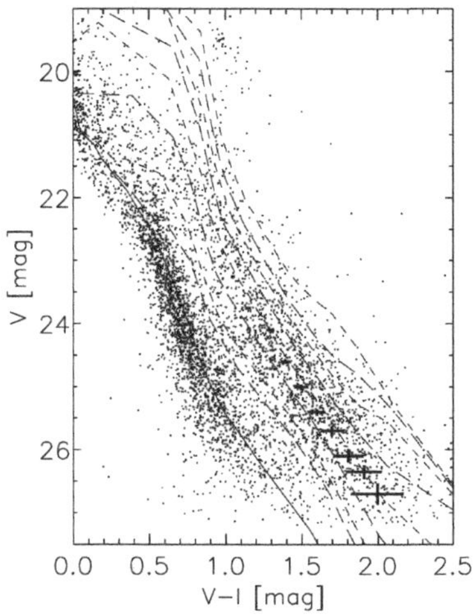

Fig. 14.2 Detail of the $V - I$, V CMD of the stars detected with ACS/WFC in the SMC association NGC 346. Simulations showed that the broadening of these stars can be explained as the result of interstellar reddening and/or continuous star formation, which indicates an age-spread [6]

we had for stellar associations in the MCs by revealing a unique rich sample of PMS stars in LH 95/N 64 (Fig. 14.1, *right*).

14.1.1 CMD Broadening of PMS Stars

The loci occupied by PMS stars in the CMD often appear to be widened areas rather than sharp strips, which could be evidence for an age spread. Such a broadening is observed with ACS/WFC also in the star-forming region NGC 346/N 66 in the SMC (Fig. 14.2). The low-mass population in subgroups within galactic OB associations exhibits evidence for age spreads on time-scales $\gtrsim$10 Myr. Still, simulations show that apart from binarity and variability, reddening seems to play the most important role in the observed CMD-widening of the PMS stars in MCs associations. However, a true age spread cannot be excluded as the reason for the observed CMD-broadening of the PMS stars in NGC 346/N 66 [6].

14.1.2 Stellar Subgroups in MCs Associations

The spatial distribution of the PMS stars in MCs associations demonstrates the existence of significant substructure ("subgroups"), as in the case of galactic OB associations. This has been observed with ACS/WFC, for example, in the NGC 346/N 66 [4]. The surface density map of the region of NGC 346/N 66 constructed from star counts of the PMS stars in the observed field is shown in Fig. 14.3. Apart from the association itself (central large concentration) there are at least five distinct "PMS clusters". Although the CMD-broadening of the PMS stars prevents an accurate estimation of their age, it was found that the northern PMS clusters represent the most recent star formation activity in the region [6].

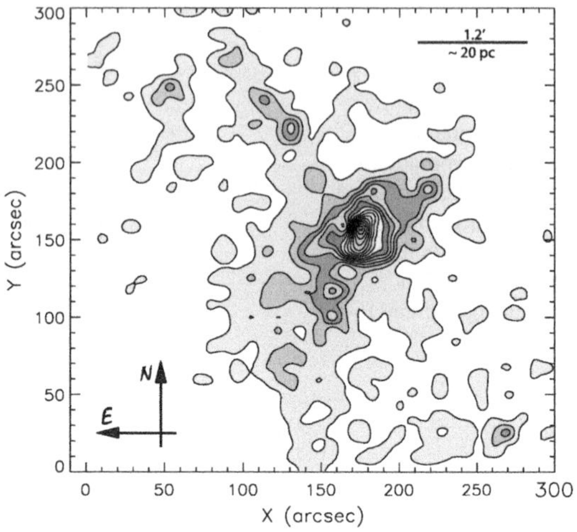

Fig. 14.3 (Color online) Isodensity contour map of the region of NGC 346/N 66 constructed from star counts of the PMS stars. The compact PMS clusters revealed in this map are suspected to be the product of star formation triggered by the action of the OB stars of the association and a nearby supernova

14.2 The IMF of MCs Associations

Young stellar systems, which host newborn PMS stars, naturally provide the testbed for a comprehensive study of the stellar IMF. A coherent sample of PMS stars is found with *Hubble* from ACS imaging in the vicinity of another SMC association, NGC 602, located in the H II region N 90 [8]. The region of NGC 602/N 90 includes no distinct subgroups, and therefore, being less complicated than NGC 346/N 66, is more suitable for the investigation of the low-mass IMF. For the construction, though, of this IMF a mass–luminosity relation derived from evolutionary models cannot be used due to the spread of the PMS stars, also apparent in the CMD of NGC 602. Instead, counting the PMS stars between evolutionary tracks, which represent specific mass ranges, seems to be the most accurate method for the construction of their mass spectrum [8].

14.3 On-going Star Formation in the MCs

The coexistence of H II regions and PMS stars in stellar associations of the MCs indicate that star formation may be still active in their vicinity (see e.g. [1], [7] and Chap. 6 by Heydari-Malayeri et al. in this book). Indeed, observations with the *Spitzer* Space Telescope of the region of NGC 602/N 90 revealed objects classified as candidate Young Stellar Objects (YSOs), and the comparison of the loci of these IR-bright sources with the *Hubble* images interestingly showed PMS stars to be their optical counterparts [5]. A variety of objects is discovered to coincide with these candidate YSOs, such as single highly embedded sources, small compact PMS clusters, as well as features similar to "Elephant Trunks", all located at the periphery of NGC 602, along the dust ridges of the molecular cloud presumably blown-away by the action of the association itself [5].

Acknowledgements D.A. Gouliermis kindly acknowledges the support of the German Research Foundation, through the individual grant 1659/1-1, and of the and the German Aerospace Center (DLR). Based on observations made with the NASA/ESA Hubble Space Telescope, obtained at the Space Telescope Science Institute, which is operated by the Association of Universities for Research in Astronomy, Inc. under NASA contract NAS 5-26555.

References

1. Chu, Y.-H., et al., 2005. Astrophys. J. Lett. 634, L189.
2. Gouliermis, D., Brandner, W., Henning, T., 2005. Astrophys. J. 623, 846.
3. Gouliermis, D., Brandner, W., Henning, T., 2006. Astrophys. J. Lett. 636, L133.
4. Gouliermis, D.A., et al., 2006. Astrophys. J. Suppl. Ser. 166, 549.
5. Gouliermis, D.A., et al., 2007. Astrophys. J. 665, 306.
6. Hennekemper, E., et al., 2008. Astrophys. J. 672, 914.
7. Meynadier, F., et al., 2004. Astron. Astrophys. 422, 129.
8. Schmalzl, M., Gouliermis, D., Dolphin, A., Henning, T., 2008. Astrophys. J. 681, 290.

Chapter 15
A Preliminary Budget for the Ionizing Photons in HII Regions of M51

Leonel Gutiérrez and John Beckman

Abstract The purpose of this work is to quantify the ionizing photon budget in the HII regions, deriving the fraction of the Lyman continuum (Lyc) absorbed by dust or escaping from the regions. An underlying aim is to explore the degree of inhomogeneity of a typical HII region.

Using broad band and narrow band (Hα) images of M51 from ACS/HST we have examined the ionizing radiation budget in a set of luminous HII regions in a spiral arm between 5 and 12 kpc from the centre. In the BVI filters we identified sets of point sources within the HII region boundaries (as defined in the Hα image), whose absolute magnitudes and colours identify them as young massive star clusters. Plotting a B-V vs V-I colour–colour diagram we estimated the internal visual extinction for a given region. Also, a mean value for the differential extinction coefficient, R_V, of 3.3, similar to the Galactic value, was derived.

We used the I band filter to continuum-subtract the Hα image, and derived absolute fluxes using the distance to M51 of 8.4 Mpc [2].

The Lyman continuum (Lyc) luminosities of the ionizing clusters were derived from their stellar contents, taking an average type for the ionizing stars as O7V [9]. We computed the equivalent O7V number for each region, and the total Lyc luminosity Qo [10]. Thence, we derived the corresponding predicted Hα luminosities from the HII regions, assuming the "Case B" of [6], and a dust free HII region opaque to Lyc. These exceeded the measured values by a factor of order 7.

L. Gutiérrez (✉) · J. Beckman
Instituto de Astrofísica de Canarias, c/Vía Láctea s/n, La Laguna, Tenerife, Spain
e-mail: leonel@iac.es

J. Beckman
e-mail: jeb@iac.es

J. Beckman
Consejo Superior de Investigaciones Científicas, Madrid, Spain

L. Gutiérrez
Universidad Nacional Autónoma de México, Ensenada, Mexico

F.D. Macchetto (ed.), *The Impact of HST on European Astronomy*,
Astrophysics and Space Science Proceedings,
DOI 10.1007/978-90-481-3400-7_15, © Springer Science+Business Media B.V. 2010

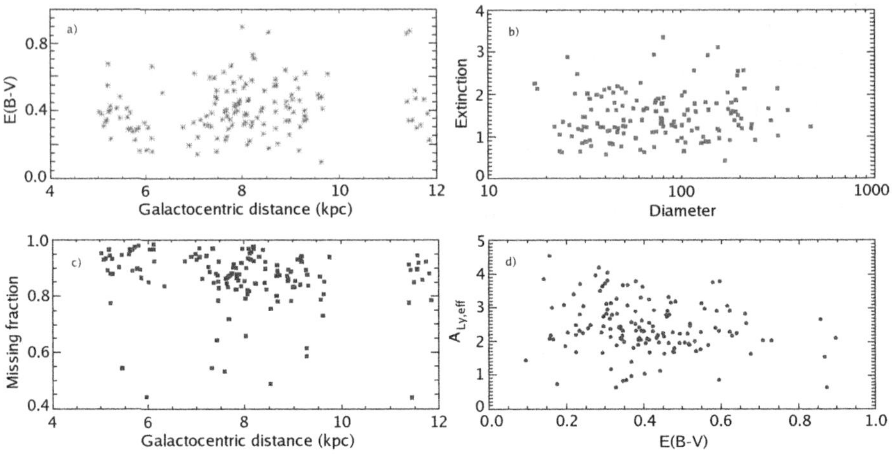

Fig. 15.1 **(a)** Plot of the reddening vs the galactocentric distance. **(b)** The extinction vs the equivalent diameter of the regions. **(c)** The missing fraction vs the galactocentric distance. **(d)** The effective Lyc extinction vs reddening

The first result to bring out is the range of extinction values: A_V between 0.42 and 3.56 with a mean value of 1.51, agreeing well with [1]. There is no correlation between reddening and galactocentric distance (Fig. 15.1a), nor between extinction and region diameter (Fig. 15.1b), indicating highly inhomogeneous regions.

A second result is the ratio of detected (extinction corrected) Hα to predicted Hα. This value is, strikingly only 0.12, i.e. 88% of the predicted Hα is not detected (Fig. 15.1c). Defining the effective Lyc extinction $A_{\mathrm{Lyc,eff}} = 2.5 \times \log(\frac{H\alpha_{\mathrm{pred}}}{H\alpha_{\mathrm{obs}}})$ as [4], and assuming no Lyc photon escape, we find values of $A_{\mathrm{Lyc,eff}}$ between 0.6 and 4.8 (Fig. 15.1d). We can see that if the gas and dust distribution in HII regions were homogeneous, this range of values is inconsistent with that of A_V (the values for $A_{\mathrm{Lyc,eff}}$ should be 5–10 times greater). In addition, if we assume that about 50% of Lyc photons escape from an HII region ([5] or [8]), the discrepancy with a homogeneous model is even bigger. This is in good qualitative agreement with the now classical [7] clumpy model for the gas and dust in HII regions (see also [3]).

References

1. Calzetti, D., Kennicutt, R.C. Jr., Bianchi, L., et al., 2005. Astrophys. J. 633, 871.
2. Feldmeier, J.J., Ciardullo, R., Jacoby, G.H., 1997. Astrophys. J. 479, 231.
3. Giammanco, C., Beckman, J.E., Cedrés, B., 2005. Astron. Astrophys. 438, 599.
4. Hill, J.K., Waller, W.H., Cornett, R.H., et al., 1997. Astrophys. J. 477, 673.
5. Oey, M.S., Kennicutt, R.C. Jr., 1997. Mon. Not. R. Astron. Soc. 291, 827.
6. Osterbrock, D., 1989. Astrophysics of Gaseous Nebulae and Active Galactic Nuclei. University Science Books, Mill Valley.
7. Osterbrock, D., Flather, E., 1959. Astrophys. J. 129, 26.
8. Relaño, M., Peimbert, M., Beckman, J., 2002. Astrophys. J. 564, 704.

9. Vacca, W.D., 1994. Astrophys. J. 421, 140.
10. Vacca, W.D., Garmany, C.D., Shull, J.M., 1996. Astrophys. J. 460, 914.

Part II
Nearby Galaxies, Bulges, Spheroids and Galaxy Formation

Chapter 16
New HST Views at Old Stellar Systems

Alvio Renzini

Abstract HST has recently revealed that many among the most massive globular clusters harbor multiple stellar populations, and — most surprisingly — some of them are extremely helium rich. How these clusters managed to generate such complex stellar populations, and what processes let to so dramatic helium enrichment, is today one of the most exciting puzzles in the astrophysics of stellar systems. HST has also been instrumental in demonstrating that both the bulge of our own Galaxy and that of M31 are dominated by old stellar populations, coeval to galactic globular clusters. Ultradeep HST imaging has also demonstrated that a major component of the M31 halo is metal rich and much younger than old globular clusters.

16.1 Introduction

HST has enormously contributed to the study of old resolved stellar populations, in the Milky Way as well as in nearby galaxies. As this Conference is meant to celebrate the impact of HST on European astronomy, it is worth saying that European astronomers have often played a prime role in the achievements with HST in this field. This brief review will focus on the most exciting HST results obtained by them in recent years, either leading the corresponding projects, or participating as co-investigators.

16.2 Helium-Rich Populations in Globular Clusters

Globular clusters (GC) have always been prime targets for HST. Yet, a wealth of extremely exciting and unexpected results have been recently obtained with ACS, which partly contradict the long-standing view of these objects as prototypical *simple stellar populations* (SSP), i.e. assemblies of coeval stars all with the same chemical composition. Early results with HST indeed confirmed such a view, showing exquisitely narrow sequences in the color–magnitude diagram (CMD) of the best studied clusters. HST superior spatial resolution was instrumental in producing such

A. Renzini (✉)
INAF, Osservatorio Astronomico di Padova, Padova, Italy
e-mail: alvio.renzini@oapd.inaf.it

F.D. Macchetto (ed.), *The Impact of HST on European Astronomy*, Astrophysics and Space Science Proceedings,
DOI 10.1007/978-90-481-3400-7_16, © Springer Science+Business Media B.V. 2010

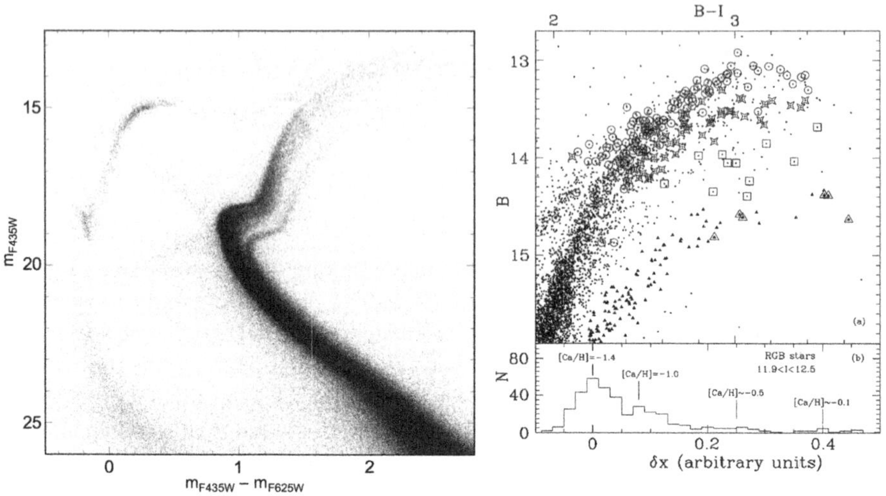

Fig. 16.1 Deep CMD of ω Cen from a 3×3 ACS mosaic [22] showing all the evolutionary phases from the main to the white dwarf sequence. The fine structure of the upper RGB is shown on the *right panel*, with the histogram giving the stellar distribution across a section of the RGB [15]

CMDs, as it allowed superb photometric accuracy *and* proper-motion decontamination from foreground/background stars (e.g., for the cluster M4 [1]). These early results reinforced the notion that sees GCs as viable SSP templates.

One exception to this paradigm was known since the early 1970s, i.e. ω Centauri, whose broad RGB indicated that stars are distributed over a wide range of metallicities. This is emphatically illustrated by a recent CMD of ω Cen from a 3×3 ACS mosaic, shown in Fig. 16.1 [22], with its multiple turnoffs, broad RGB, and complex HB morphology. Being the most massive GC in the Galaxy, ω Cen was not felt as a too embarrassing exception: thanks to its deeper potential well it may have retained enriched gas out of which successive stellar generation were formed. Perhaps, it also started much more massive than at present, possibly a compact dwarf galaxy in itself.

A first surprise came from a particularly accurate CMD of ω Cen obtained with WFPC2 [2], shown here in Fig. 16.2. The main sequence (MS) appears in fact split in two parallel sequences, indicating that at least two distinct star formation episodes had to take place, rather than a continuous star formation process. A third distinct population is also evident in Fig. 16.1, from its faint turnoff and subgiant branch (SGB) and very red RGB. So, ω Cen harbors at least three distinct populations. Now, the majority of stars in this cluster are relatively metal poor, as indicated e.g., by the blue side of the RGB being the most populated (see Fig. 16.1). One would then expect that the least populated of the two MSs in Fig. 16.2 would correspond to the minority, metal rich population. However, [2] noted that the metal rich MS should lie to the red of the metal poor one, instead it lies to the blue! This is just contrary to what well understood stellar models predict. Among the possible solutions of the conundrum Bedin et al. mention an enhanced helium abun-

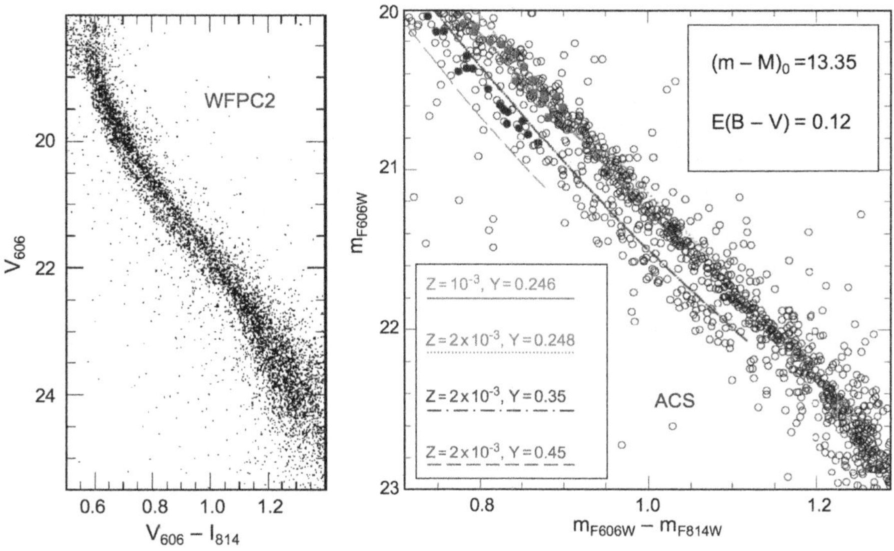

Fig. 16.2 (Color online) *Left panel*: the main sequence of ω Cen splits in two parallel sequences in this CMD from WFPC2 data [2]. *Right panel*: the double main sequence of ω Cen is even better resolved using ACS data, with overlapped theoretical sequences with appropriate metallicities and different helium abundances [16]

dance up to $Y \sim 0.3$ or more in the metal rich population, a solution further convincingly explored by [13]. That the blue MS is indeed the metal rich one was then demonstrated spectroscopically using the FLAMES multiobject spectrograph at the VLT [16], and at this point it became virtually inescapable to conclude that the cluster contains a minority population with an helium abundance as high as $Y \simeq 0.35$.

Figure 16.3 shows a blow up of the turnoff and SGB region of the CMD in Fig. 16.1. It is clear that at least four distinct populations coexist (labelled A, B, C, and D), and there is a rather convincing trace of a fifth one, intermediate between C and D. Indeed, a section of the RGB shows the presence of five populations, each with a different metallicity [18]. Thus, in ω Cen one can distinguish at least three MSs (the third being relative to the D population seen in Fig. 16.3), four or five SGBs, five RGBs, and a complex, multimodal HB. The real puzzle is how to connect the various parts of the CMD, recognizing each of the five MS–SGB–RGB–HB sequences, and estimate age, metallicity and helium (t, Z, Y) for each of the corresponding populations.

To help composing this puzzle, FLAMES spectroscopy at the VLT was then undertaken focusing on the SGB components [19, 22]. The conclusions of these two studies differ somewhat, with one favoring nearly equal ages (within 1–2 Gyr uncertainty) for the five populations, but each with a distinct helium abundance [19], while the other argues for at least four age/metallicity groups with a ~30% age range, having assumed just two helium abundances [22]. The t, Z, Y combinations of the five populations remain discrepant from one study to another (see also [11]),

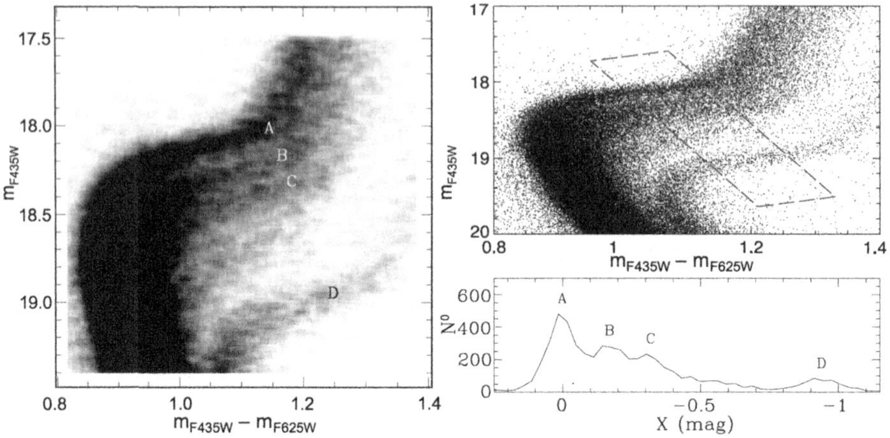

Fig. 16.3 *Left*: the Hess diagram of the turnoff/SGB portion of the ω Cen CMD in Fig. 16.1, showing four (perhaps) five distinct SGBs. *Right*: the same portion of the CMD with indicated the cut used for the number counts shown in the *lower panel* [22]

but only narrowing down these discrepancies one will understand the formation of such a complex cluster.

Whereas ω Cen had for a long time been regarded as a unique exception, suddenly it started to turn out that it was not at all so. Another cluster, NGC 2808 was known for having a multimodal HB, somehow analog to that of ω Cen [20]. Thus, it was speculated that the multimodal HB of this cluster could also be due to a multimodal distribution of helium abundances among the cluster stars [9]. Rarely speculations receive such a fine observational confirmation as was the case for this one: Fig. 16.4 shows the *triple* MS of this cluster, for which no metallicity differences appear to exist [17]. Thus, with three distinct MSs, three groups of RGB stars with different [O/Fe] ratios [6], and an HB made of four separated clumps, NGC 2808 hosts at least three distinct populations, each with different helium, from $Y \simeq 0.24$ up to $Y \simeq 0.37$.

For two other clusters, namely NGC 6388 and NGC 6441, helium-rich subpopulations have been suggested in order to account for their unique HB morphology and periods of the RR Lyrae variables [5, 12]. At this point it soon became apparent that all four GCs with multiple, helium-enhanced populations were very massive, i.e. among the 11 GCs in the Galaxy which are more massive than $10^6\ M_\odot$. Of the remaining most massive clusters, 47 Tuc does not show evidence for multiple populations, but NGC 1851 does. Among this set of supermassive globulars, NGC 6715 is being observed in Cycle 15 and NGC 6093, 6388, 7078, and 7089 will be observed in Cycle 16 (PI G. Piotto), in all cases aiming at checking whether there is evidence for multiple MSs. So, we shall soon know.

These exciting discoveries still ask more questions than give answers:

- ⋆ How did the most massive GCs manage to form/accrete their multiple stellar populations?

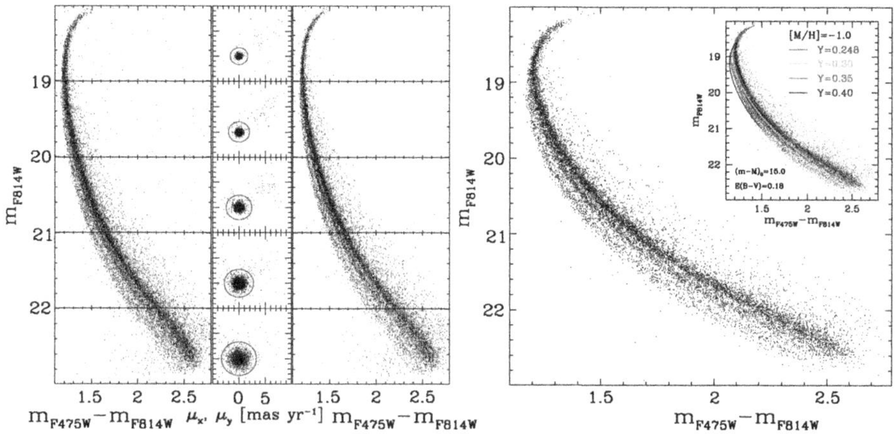

Fig. 16.4 (Color online) *Left panels*: the ACS CMD of NGC 2808 before and after proper motion decontamination, and, in the *right panel*, the same CMD corrected for differential reddening, clearly showing its triple main sequence [17]. In the *insert* theoretical sequences with different helium abundances are overplotted

- ⋆ Where did the huge amount of helium come from? From 3–8 $M_\odot$ AGB stars? Or from where else?
- ⋆ Is ω Cen the remnant core of a tidally disrupted galaxy? And if so, what about the other heavyweight GCs?
- ⋆ Are super-helium populations confined to massive GCs? I.e., can we exclude their presence in massive elliptical galaxies?

16.3 The Milky Way Bulge and Its Globulars

The Galactic bulge harbors a fair fraction of the total populations of GCs, and unlike the metal poor GCs in the halo many of them approach solar metallicity. Observations of two such metal rich GCs taken in the very first Cycle with WFC2 showed that they are as old as Halo GCs, demonstrating that the bulge underwent rapid chemical enrichment and is virtually coeval to the Halo [14]. Figure 16.5 shows a comparison between the near-IR CMD of a bulge field obtained with NTT/SOFI, with the CMD of the bulge GC NGC 6528 obtained with HST/NICMOS [23]. The near identity of the two CMDs, and in particular of the luminosity difference between the HB and the MS turnoff, ensure that the bulge as a whole is as old as the cluster, with no trace of an intermediate age population. However, the bulge CMD was cleaned by the disk contamination in a statistical fashion, a procedure that might have also removed a trace young population. A better way of removing the disk contamination was by picking only HST proper motion bulge members [10], which indeed conclusively demonstrated the virtual absence of an intermedi-

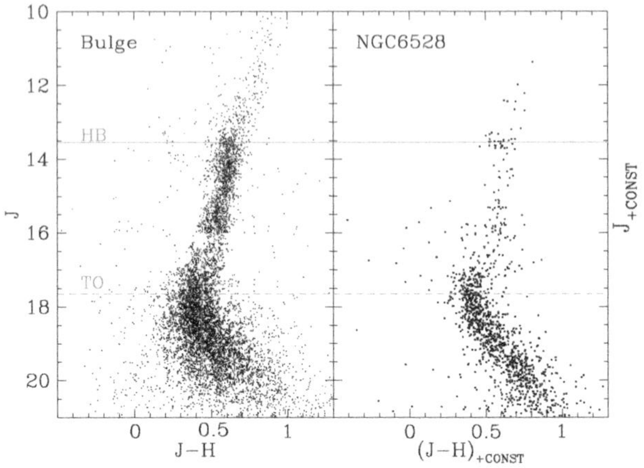

Fig. 16.5 (Color online) The near-IR CMD of a bulge field from NTT/SOFI data (*left*) on the same scale of the HST/NICMOS CMD of the old bulge globular NGC 6528 (*right*) [23]

ate age population in the bulge. The question then arose as to whether among bulges our own is typical or atypical in this respect.

16.4 HST Visits to Andromeda

Next bulge worth checking is that of M31. There had been early claims for the M31 bulge being dominated by an intermediate age population, based on ground based near-IR photometry. This showed an ubiquitous population of very bright red *giants*, then interpreted as intermediate age AGB stars, but there were good reasons to doubt such interpretation, given the extreme crowding of the observed fields. That indeed the apparent bright AGB stars were clumps of fainter RGB stars was beautifully demonstrated by HST/NICMOS imaging [21], as illustrated here in Fig. 16.6. This study also showed that the near-IR luminosity function of M31 bulge is indistinguishable from that of the Galactic bulge, hence both bulges ought to be equally old.

With the advent of ACS it became affordable trying to reach the old MS turnoff in M31. This was first accomplished on an inner halo field, ~11 kpc from the nucleus on the minor axis, investing some 120 HST orbits [3]. Surprisingly, the result was that, besides the old/metal poor population, the field includes also an intermediate age (~7 Gyr old), metal rich population making up to 40% of the total. Subsequent HST/ACS projects by the same team probed a disk, a stream, and more outer halo fields in the attempt of mapping the star formation/assembly history of our companion spiral that looks so similar, yet turns out to be so different from ours. There is no room left to make justice of these HST results, and to illustrate the power of HST/ACS should suffice to show in Fig. 16.7 the image of a GC in M31 [4].

With the Milky Way and the M31 bulges being dominated by stars at least ~10 Gyr old, this is like saying that these bulges had to form the bulk of their stars at $z \gtrsim 1.5$, and evolve passively since then. If so, we should see passive galaxies at

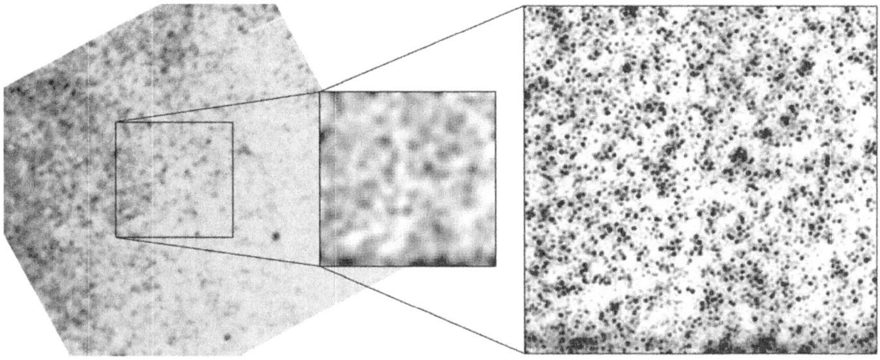

Fig. 16.6 A ground based K-band image of a field in the M31 bulge, and on the *right* the same field as seen by HST/NICMOS [21]. What appear as individual bright "stars" on the ground image, are resolved into stochastic clumps of many fainter stars on the HST image

Fig. 16.7 (Color online) A globular cluster in M31 looks so close in this HST/ACS image [4]

such high redshifts in deep spectroscopic surveys, and indeed we do, and once more HST has been instrumental in reveling their *spheroidal* morphology [7, 8].

In summary, there has been a very strong and successful use of HST by European astronomers in the field of resolved stellar populations (globular clusters, MW & M31 bulges and more) as well as on high redshift galaxies. It is also important to stress that the HST+VLT synergy has been very effectively exploited for many programs. Much more is now expected to come in the final years of HST (2008–2014), hopefully with a telescope more powerful than ever (thanks to WFC3/COS/ACS/STIS/NICMOS/FGS).

Acknowledgements I would like to thank again Duccio Macchetto, for having invited me to this exciting conference, and for having done so much for HST and for attracting the European astronomers to its scientific use.

References

1. Bedin, L.R., Anderson, J., King, I.R., Piotto, G., 2001. Astrophys. J. 560, L75.
2. Bedin, L.R., Piotto, G., Anderson, J., Cassisi, S., King, I.R., Momany, Y., Carraro, G., 2004. Astrophys. J. 605, L125.
3. Brown, T.M., Ferguson, H.C., Smith, E., Kimble, R.A., Sweigart, A.V., Renzini, A., Rich, R.M., VandenBerg, D.A., 2003. Astrophys. J. 592, L17.
4. Brown, T.M., et al., 2004. Astrophys. J. 613, L125.
5. Caloi, V., D'Antona, F., 2007. Astron. Astrophys. 463, 949.
6. Carretta, E., Bragaglia, A., Gratton, R.G., Leone, F., Recio-Blanco, A., Lucatello, S., 2006. Astron. Astrophys. 450, 523.
7. Cimatti, A., Daddi, E., Renzini, A., Cassata, P., Vanzella, E., Pozzetti, L., Cristiani, S., Fontana, A., Rodighiero, G., et al., 2004. Nature 430, 184.
8. Daddi, E., Renzini, A., Pirzkal, N., Cimatti, A., Malhotra, S., Stiavelli, M., Xu, C., Pasquali, A., Rhoads, J.E., Brusa, M., et al., 2005. Astrophys. J. 626, 680.
9. D'Antona, F., Caloi, V., 2004. Astrophys. J. 611, 871.
10. Kuijken, K., Rich, R.M., 2002. Astron. J. 123, 2054.
11. Lee, Y.-W., Joo, S.-J., Han, S.-I., Chung, C., et al., 2005. Astrophys. J. 621, L57.
12. Möhler, S., Sweigart, A.V., 2006. Astron. Astrophys. 455, 943.
13. Norris, J.E., 2004. Astrophys. J. 612, L25.
14. Ortolani, S., Renzini, A., Gilmozzi, R., Marconi, G., Barbuy, B., Bica, E., Rich, R.M., 1995. Nature 377, 701.
15. Pancino, E., Ferraro, F.R., Bellazzini, M., Piotto, G., Zoccali, M., 2000. Astrophys. J. 534, L83.
16. Piotto, G., Villanova, S., Bedin, L.R., Gratton, R., Cassisi, S., Momany, Y., Recio-Blanco, A., Lucatello, S., Anderson, J., et al., 2005. Astrophys. J. 621, 777.
17. Piotto, G., Bedin, L.R., Anderson, J., King, I.R., Cassisi, S., Milone, A.P., Villanova, S., Pietrinferni, A., Renzini, A., 2007. Astrophys. J. 661, L53.
18. Sollima, A., Ferraro, F.R., Pancino, E., Bellazzini, M., 2005. Mon. Not. R. Astron. Soc. 357, 265.
19. Sollima, A., Pancino, E., Ferraro, F.R., Bellazzini, M., Straniero, O., Pasquini, L., 2005. Astrophys. J. 634, 332.
20. Sosin, C., Dorman, B., Djorgovski, S.G., Piotto, G., et al., 1997. Astrophys. J. 480, L35.
21. Stephens, A.W., Frogel, J.A., DePoy, D.L., Freedman, W., Gallart, C., Jablonka, P., Renzini, A., Rich, R.M., Davies, R., 2003. Astron. J. 125, 2473.
22. Villanova, S., Piotto, G., King, I.R., Anderson, J., Bedin, L.R., Gratton, R.G., Cassisi, S., Momany, Y., Bellini, A., Cool, A.M., et al., 2007. Astrophys. J. 663, 296.
23. Zoccali, M., Renzini, A., Ortolani, S., Greggio, L., Saviane, I., Cassisi, S., Rejkuba, M., Barbuy, B., Rich, R.M., Bica, E., 2003. Astron. Astrophys. 399, 931.

Chapter 17
Young Massive Star Clusters in the Era of the Hubble Space Telescope

Richard de Grijs

Abstract The *Hubble Space Telescope* (*HST*) has been instrumental in the discovery of large numbers of extragalactic young massive star clusters (YMCs), often assumed to be proto-globular clusters (GCs). As a consequence, the field of YMC formation and evolution is thriving, generating major breakthroughs as well as controversies on annual (or shorter) time-scales. Here, I review the long-term survival chances of YMCs, hallmarks of intense starburst episodes often associated with violent galaxy interactions. In the absence of significant external perturbations, the key factor determining a cluster's long-term survival chances is the shape of its stellar initial mass function (IMF). It is, however, not straightforward to assess the IMF shape in unresolved extragalactic YMCs. I also discuss the latest progress in worldwide efforts to better understand the evolution of entire cluster populations, predominantly based on *HST* observations, and conclude that there is an increasing body of evidence that GC formation appears to be continuing until today; their long-term evolution crucially depends on their environmental conditions, however.

17.1 Introduction

Young, massive star clusters (YMCs) are the hallmarks of violent star-forming episodes triggered by galaxy collisions and close encounters. This field has seen major progress and a flurry of renewed interest ever since such YMCs were first reported in the starburst galaxy NGC 1275 by [14] using pre-COSTAR HST images. The question remains, however, whether or not at least a fraction of the compact YMCs seen in abundance in extragalactic starbursts, are potentially the progenitors of ($\geq$10 Gyr) old globular cluster (GC)-type objects — although of higher metallicity than the present-day GCs. If we could settle this issue convincingly, one way or the other, such a result would have far-reaching implications for a wide range of astrophysical questions, including our understanding of the process of galaxy

R. de Grijs (✉)
Department of Physics & Astronomy, University of Sheffield, Hicks Building, Hounsfield Road, Sheffield S3 7RH, UK
e-mail: R.deGrijs@sheffield.ac.uk

R. de Grijs
National Astronomical Observatories, Chinese Academy of Sciences, 20A Datun Road, Chaoyang District, Beijing 100012, China

F.D. Macchetto (ed.), *The Impact of HST on European Astronomy*, Astrophysics and Space Science Proceedings,
DOI 10.1007/978-90-481-3400-7_17, © Springer Science+Business Media B.V. 2010

formation and assembly, and the process and conditions required for star (cluster) formation. Because of the lack of a statistically significant sample of YMCs in the Local Group, however, we need to resort to either statistical arguments or to the painstaking approach of case-by-case studies of individual objects in more distant galaxies.

17.2 Individual YMC Evolution

The evolution to old age of young clusters depends crucially on their stellar initial mass function (IMF). If the IMF slope is too shallow, i.e., if the clusters are significantly deficient in low-mass stars compared to, e.g., the solar neighbourhood, they will likely disperse within about a Gyr of their formation (see e.g. [2, 3, 11, 17, 19]). As a case in point, [11] simulated the evolution of $\sim 10^4$–10^5 $M_\odot$ YMCs similar to those observed in the Large Magellanic Cloud (LMC), with IMF slopes $\alpha = 2.35$ (see [18], where the IMF is characterised as $\phi(m_*) \propto m_*^{-\alpha}$, as a function of stellar mass, m_*) and $\alpha = 1.50$, i.e., roughly covering the range of (present-day) mass function slopes observed in LMC clusters at the time he performed his N-body simulations. Following [3], and based on a detailed comparison between the initial conditions for the LMC YMCs derived in [11] and the survival chances of massive star clusters in a Milky Way-type gravitational potential (see [10]), reference [11] concluded that — for Galactocentric distances $\geq$12 kpc — some of his simulated LMC YMCs should be capable of surviving for a Hubble time if $\alpha \geq 2$ (or even $\geq$3; [17]), but not for shallower IMF slopes for any reasonable initial conditions (cf. [2, 3]). More specifically, [3] and [20], based on numerical cluster simulations employing the Fokker–Planck approximation, suggest that the most likely survivors to old age are, additionally, characterised by King model concentrations, $c \geq 1.0$–1.5. Reference [17] (see their Fig. 9) use these considerations to argue that their sample of YMCs observed in the Antennae interacting system might survive for at least a few Gyr, but see de [1, 7, 12] for counterarguments related to environmental effects and to variations in the clusters' star-formation efficiencies, respectively.

In addition, YMCs are subject to a variety of internal and external drivers of cluster disruption. These include internal two-body relaxation effects, the nature of the stellar velocity distribution function, the effects of stellar mass segregation, disk and bulge shocking, and tidal truncation (e.g. [2, 9]). All of these act in tandem to accelerate cluster expansion, thus leading to cluster dissolution — since expansion will lead to greater vulnerability to tidally-induced mass loss.

With the ever increasing number of large-aperture ground-based telescopes equipped with state-of-the-art high-resolution spectrographs and the wealth of observational data provided by the *HST*, we may now finally be getting close to resolving the issue of potential YMC longevity conclusively. To do so, one needs to obtain (i) high-resolution spectroscopy, in order to obtain dynamical mass estimates, and (ii) high-resolution imaging to measure their sizes (and luminosities). One could then construct diagnostic diagrams of YMC mass-to-light (M/L) ratio versus age, and compare the YMC loci in this diagram with simple stellar population models using a variety of IMF descriptions. In [6] we present an updated version of the M/L ratio versus age diagram, including all of the YMCs for which the required observables are presently available.

Despite some outstanding issues (particularly for the youngest clusters), it appears that most of the YMCs for which high-resolution spectroscopy is available are characterised by "standard" Salpeter [18] or Kroupa [16] IMFs. As such, a fraction of the YMCs seen today may potentially evolve to become old GCs, depending on their environmental conditions.

17.3 The Evolution of Star Cluster Systems

Following the violent relaxation induced by the supernova-driven expulsion of the left-over star-forming gas, star clusters — at least those that survive the "infant mortality" phase (i.e., roughly the first 10–30 Myr of their lives) — settle back into virial equilibrium by the time they reach an age of about 40–50 Myr [1, 12]. Subsequently, the initial conditions characterising these gas-free bound star clusters are modified as secular evolution proceeds. Internal (two-body relaxation) and external effects (due to interactions with the tidal field associated with the underlying galactic gravitational potential) lead to tidal stripping and the evaporation of a fraction of the low-mass cluster stars, thus resulting in the gradual dissolution of star clusters.

One of the most important diagnostics used to infer the formation history, and to follow the evolution of a star cluster population is the CMF (i.e., the number of clusters per constant logarithmic cluster mass interval, $\mathrm{d}N/\mathrm{d}\log m_{\mathrm{cl}}$). Of particular importance is the *initial* cluster mass function (ICMF), since this holds clues to the star and cluster formation processes. The debate regarding the shape of the ICMF, and of the CMF in general, is presently very much alive, both observationally and theoretically. This is so because it bears on the very essence of the star-forming process, as well as on the formation, assembly history and evolution of the clusters' host galaxies over cosmic time. Yet, the observable property one has access to is the CLF (i.e., the number of objects per unit magnitude, $\mathrm{d}N/\mathrm{d}M_V$).

In this respect, a remaining contentious issue is whether the observed CMF of YMCs will eventually evolve into that of the ubiquitous old GCs. The GCMF is a Gaussian with a mean $\langle\log(m_{\mathrm{cl}}[\mathrm{M}_\odot])\rangle \sim 5.2$–$5.3$ and a standard deviation of $\sigma_{\log m_{\mathrm{cl}}} \simeq 0.5$–$0.6$ dex. It seems to be almost universal, both *among* and *within* galaxies. On the other hand, many CMFs of YMCs appear to be featureless power laws with a spectral index $\alpha \sim -2$ down to a few $\times 10^3$ $\mathrm{M}_\odot$ (e.g. [5, 15] for the LMC). Some cluster systems exhibit differently shaped ICMFs, however (M82 B [7]; NGC 1316 [13]; NGC 5253 [4]). Evolving such an initial power-law CMF into the near-invariant Gaussian GCMF regardless both of the host galaxy properties and of the details of the cluster loci turns out to be most challenging and requires significant fine-tuning of the models, which is not necessarily compatible with the available observational constraints (see, e.g., [8] vs. [22]; see also [21]).

In order to settle the issues of cluster evolution and ICMF shape more conclusively, major improvements are required in the near future, both observationally and theoretically. Observations reaching low-mass clusters, and with sufficiently accurate photometry, in order to derive reliable cluster ages, are required to follow the temporal evolution of the CMF. From a modeling point of view, a better treatment of the initially loosely bound clusters (i.e., the low-concentration clusters) is required, since these may account for the missing link between the Gaussian GCMF and the power laws seen for YMC systems [21]. In addition, the inclusion of a time-dependent host galaxy gravitational potential will enable us to better follow the early evolution of both old GCs and YMCs formed in interacting and merging galaxies.

References

1. Bastian, N., Goodwin, S.P., 2006. Mon. Not. R. Astron. Soc. 369, L9.
2. Chernoff, D.F., Shapiro, S.L., 1987. Astrophys. J. 322, 113.
3. Chernoff, D.F., Weinberg, M.D., 1990. Astrophys. J. 351, 121.
4. Cresci, G., Vanzi, L., Sauvage, M., 2005. Astron. Astrophys. 433, 447.
5. de Grijs, R., Anders, P., 2006. Mon. Not. R. Astron. Soc. 366, 295.
6. de Grijs, R., Parmentier, G., 2007. Chin. J. Astron. Astrophys. 7, 155.
7. de Grijs, R., Parmentier, G., Lamers, H.J.G.L.M., 2005. Mon. Not. R. Astron. Soc. 364, 1054.
8. Fall, S.M., Zhang, Q., 2001. Astrophys. J. 561, 751.
9. Gnedin, O.Y., Ostriker, J.P., 1997. Astrophys. J. 474, 223.
10. Goodwin, S.P., 1997. Mon. Not. R. Astron. Soc. 284, 785.
11. Goodwin, S.P., 1997. Mon. Not. R. Astron. Soc. 286, 669.
12. Goodwin, S.P., Bastian, N., 2006. Mon. Not. R. Astron. Soc. 373, 752.
13. Goudfrooij, P., Gilmore, D., Whitmore, B.C., Schweizer, F., 2004. Astrophys. J. 613, L121.
14. Holtzman, J., et al., 1992. Astron. J. 103, 691.
15. Hunter, D.A., Elmegreen, B.G., Dupuy, T.J., Mortonson, M., 2003. Astron. J. 126, 1836.
16. Kroupa, P., 2001. Mon. Not. R. Astron. Soc. 322, 231.
17. Mengel, S., Lehnert, M.D., Thatte, N., Genzel, R., 2002. Astron. Astrophys. 383, 137.
18. Salpeter, E.E., 1955. Astrophys. J. 121, 161.
19. Smith, L.J., Gallagher, J.S. III, 2001. Mon. Not. R. Astron. Soc. 326, 1027.
20. Takahashi, K., Portegies Zwart, S.F., 2000. Astrophys. J. 535, 759.
21. Vesperini, E., Zepf, S.E., 2003. Astrophys. J. 587, L97.
22. Vesperini, E., Zepf, S.E., Kundu, A., Ashman, K.M., 2003. Astrophys. J. 593, 760.

Chapter 18
The Central Regions of Early-Type Galaxies

Andrés Jordán

Abstract The ACS Virgo and Fornax cluster surveys present homogeneous HST/ACS observations of 143 early-type galaxies and their globular cluster systems in the Virgo and Fornax clusters. I briefly describe these surveys and present some of their recent results on the properties of the central regions of galaxies. In particular, we have shown that on nuclear scales, galaxies are found to exhibit a gradual progression from a light "deficit" (cores) to a light "excess" (stellar nuclei), while on larger scales the brightness profiles of galaxies are accurately described by Sérsic models. Our observations suggest that a generic by-product of galaxy formation might be the creation of a central massive object — either a supermassive black hole or a compact stellar nucleus — that contains a mean fraction $\sim$0.2% of the total galactic mass.

18.1 Introduction

Much of our understanding of galaxy formation and evolution is based on observations of galaxies in cluster environments. The Virgo and Fornax clusters are among the most thoroughly studied clusters of galaxies by virtue of being near to our Galaxy and observations of them have had an important role in the study of several astrophysical problems. In Cycle 11 of the Hubble Space Telescope we initiated the ACS Virgo Cluster Survey (ACSVCS; [4]), an imaging survey of 100 galaxies in the Virgo cluster carried out with the Advanced Camera for Surveys (ACS) on board the Hubble Space Telescope (HST). In Cycle 13 we initiated a similar survey of 43 galaxies in the poorer Fornax cluster: the ACS Fornax Cluster Survey (ACSFCS; [18]). In this contribution we briefly describe the ACS Virgo and Fornax cluster surveys and present recent results on the properties of the central region of galaxies.

18.2 The ACS Virgo and Fornax Cluster Surveys

We selected a sample of 100 early-type galaxies from the Virgo Cluster Catalog of [1] and 43 early-type galaxies from the Fornax Cluster Catalog of [8]. The sample

A. Jordán (✉)
European Southern Observatory, Karl-Schwarzschild-Str. 2, 85748 Garching bei München, Germany
e-mail: ajordan@eso.org

F.D. Macchetto (ed.), *The Impact of HST on European Astronomy*, Astrophysics and Space Science Proceedings,
DOI 10.1007/978-90-481-3400-7_18, © Springer Science+Business Media B.V. 2010

includes morphological types E, S0, dE, dE,N and dS0. Each galaxy is observed in a single orbit in the F475W (≈ Sloan g) and F850LP (≈ Sloan z) bands. The choice of the observational setup and the data reduction procedures are documented in [4, 14, 18].

The data from the ACSVCS have been used to address several topics. In particular, papers based on the survey data have discussed the connection between globular clusters (GCs) and low-mass X-ray binaries [15, 30], the measurement and calibration of surface brightness fluctuations magnitudes and distances [22–24], the morphology, isophotal parameters and surface brightness profiles for early-type galaxies [10], the connection between GCs and ultracompact dwarf galaxies [13], the nuclei of early-type galaxies [5], the color distribution of GCs [26], the half-light radii of GCs and their use as a distance indicator [16], diffuse star clusters in early-type galaxies [27], the connection between supermassive black holes and central stellar nuclei in early-type galaxies [11], and the luminosity function and color-magnitude relations for GCs in early-type galaxies [17, 19, 25].

Below we present some results of the ACSVCS regarding the properties of the central region of galaxies. While we do not mention the ACSFCS explicitly, all the results presented below are confirmed in our ACSFCS sample [6].

18.3 The Central Regions of Early-Type Galaxies

High-resolution imaging from HST has made it possible to probe the innermost regions of nearby galaxies allowing the systematic study of brightness profiles down to scales of tens of parsecs. An interesting result of the ACSVCS was the realization that previous ground-based studies of early-type galaxies had significantly underestimated the number of galaxies which contain compact stellar nuclei at, or near, their photocenters. It was found that ≈70–80% of the ACSVCS sample galaxies contained such nuclei, roughly three times higher than previously believed [5]. This incidence of nucleation is similar to that found for late-type galaxies, which often contain a "nuclear star cluster" at, or near, their photocenters (e.g., [2, 3, 21, 29, 31]).

The ACSVCS has allowed an exploration of the connection between core structure, stellar nuclei, and supermassive black holes (SBHs) in new levels of detail. On subarcsecond scales (i.e., $r \sim 0.''1$–$1''$, or ∼10–100 pc), the surface brightness profiles were found to vary systematically as one moves down the luminosity function [5, 10]. Bright galaxies ($M_B \lesssim -20$), which in agreement with previous HST studies (e.g. [7, 9, 20]) exhibit a nearly constant surface brightness cores, have surface brightness profiles that fall below the inward extrapolation of the Sérsic model fitted beyond a few arcseconds (see also [12]). Meanwhile, progressively fainter galaxies show increasingly steep upturns over the Sérsic models that fit the galaxies on scales greater than $\sim 0.''1$–$1''$. In other words, on small (subarcsecond) angular scales, galaxies were found to exhibit a gradual progression from a light "deficit" to a light "excess" (see Fig. 18.1), while on larger scales (i.e., $\gtrsim$ a few arcseconds), the

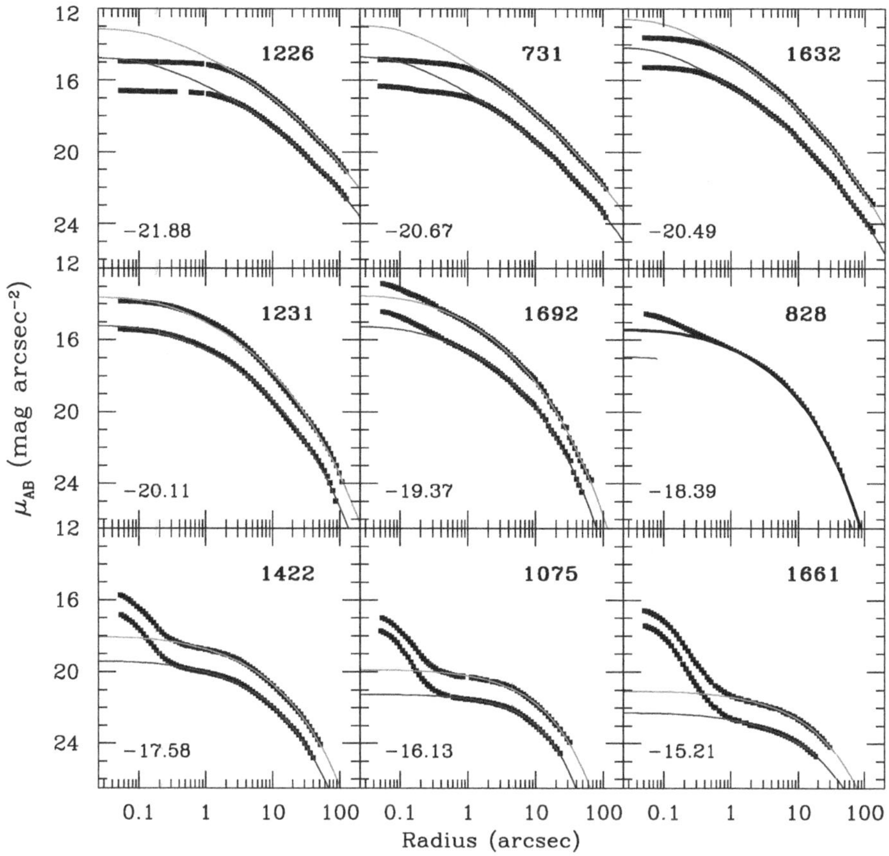

Fig. 18.1 (Color online) Representative surface brightness profiles for nine early-type galaxies from the ACSVCS spanning a factor of ~460 in blue luminosity; the B-band magnitude of each galaxy is listed in the corresponding *panel*. In each *panel*, we show the azimuthally-averaged brightness profile in the g and z bands plotted as a function of mean geometric radius (*lower and upper profiles*, respectively). The *solid curves* show Sérsic models fitted to the profiles beyond $\sim 0.''2$–$2''$. Note the gradual progression from a central light "deficit" to "excess", with a transition at $M_B \sim -20$

curved brightness profiles of real galaxies (both giants and dwarfs) are accurately captured by Sérsic models.

Stellar nuclei are consistently absent only in the brightest galaxies — the same galaxies which are believed to host SBHs. In addition, it was found that nuclei in the low- and intermediate-mass galaxies contribute a mean fraction, $\eta \sim 0.2\%$, of the total galaxy luminosity [5, 11]. This fraction is, to within the errors, the same as the fractional *mass* contribution of the central SBHs in massive early-type galaxies (see also [28] for similar conclusions regarding nuclear star clusters in late-type galaxies). Long-slit, integrated-light spectra for several dozen galaxies in the ACSVCS were used to derive dynamical galaxy masses $\mathcal{M}_{\rm gal}$. Combining these masses with masses for the nuclei derived from the brightness profiles and stellar population models, it was found that a single $\mathcal{M}_{\rm CMO}$–$\mathcal{M}_{\rm gal}$ relation extends smoothly from SBHs to nuclei as one moves down the mass function for early-type galaxies [11] (see also [32]).

Our observations suggest that a generic by-product of galaxy formation might be the creation of a central massive object — either a supermassive black hole or a compact stellar nucleus — that contains a mean fraction $\sim$0.2% of the total galactic mass. Galaxy formation models will need to account for the generic formation of central massive objects, which offer new constraints on the physical processes at the centers of galaxies.

References

1. Binggeli, B., Sandage, A., Tammann, G.A., 1985. Astron. J. 90, 1681.
2. Böker, T., et al., 2004. Astron. J. 127, 105.
3. Carollo, C.M., Stiavelli, M., Mack, J., 1998. Astron. J. 116, 68.
4. Côté, P., et al., 2004. Astrophys. J. Suppl. Ser. 153, 223.
5. Côté, P., et al., 2006. Astrophys. J. Suppl. Ser. 165, 57.
6. Côté, P., et al., 2007. Astrophys. J. 671, 1456.
7. Crane, P., et al., 1993. Astron. J. 106, 1371.
8. Ferguson, H.C., 1989. Astron. J. 98, 367.
9. Ferrarese, L., et al., 1994. Astron. J. 108, 1598.
10. Ferrarese, L., et al., 2006. Astrophys. J. Suppl. Ser. 164, 334.
11. Ferrarese, L., et al., 2006. Astrophys. J. 644, L21.
12. Graham, A.W., et al., 2003. Astron. J. 125, 2951.
13. Haşegan, M., et al., 2005. Astrophys. J. 627, 203.
14. Jordán, A., et al., 2004. Astrophys. J. 613, 279.
15. Jordán, A., et al., 2004. Astrophys. J. Suppl. Ser. 154, 509.
16. Jordán, A., et al., 2005. Astrophys. J. 634, 1002.
17. Jordán, A., et al., 2006. Astrophys. J. 651, L25.
18. Jordán, A., et al., 2007. Astrophys. J. Suppl. Ser. 169, 213.
19. Jordán, A., et al., 2007. Astrophys. J. Suppl. Ser. 171, 101.
20. Lauer, T.R., et al., 1995. Astron. J. 110, 2622.
21. Matthews, L.D., et al., 1999. Astron. J. 118, 208.
22. Mei, S., et al., 2005. Astrophys. J. Suppl. Ser. 156, 113.
23. Mei, S., et al., 2005. Astrophys. J. 625, 121.
24. Mei, S., et al., 2007. Astrophys. J. 655, 144.
25. Mieske, S., et al., 2006. Astrophys. J. 653, 193.
26. Peng, E.W., et al., 2006. Astrophys. J. 639, 95.
27. Peng, E.W., et al., 2006. Astrophys. J. 639, 838.
28. Rossa, J., et al., 2006. Astron. J. 132, 1074.
29. Seth, A.C., et al., 2006. Astron. J. 132, 2539.
30. Sivakoff, G.R., et al., 2007. Astrophys. J. 660, 1246.
31. Walcher, C.J., et al., 2005. Astrophys. J. 618, 237.
32. Wehner, E.H., Harris, W.E., 2006. Astrophys. J. 644, L17.

Chapter 19
Nuclear Star Clusters Across the Hubble Sequence

Torsten Böker

Abstract Over the last decade, HST imaging studies have revealed that the centers of most galaxies are occupied by compact, barely resolved sources. Based on their structural properties, position in the fundamental plane, and spectra, these sources clearly have a stellar origin. They are therefore called "nuclear star clusters" (NCs) or "stellar nuclei". NCs are found in galaxies of all Hubble types, suggesting that their formation is intricately linked to galaxy evolution. In this contribution, I briefly review the results from recent studies of NCs, touch on some ideas for their formation, and mention some open issues related to the possible connection between NCs and supermassive black holes.

19.1 Introduction

The nuclei of galaxies are bound to provide "special" physical conditions because they are located at the bottom of the potential well of their host galaxies. This unique location manifests itself in various distinctive phenomena such as active galactic nuclei (AGN), central starbursts, or extreme stellar densities. The evolution of galactic nuclei is closely linked to that of their host galaxies, as inferred from a number of global-to-nucleus scaling relations discovered in the last decade.

Recently, observational and theoretical interest has been refocused onto the compact and massive star clusters found in the nuclei of galaxies of all Hubble types. Historically, the nuclei of dE,N galaxies have been best studied, but it has become clear from recent HST studies that similar objects exist also in normal spirals and ellipticals. At face value, these "nuclear star clusters[1]" are an intriguing environment for the formation of massive black holes because of their extreme stellar

T. Böker (✉)
European Space Agency, Dept. RSSD, Keplerlaan 1, 2200 AG Noordwijk, Netherlands
e-mail: tboeker@rssd.esa.int

[1] A word on terminology: historically, the compact light excess found in the centers of "nucleated" early-type galaxies is often referred to as a "stellar nucleus". In most studies of disk-dominated galaxies, however, the term "nuclear star cluster" is used. Because we now know that in terms of size and luminosity, both types of objects are indistinguishable, and likely differ only their in evolutionary stage, it seems reasonable to adopt a single term for them. Since "star cluster" seems more descriptive of the nature of these objects than "nucleus" (after all, every galaxy has a nucleus, i.e. a center), we will refer to them as "nuclear star clusters" or NCs for all galaxies types.

F.D. Macchetto (ed.), *The Impact of HST on European Astronomy*, Astrophysics and Space Science Proceedings,
DOI 10.1007/978-90-481-3400-7_19, © Springer Science+Business Media B.V. 2010

density. NCs may also constitute the progenitors of at least some halo globular clusters via "NC capture" following the tidal disruption of a satellite galaxy. Finally, their formation process is influenced by (and important for) the central potential, which in turn governs the secular evolution of their host galaxies. In what follows, I briefly summarize what has been learned about NCs over the last few years.

19.2 Properties of Nuclear Star Clusters

Extragalactic star clusters are compact sources, and in general, their study requires space-based resolution. It doesn't come as a surprise, therefore, that the HST has been instrumental for recent progress in the understanding of NCs. Over the last decade, a number of HST studies — both via imaging and spectroscopic observations — have contributed to the following picture of NCs:

(1) NCs are common: the fraction of galaxies with an unambiguous NC detection is 75% in late-type (Scd-Sm) spirals [1], 50% in earlier-type (Sa-Sc) spirals [4], and 70% in spheroidal (E & S0) galaxies [5]. All these numbers are likely lower limits, although for different reasons. In the latest-type disks, it is sometimes not trivial to locate the galaxy center unambiguously so that no particular source can be identified with it. In contrast, many early-type galaxies have very steep surface brightness profiles (SBPs) that make it difficult to detect even luminous clusters against this bright background.
(2) NCs are much more luminous than "normal" globular clusters (GCs). With typical absolute I-band magnitudes between -14 and -10 [1, 5], they are roughly 40 times more luminous than the average MW globular cluster [10].
(3) However, NCs are as compact as MW GCs. Their half-light radius typically is 2–5 pc, independent of galaxy type [2, 5, 9].
(4) Despite their compactness, NCs are very massive: their typical dynamical mass is 10^6–10^7 $M_\odot$ [19] which is at the extreme high end of the GC mass function.
(5) Their mass density clearly separates NCs from compact galaxy bulges. This is demonstrated in Fig. 19.1 which compares the mass and mass density of NCs to that of other spheroidal stellar systems. The clear gap between bulges/ellipticals on the one hand, and NCs on the other hand makes a direct evolutionary connection between the two classes of objects unlikely.
(6) The star formation history of NCs is complex, as evidenced by the fact that most NCs have stellar populations comprised of multiple generations of stars [13, 20]. The youngest generation is nearly always younger than 100 Myr which is strong evidence that NCs experience frequent and repetitive star formation episodes [20].
(7) Due to three recent and independent studies of NCs in different galaxy types [5, 13, 21], it has become clear that NCs obey similar scaling relationships with host galaxy properties as do supermassive black holes (SMBHs). As an example, Fig. 19.2 shows the NC mass as a function of bulge luminosity. While the

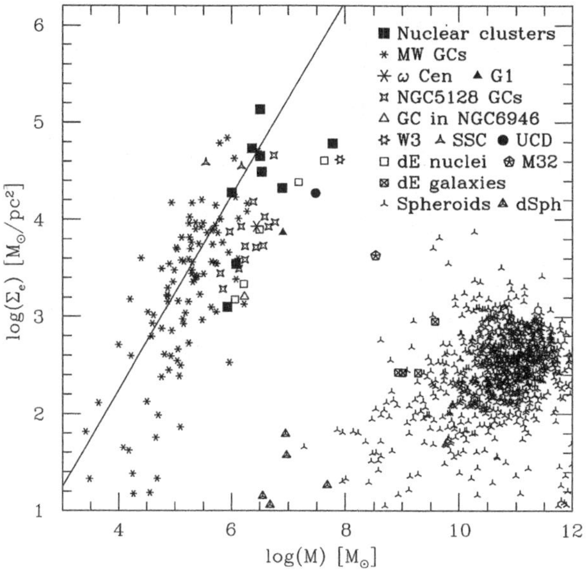

Fig. 19.1 Mean projected mass density of various stellar systems inside their effective radius r_e, plotted against their total mass. This is similar to a face-on view of the fundamental plane. NCs occupy the high end a region populated by other types of massive stellar clusters, and are well separated from elliptical galaxies and spiral bulges. The *solid line* represents a constant cluster size, i.e. $r_e = 3$ pc (from [19])

Fig. 19.2 Relation between NC mass in spiral galaxies and bulge luminosity $\log(L_B)[L_\odot]$ of the host galaxy (from [13]). *Open symbols* denote early-type spirals, and *filled symbols* denote late-type spirals. There is a strong correlation in the sense that galaxies with more luminous bulges have more massive NCs. The *solid line* indicates the best linear fit to the data, while the *dashed line* indicates the relation between the SMBH mass and bulge luminosity for the sample of [11]

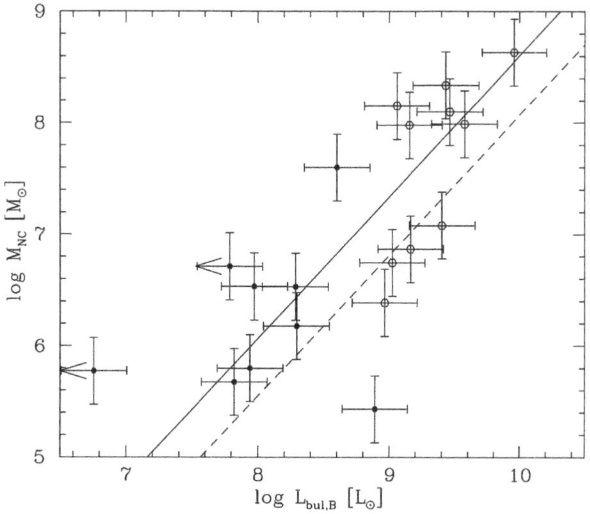

implications of this result are not yet clear (see Sect. 19.4), these studies have renewed interest in NCs because of the potentially important role that NCs play in the evolution of their host galaxies.

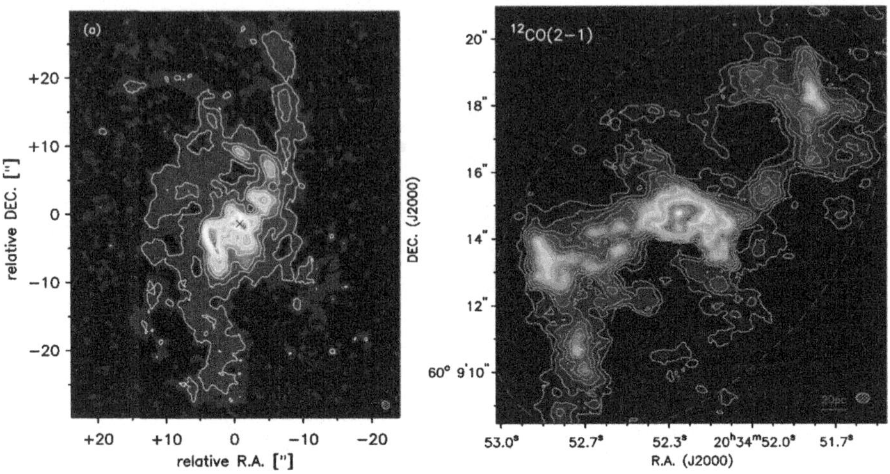

Fig. 19.3 (Color online) *Left*: intensity map of the ^{12}CO(1 − 0) line in the central 1′ (1.6 kpc) in the nearby spiral galaxy NGC 6946 [16]. *Right*: higher resolution (0.4″) map of the ^{12}CO(2 − 0) emission in the central 300 pc. Note the S-shaped spiral that appears to "funnel" molecular gas towards the central 10 pc where about $1.6 \cdot 10^7$ $M_\odot$ of molecular gas have accumulated [17]

19.3 How (and When) Do Nuclear Clusters Form?

The processes that funnel gas onto NCs in the local universe have recently been studied in some detail, enabled by significant improvements to the sensitivity and spatial resolution of mm-interferometers. Figure 19.3 shows the molecular gas distribution in the nearby spiral NGC 6946. Both the morphology and the kinematics of the gas can be well explained by the effects of a small-scale stellar bar. The S-shaped flow pattern onto the nucleus and the large ($1.6 \cdot 10^7$ $M_\odot$) gas concentration in the inner $\approx$10 pc lend credibility to the "repetitive burst" scenario for NC growth.

Less clear, however, are the reasons for why gas accumulates in the nucleus of a shallow disk galaxy *in the absence* of a prominent central mass concentration, i.e. how the "seed clusters" form initially. A few studies have attempted to provide an explanation for this puzzle. For example, [12] suggests the magneto-rotational instability in a differentially rotating gas disk as a viable means to transport gas towards the nucleus and to support (semi)continuous star formation there.

More recently, [6] have pointed out that the tidal field becomes compressive in shallow density profiles, causing gas to collapse onto the nucleus of a disk galaxy. If correct, then NC formation is indeed expected to be a natural consequence of galaxy formation, which would go a long way towards explaining at least some of the observed scaling relations between NCs and their host galaxies.

The question of when a particular NC (i.e. its "seed" cluster) has formed is equivalent to asking how old its oldest stars are. This question is extremely difficult to answer in all galaxy types, albeit for different reasons. In late-type spirals, for example, the NC nearly always contains a young stellar population which dominates the spectrum and thus makes the detection of an underlying older population challenging, not to mention its accurate age determination.

Early-type, spheroidal galaxies, on the other hand, lack the large gas reservoirs of spirals, and thus should experience less frequent nuclear starbursts. One therefore would expect their NCs to contain fewer and older stellar populations. However, early-type galaxies have much steeper surface brightness profiles, and therefore a low contrast between NC and galaxy body. This makes spectroscopic studies of NCs in E's and S0's exceedingly difficult. The few published studies have focused on the NCs of dE,N galaxies, and have shown that even these can have rather young (few hundred Myrs) stars, as demonstrated e.g. by [3] in the case of NGC 205. Generally speaking, however, most NCs in dE,N galaxies have integrated colors that — while different from those of their host galaxies — are generally consistent with evolved stellar populations at least 1 Gyr old [18]. Considering that there may be even older stellar populations "hiding", this age should only be considered a lower limit for the oldest stellar population in dE,N nuclei.

In fact, it is not implausible that the "seed clusters" for present-day NCs were in place very early in the universe. The average star formation rate over the last 100 Myr in NCs of late-type spirals is $2 \cdot 10^{-3}$ $M_{\odot}$/yr [20]. Assuming this SFR was constant for the past 10 Gyr, one would expect a stellar mass of $\approx 2 \cdot 10^{7}$ $M_{\odot}$ which is within a factor of 4 from the typical NC mass of $5 \cdot 10^{6}$ $M_{\odot}$ [19]. Turning the argument around, if NCs indeed build up their entire present-day mass via a series of repetitive starbursts, then they must have been in place at least 3 Gyr ago, unless their SFRs were significantly higher in the past than over the last 100 Myr. Given that observations today are somewhat biased towards more luminous NCs which likely have a time-averaged SFR higher than the "typical" NC, this estimate might even be too low.

19.4 A New Paradigm?

It has recently been proposed by [7] that NCs extend the well-known scaling relation between the mass of a galaxy and that of its central super-massive black hole (SMBH) to lower masses. This has triggered speculation about a common formation mechanism of NCs and SMBHs, being governed mostly by the mass of the host galaxy. The idea put forward is that NCs and SMBHs are two incarnations of a "central massive object" (CMO) which forms in every galaxy. In galaxies above a certain mass threshold ($\approx 10^{10}$ $M_{\odot}$), galaxies form predominantly SMBHs while lower mass galaxies form NCs.

While tantalizing, this apparent connection opens more questions than it answers. For example, we know that some galaxies contain *both* an NC and a SMBH. A well-known example is the "mini-Seyfert" NGC 4395 [8], but others have been found recently [14, 15]. Why then do some NCs contain SMBHs, but not all? Why do some galaxies apparently contain *neither* NC nor SMBH? Is an NC possibly a prerequisite for the formation of a SMBH? Is the formation of a BH (not necessarily a super-massive one) a logical consequence of the high stellar densities present in

NCs? Progress along these lines will require a better understanding of the formation of "pure" disk galaxies in the early universe, as well as improved models for the evolution of extremely dense stellar systems.

References

1. Böker, T., et al., 2002. Astron. J. 123, 1389.
2. Böker, T., et al., 2004. Astron. J. 127, 105.
3. Butler, D.J., Martínez-Delgado, D., 2005. Astron. J. 129, 2217.
4. Carollo, C.M., et al., 1997. Astron. J. 114, 2366.
5. Côté, P., et al., 2006. Astrophys. J. Suppl. Ser. 165, 57.
6. Emsellem, E., van de Ven, G., 2008. Astrophys. J. 674, 653.
7. Ferrarese, L., et al., 2006. Astrophys. J. 644, L21.
8. Filippenko, A.V., Ho, L.C., et al., 2003. Astrophys. J. 588, L13.
9. Geha, M., Guhathakurta, P., van der Marel, R.P., 2002. Astron. J. 124, 3073.
10. Harris, W.E., 1996. Astron. J. 112, 1487.
11. Marconi, A., Hunt, L.K., 2003. Astrophys. J. 589, L21.
12. Milosavlević, M., 2004. Astrophys. J. 605, L13.
13. Rossa, J., et al., 2006. Astron. J. 132, 1074.
14. Satyapal, S., et al., 2007. Astrophys. J. 663, L9.
15. Shields, J.C., Walcher, C.J., Boeker, T., et al., 2008. Astrophys. J. 682, 104.
16. Schinnerer, E., et al., 2006. Astrophys. J. 649, 181.
17. Schinnerer, E., et al., 2007. Astron. Astrophys. 462, L27.
18. Stiavelli, M., et al., 2001. Astron. J. 121, 1385.
19. Walcher, C.J., et al., 2005. Astrophys. J. 618, 237.
20. Walcher, C.J., et al., 2006. Astrophys. J. 649, 692.
21. Wehner, E.H., Harris, W.E., 2006. Astrophys. J. 644, L17.

Chapter 20
Stellar Populations in the Outskirts of M31: The View from HST

Annette M.N. Ferguson

Abstract M31, our nearest large galactic neighbour, offers an opportunity to study in superb detail the processes which drive the formation and evolution of spiral galaxies. Over the last decade, our view of this oft-considered quiescent system has been revolutionized as a result of wide-field photometric and spectroscopic surveys from the ground, as well as deep pencil beam studies with HST. These studies, which have probed the galaxy to unprecedented surface brightness levels and established the outer stellar boundary to be at least ~150 kpc, reveal that M31 has had a much more violent history than previously thought, having devoured at least one moderate-sized luminous satellite in the recent past. I review here some key results, focusing on the particular role that HST has played.

20.1 Introduction

The study of the faint outskirts of galaxies has become increasingly popular in recent years. From a theoretical perspective, it has been realised that many important clues about the galaxy assembly process should lie buried in these parts. Cosmological simulations of disk galaxy formation yield predictions for the large-scale structure and stellar content at large radii — for example, the abundance and nature of stellar substructure and the ubiquity, structure and content of stellar halos and thick disks. These models generally predict a wealth of (sub)structure at levels of $\mu_V \gtrsim 30\ \mathrm{mag/arcsec^2}$ (e.g. [5]); their verification thus requires studies of galaxies to ultra-faint surface brightness levels.

Our nearest large neighbour, M31, provides an excellent laboratory in which to test such ideas. Located at 785 kpc, M31 is close enough to allow luminous giant stars to be resolved from the ground, enabling its outermost parts to be surveyed via resolved star counts as opposed to the much more challenging technique of diffuse light photometry. The most significant problem posed is the huge solid angle that M31 subtends on the sky: the area inside the optical disk subtends $\approx 2.5\ \mathrm{deg^2}$ while the area inside the virial radius subtends a staggering $\approx 350\ \mathrm{deg^2}$. Detailed study of the outer structure of M31 thus demands deep wide-field observations which

A.M.N. Ferguson (✉)
Institute for Astronomy, University of Edinburgh, Blackford Hill, Edinburgh, Scotland EH9 3HJ, UK
e-mail: ferguson@roe.ac.uk

F.D. Macchetto (ed.), *The Impact of HST on European Astronomy*, Astrophysics and Space Science Proceedings,
DOI 10.1007/978-90-481-3400-7_20, © Springer Science+Business Media B.V. 2010

can currently only be obtained from the ground. The role of HST is to provide an exquisitely detailed view of the stellar populations at selected locations, and thus shed light on how the various structural components present have been assembled.

20.2 The Stellar Outskirts of M31: The View in 2000

Since the first ground-based work on the outer field populations in M31, it has generally been believed that the M31 halo is radically different from that of the Milky Way (MW) — a result that has caused considerable unease within the stellar populations community. Reference [24] presented a colour–magnitude diagram (CMD) of a field located ~7 kpc along the southern minor axis of M31 and demonstrated that the mean metallicity of the population was ~ -0.6 dex and had a significant spread; this value is approximately a dex higher than the mean in the MW's stellar halo.

WFPC2 studies of the outer regions of M31 were crucial in confirming and extending this result. References [12, 26] not only derived similarly high mean metallicities and metallicity spreads at other locations in the outskirts of the galaxy but they also revealed that the horizontal branch morphology is dominated by a prominent red clump, providing further evidence of the metal rich nature of the population (see also [1]). The outer regions are not entirely metal rich, however; studies also noted a weak extended blue horizontal branch indicating that some fraction (25–50%) of the population is ancient and metal poor.

Other differences in the outer populations of M31 and the MW were noted in early studies. Reference [25] used star counts out to ~20 kpc along the southern minor axis of M31 to show that an $R^{1/4}$ law provides a much better description of the halo surface brightness profile than the shallow power-law description that best fits the MW. In addition, M31 was shown to have a much larger population of globular clusters than the MW and, unlike the case for the MW, the halo globulars were shown to be significantly more metal poor than the halo field stars.

20.3 The Stellar Outskirts of M31: The View in 2008

Significant progress has been made in understanding the apparent differences between outskirts of the MW and M31 in the last five or so years. The first digital panoramic ground-based survey of M31 and its immediate environs was conducted using the Isaac Newton Telescope's Wide Field Camera [10, 15, 18]. This survey has mapped the surface density of red giant branch (RGB) stars over a $\approx$50 deg^2 area and revealed a wealth of previously-unknown substructure in the inner halo as well as two new dwarf spheroidal (dSph) galaxies, And IX [6, 29] and And XVII [19] and more than 40 new globular star clusters [13, 14]. The left panel of Figure 20.1 shows a map of the RGB population (ages $\gtrsim$1–2 Gyr) while the right panel outlines regions of prominent stellar substructure. The faintest features visible by eye in the

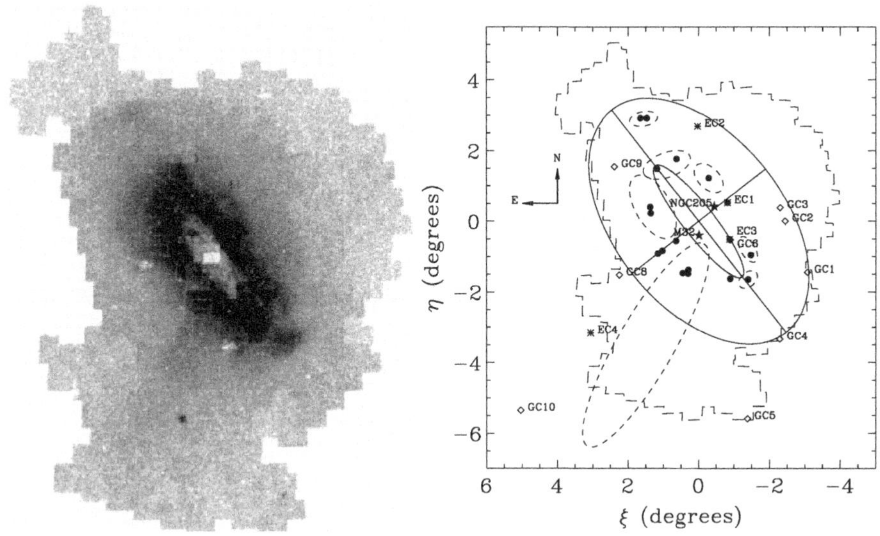

Fig. 20.1 The INT/WFC RGB star count map of M31 (*left*) with regions of prominent stellar substructure identified (*right*). The star count map spans $\sim 125 \times 95$ kpc^2, corresponding to $\sim$50 deg^2 at the distance of M31. *Filled circles* in the *right-hand panel* indicate the deep ACS pointings analysed by [27]. Globular clusters studied by [21, 22] are indicated by *asterixes* (extended clusters) and *open diamonds* (classical clusters). M32 and NGC 205 are shown as *filled stars*

map have effective V-band surface brightnesses of $\mu_V \sim 28$–31 mag/arcsec^2. It can clearly be seen that much of the inner halo — i.e. the region which most previous studies have concentrated on — is littered with tidal debris. A follow-up survey with Megacam on the Canada France Hawaii Telescope extends the INT coverage out to a projected radius of $\sim$150 kpc and has so far led to the discovery of five new dSphs (And XI, XII, XIII, And XV and And XVI) and several new globulars, including the most remote currently known [17, 23]. This survey has also uncovered various faint tidal streams in the outer halo, including a series of azimuthal shell features along the minor axis to $\sim$120 kpc.

20.3.1 Stellar Substructure

Figure 20.1 shows various distinct features in the inner halo of M31 identified from the INT survey, including the giant stream in the south-east, stellar overdensities at large radii along the major axis, a diffuse structure in the north-east and a loop of stars projected near NGC 205 [10, 15]. It is of obvious importance to establish the origin of these substructures: are they tidal debris from the accretion and disruption of one or many luminous satellites? How much of the material can be explained by a perturbed thin disk? A detailed knowledge of the constituent stellar populations and their kinematics is required in order to address these issues.

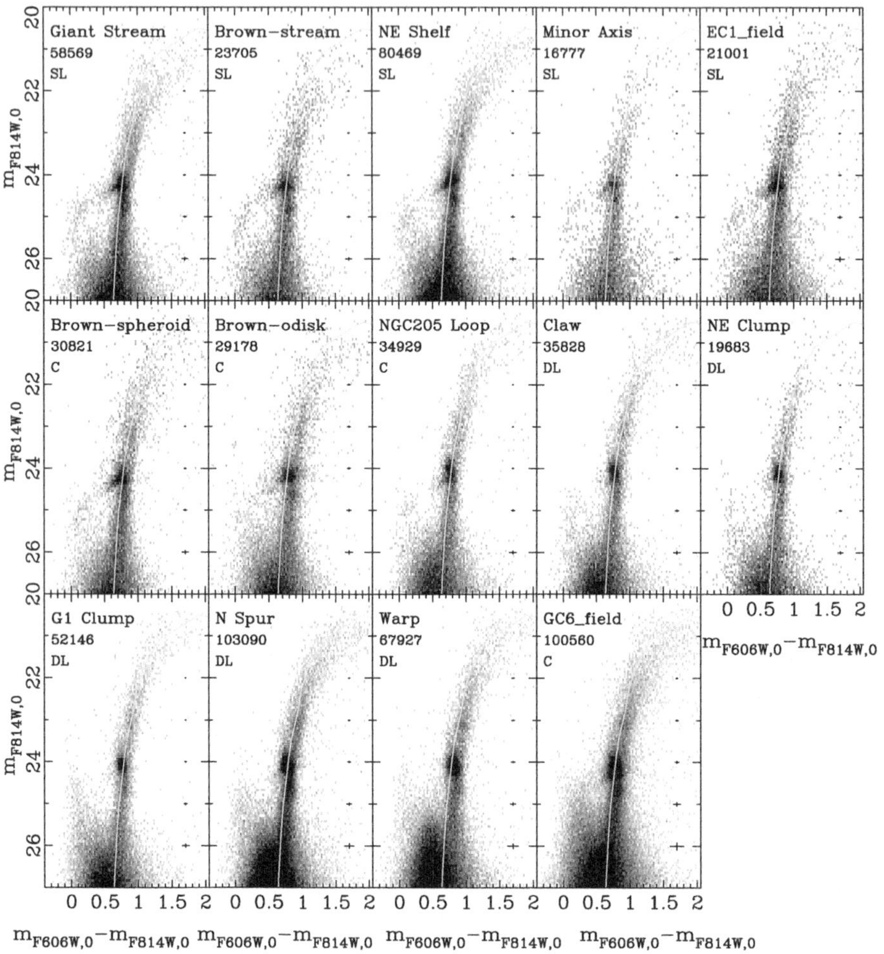

Fig. 20.2 (Color online) Hess CMDs constructed from deep HST/ACS observations of 14 fields in the outskirts of M31. The *number* of stars in each *diagram* is given below the *name*. The ridge line of 47 Tuc, which has [Fe/H] = −0.7 and age = 12.5 Gyr, has been shifted to the distance of M31 and over-plotted. The CMDs are more than 90% complete at $m_{F814W,0} = 27.0$ mag (~80% for EC1_field and GC6_field)

Reference [27] have carried out the largest and most detailed survey to date of the stellar populations in the outskirts of M31 based on the homogeneous analysis of 14 deep HST/ACS pointings, spanning the range 11.5–45 kpc. Despite the fact that many of these pointings probe complex and diverse substructures (see also [9, 11], the CMD morphologies (Fig. 20.2) can be rather simply classified into two main categories: 'stream-like' fields (SL), so named for their similarity to the CMD of the giant stellar stream, are characterized by a red clump that slants bluewards at fainter magnitudes and an extended horizontal branch. On the other hand, 'disk-like' (DL) fields exhibit red clumps with significant luminosity width, lack an obvious horizontal branch and show evidence for recent star formation (~0.25–2 Gyr ago). A third class of 'composite' fields (C) show signatures of both SL and DL populations.

Reference [27] also show that the projected and line-of-sight distribution of SL fields is in excellent agreement with the predictions of recent orbital calculations for the stream progenitor [8]. These fields are found across much of the inner halo of M31, and attest to the high degree of pollution caused by this recent accretion event. Indeed, it now seems likely that several of the M31 ultra-deep pointings formerly presented as pure outer disk and pure halo in the literature [3, 4] are significantly contaminated by this material, rendering their interpretation difficult. It is interesting to note that the currently-favoured progenitor orbit does not easily implicate the more luminous satellites (e.g. M32, NGC 205) in the accretion event; in fact, the progenitor, if still intact, has yet to be identified. Disk-like material generally resides in the extended disk structure of M31 and is detected to radii of 44 kpc here; the uniform populations in these fields, including the ubiquitous presence of young populations and the strong rotational signature reported by [16], are most consistent with this structure having formed through heating and disruption of an existing thin disk, perhaps due to the impact of the giant stream progenitor.

20.3.2 The Smooth Underlying Halo

Using data from the INT/WFC survey, reference [18] combined diffuse light photometry with resolved star counts to probe the minor axis profile of M31 to ~55 kpc ($\mu_V \sim 32$ mag/arcsec^2). The profile shows an unexpected flattening (relative to the inner $R^{1/4}$ decline) at a radius of ~30 kpc, beyond which it can be described by a shallow power-law (index ≈ -2.3). Reference [17] used their CFHT/Megacam survey to trace this power-law halo to a radius of ~150 kpc, arguing for a somewhat shallower radial fall-off with index -1.9. These values compare favourably with the MW outer halo, which exhibits a power-law index of ~ -3 in volume density (e.g. [28]). M31 halo stars at such immense radial distances have also been reported in the spectroscopic study of [20]. The discovery of a power-law component which dominates the light at large radius in M31 has profound implications for the interpretation of early studies of the M31 "stellar halo". Since these studies generally targetted regions lying within 30 kpc along the minor axis, they most likely probed the debris-littered extended disk/bulge region of the galaxy and not the true stellar halo.

Keck/DEIMOS spectroscopy has been used to study the kinematics and metallicities of RGB stars in the far outer regions of M31. By windowing out stars which corotate with the HI disk, [7] have detected an underlying metal-poor ([Fe/H] ~ -1.4), high velocity dispersion ($\sigma \sim 100$ km/s) component. They do not find evidence for a strong metallicity gradient in the halo, consistent with the very similar radial profiles for metal-rich and poor RGB stars presented by [17] as well as HST/ACS CMDs of the outer halo which indicate the presence of a dominant red clump to at least R ~ 60 kpc. Although there is clearly a need for more study,

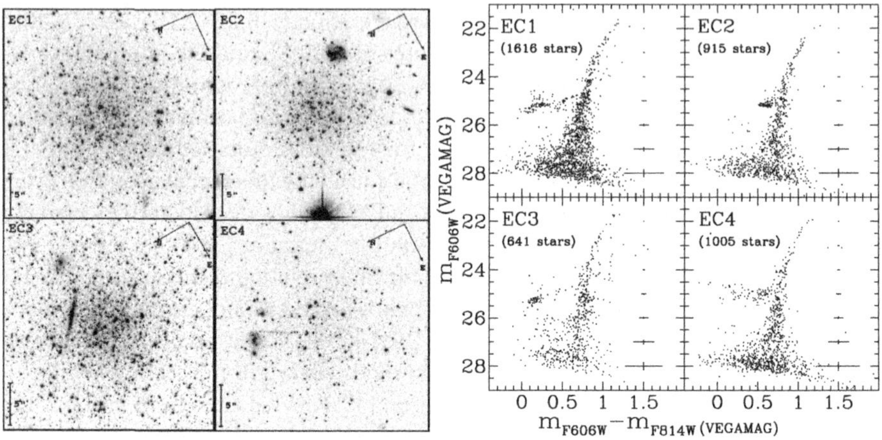

Fig. 20.3 Drizzled ACS/WFC F606W images of four extended clusters in the outskirts of M31 (*left*). Each image spans 25″ on a side. The *right-hand panel* shows ACS CMDs for these clusters

it appears as though M31 does indeed have a stellar component which is similar to that of the Milky Way's halo in terms of structure, metallicity and kinematics.

20.4 Star Clusters in the Outskirts of M31

We have also used our imaging surveys to conduct a search for star clusters in the outskirts of M31 and more than 40 new systems have been identified to date, out to ~120 kpc [13, 14, 23]. Of particular interest has been the discovery of unusually extended clusters with half-light radii of ≈30 pc yet luminosities ($M_V \approx -7$) which place them near the peak of the globular cluster luminosity function. Such a combination of luminosity and size has rarely been observed before within the star cluster population. Indeed, when placed on the M_V–R_h plane, these extended clusters encroach on the gap in parameter space between classical globulars and dSphs (e.g. [2]) and bear some resemblance to enigmatic Galactic objects, such as Willman I (though these are far fainter).

Reference [21] present deep HST/ACS CMDs of the four most luminous extended clusters (see Figs. 20.1 and 20.3). In all cases, a narrow steep RGB and a clear horizontal branch are seen. Three of the clusters possess extended blue horizontal branches and display broadened regions at intermediate colour, suggestive of the presence of RR Lyraes. The derived metallicities are low ([Fe/H] ~ -1.8 to -2.2 dex) and there is no evidence for a strong age spread. Thus, in spite of their unusual structures and special location on the M_V–R_h plane, these objects appear consistent with being genuine globular clusters as opposed to the stripped cores of dSphs. However, it would be premature to conclude that the extended clusters are completely understood. Preliminary analysis of Keck DEIMOS spectra of one of our ACS targets hints at a non-negligible mass-to-light ratio and a high proportion of the extended clusters discovered to date are projected on or near stellar substructure. For example, two extended clusters lie very close to the main body of the newly-discovered dSph And XVII, an alignment which has only ≈0.2% chance of being random [19].

We have also obtained deep HST/ACS observations of a sample of newly-discovered classical globular clusters in the outskirts of M31 ($15 \leq R_{\rm proj} \leq 100$ kpc, see Fig. 20.1). All but one of these objects appear old and metal-poor ($-2.2 \leq$ [Fe/H] ≤ -1.8). Curiously, the outermost clusters ($R_{\rm proj} \geq 40$ kpc) are particularly compact ($R_h \approx 5$ pc) and luminous ($M_V \approx -8.5$) when compared to their outer MW counterparts; this may provide a clue about the early evolution and subsequent accretion histories of the two galaxies.

20.5 Summary

Quantitative study of the faint outskirts of our nearest large neighbour, M31, is providing important new insight into the galaxy assembly process. While the vast area subtended on the sky by M31 requires wide-field ground-based imagery to map its global structure, HST is playing a key role in providing superbly detailed information on the stellar populations at selected locations. CMD analyses can be combined with structural and kinematic data to build a picture of how the distinct components of M31 — i.e. bulge, thin/thick disks, stellar halo — have formed and evolved. Ultimately, this information can be used to test and refine models of galaxy formation within a cosmological framework. M31 plays an especially important role in such endeavours, as it provides a means to test the universality of models that have been fine-tuned to explain the properties of the Milky Way. With the era of ESA's GAIA mission approaching, an understanding of just how typical our home galaxy really is becomes all the more important.

Acknowledgements I thank Scott Chapman, Daniel Faria, Rodrigo Ibata, Mike Irwin, Avon Huxor, Rachel Johnson, Kathryn Johnston, Geraint Lewis, Dougal Mackey, Nicolas Martin, Alan McConnachie, Jenny Richardson and Nial Tanvir for their collaboration. Support from a Marie Curie Excellence Grant is gratefully acknowledged.

References

1. Bellazzini, M., Cacciari, C., Federici, L., Fusi Pecci, F., Rich, M., 2003. Astron. Astrophys. 405, 867.
2. Belokurov, V., et al., 2007. Astrophys. J. 654, 897.
3. Brown, T.M., Ferguson, H.C., Smith, E., Kimble, R.A., Sweigart, A.V., Renzini, A., Rich, R.M., VandenBerg, D.A., 2003. Astrophys. J. Lett. 592, L17.
4. Brown, T.M., Smith, E., Ferguson, H.C., Rich, R.M., Guhathakurta, P., Renzini, A., Sweigart, A.V., Kimble, R.A., 2006. Astrophys. J. Lett. 652, 323.
5. Bullock, J.S., Johnston, K.V., 2005. Astrophys. J. 635, 931.

6. Chapman, S.C., Ibata, R., Lewis, G.F., Ferguson, A.M.N., Irwin, M., McConnachie, A., Tanvir, N., 2005. Astrophys. J. Lett. 632, L87.
7. Chapman, S.C., Ibata, R., Lewis, G.F., Ferguson, A.M.N., Irwin, M., McConnachie, A., Tanvir, N., 2006. Astrophys. J. 653, 255.
8. Fardal, M.A., Guhathakurta, P., Babul, A., McConnachie, A.W., 2007. Mon. Not. R. Astron. Soc. 380, 15.
9. Faria, D., Johnson, R., Ferguson, A.M.N., Irwin, M.J., Ibata, R.A., Johnston, K., Lewis, G.F., Tanvir, N.R., 2007. Astron. J. 133, 1275.
10. Ferguson, A.M.N., Irwin, M.J., Ibata, R.A., Lewis, G.F., Tanvir, N.R., 2002. Astron. J. 124, 1452.
11. Ferguson, A.M.N., Johnson, R.A., Faria, D.C., Irwin, M.J., Ibata, R.A., Johnston, K.V., Lewis, G.F., Tanvir, N.R., 2005. Astrophys. J. Lett. 622, L109.
12. Holland, S., Fahlman, G.G., Richer, H.B., 1996. Astron. J. 112, 1035.
13. Huxor, A.P., Tanvir, N.R., Irwin, M.J., Ibata, R., Collett, J.L., Ferguson, A.M.N., Bridges, T., Lewis, G.F., 2005. Mon. Not. R. Astron. Soc. 360, 1007.
14. Huxor, A.P., Tanvir, N.R., Ferguson, A.M.N., Irwin, M.J., Ibata, R., Bridges, T., Lewis, G.F., 2008. Mon. Not. R. Astron. Soc. 385, 1989.
15. Ibata, R., Irwin, M., Lewis, G., Ferguson, A.M.N., Tanvir, N., 2001. Nature 412, 49.
16. Ibata, R., Chapman, S., Ferguson, A.M.N., Lewis, G., Irwin, M., Tanvir, N., 2005. Astrophys. J. 634, 287.
17. Ibata, R., Martin, N.F., Irwin, M., Chapman, S., Ferguson, A.M.N., Lewis, G.F., McConnachie, A.W., 2007. Astrophys. J. 671, 1591.
18. Irwin, M.J., Ferguson, A.M.N., Ibata, R.A., Lewis, G.F., Tanvir, N.R., 2005. Astrophys. J. Lett. 628, L105.
19. Irwin, M.J., Ferguson, A.M.N., Huxor, A.P., Tanvir, N.R., Ibata, R.A., Lewis, G.F., 2008. Astrophys. J. Lett. 676, L17.
20. Kalirai, J.S., et al., 2006. Astrophys. J. 648, 389
21. Mackey, A.D., et al., 2006. Astrophys. J. Lett. 653, L105.
22. Mackey, A.D., et al., 2007. Astrophys. J. Lett. 655, L85.
23. Martin, N.F., Ibata, R.A., Irwin, M.J., Chapman, S., Lewis, G.F., Ferguson, A.M.N., Tanvir, N., McConnachie, A.W., 2006. Mon. Not. R. Astron. Soc. 371, 1983.
24. Mould, J., Kristian, J., 1986. Astrophys. J. 305, 591.
25. Pritchet, C.J., van den Bergh, S., 1994. Astron. J. 107, 1730.
26. Rich, R.M., Mighell, K.J., Freedman, W.L., Neill, J.D., 1996. Astron. J. 111, 768.
27. Richardson, J.C., et al., 2008. Astron. J. 135, 1998.
28. Vivas, A.K., Zinn, R., 2006. Astron. J. 132, 714.
29. Zucker, D.B., et al., 2004. Astrophys. J. Lett. 612, L121.

Chapter 21
Variable Stars in "Nearby" Galaxies with HST

Gisella Clementini

Abstract Results are presented on the study of the variable star population in a number of extragalactic stellar systems that range from the globular clusters (GCs) of the Small Magellanic Cloud (SMC), the Fornax dwarf spheroidal (dSph) and the Andromeda spiral galaxy, to one of the most remote and metal poor systems resolved in stars: the blue compact dwarf galaxy IZw 18. All these studies are largely based on proprietary observations with the Wide Field Planetary Camera 2 (WFPC2) and the Advanced Camera for Surveys (ACS) on board to the Hubble Space Telescope (HST) and on HST archival data.

21.1 Introduction

Pulsating variables are powerful tools in astrophysics. The light variation caused by the periodic expansion/contraction of the surface layers makes them much easier to recognize than normal stars in very distant stellar systems and when stellar crowding is high. Their main parameter, the pulsation period, is measured at great precision, is unaffected by distance and reddening, and is directly related to stellar intrinsic parameters such as the star mass, radius, and luminosity. Among the pulsating variables the Cepheids and the RR Lyrae stars are standard candles, and serve as primary distance indicators in establishing the cosmic distance scale, the Cepheids because they obey to a period–luminosity (P–L) relationship, and the RR Lyrae stars because they follow a luminosity-metallicity relation in the visual-band and a period–luminosity relation in the K-band. Moreover, since pulsating variables of different types are in different evolutionary phases, they allow to trace the stellar components of different ages in the host systems. Their role becomes crucial when the stellar population is mixed, and stars of different age and metal abundance share the same region of the CMD, as in a galaxy. Among different types of pulsating variables the RR Lyrae and the SX Phoenicis stars are as old as $t \geq 10$ Gyr and trace the oldest component first to form in galaxies. They eyewitnessed the first epochs of galaxy's formation thus providing hints on the mechanism that led to the galaxy's formation and assembling. Classical Cepheids are instead relatively young stars ($t < 200$ Myr), found in large numbers in spiral galaxies and gas-reach systems, and are among the brightest stellar standard candles.

G. Clementini (✉)
INAF, Osservatorio Astronomico di Bologna, via Zamboni 33, 40126 Bologna, Italy
e-mail: gisella.clementini@obao.inaf.it

F.D. Macchetto (ed.), *The Impact of HST on European Astronomy*, Astrophysics and Space Science Proceedings,
DOI 10.1007/978-90-481-3400-7_21, © Springer Science+Business Media B.V. 2010

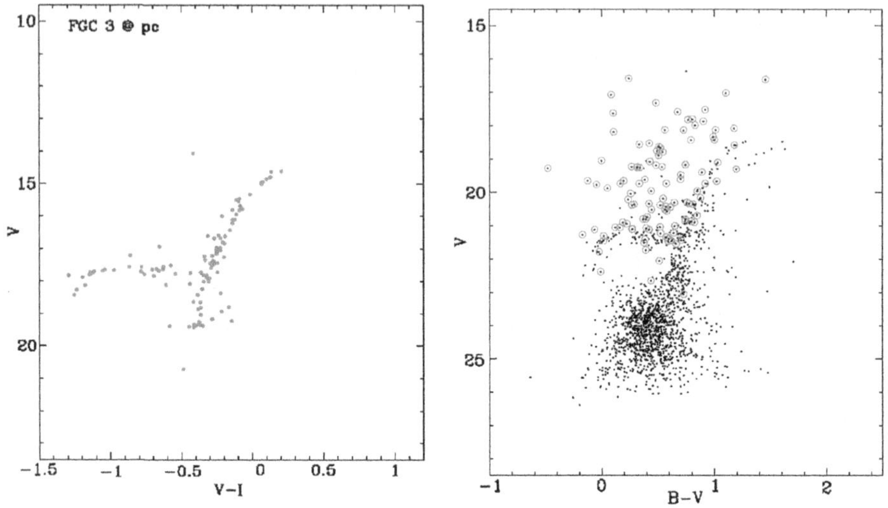

Fig. 21.1 (Color online) *Right panel*: CMD of the GC Fornax 3 from ground-based observations. Stars within 16 arcsec from the cluster center (*magenta circles*) are heavily contaminated by companions, but they are fully resolved in the WFPC2@HST dataset (*left panel*)

The potential of the pulsating stars is enhanced when they are observed with the HST. The following few examples illustrate how powerful is the combination of pulsating variable stars and the HST.

21.2 Variable Stars in Extragalactic Globular Clusters

The Sagittarius dSph currently disrupting into the Milky Way (MW) [9] and the M31 giant tidal stream [10] are among the most spectacular evidences that merging processes play a fundamental role in the formation of large galaxies. The properties of the RR Lyrae stars in extragalactic globular clusters (GCs) can provide essential insight to reconstruct the merging events that led to building the halos of the MW and M31. Indeed, in the MW most of the GCs that host RR Lyrae stars divide into two separate groups according to the average period of their fundamental mode RR Lyrae stars. This phenomenon is called Oosterhoff dichotomy [13]. If the halos of the two large spiral galaxies of the Local Group (LG) originated by past accretions of protogalactic fragments resembling their present-day satellite companions, then the GCs in the MW and M31 satellites should conform to the Oosterhoff division observed among the MW GCs.

Among the MW dSph companions the Fornax dSph hosts five GCs. We have combined proprietary ground-based time-series observations obtained at the 2.2 m ESO and the 6.5 m Magellan telescopes, and WFPC2@HST archive data of the Fornax GCs, to study the properties of the RR Lyrae stars in these extragalactic GCs. The left panel of Fig. 21.1 shows that the WFPC2@HST allows to resolve

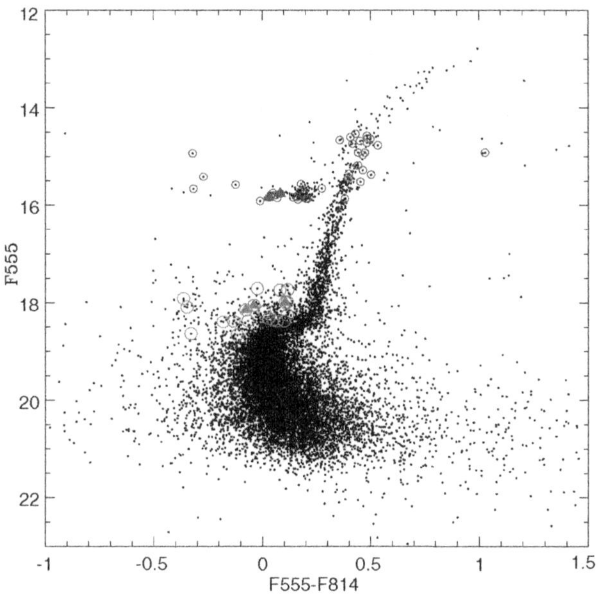

Fig. 21.2 (Color online) CMD of NGC 121 from WFPC2 data with candidate variable stars marked by different symbols. *Open circles*: candidate RR Lyrae (*blue*) and DCs (*red*); *filled triangles*: confirmed RR Lyrae stars (*blue*) and DCs (*red*)

stars within 16 arcsec from the center of the GC Fornax 3 which are unresolved in the ground-based photometry, and correctly places them on the horizontal and red giant branches of the cluster color magnitude diagram (CMD). Many of these stars are RR Lyrae variables. According to the pulsation properties of their RR Lyrae stars the Fornax GCs are found to belong to an Oosterhoff-intermediate class and to fill the Oosterhoff-gap defined by the Galactic GCs [7, 8, 11]. This finding allow us to exclude that the MW halo can have assembled by stripping of Fornax dSph-like protogalactic fragments.

As part of an ongoing project aimed to study the stellar populations, the structure and the evolution of the SMC, new candidate variable stars have been identified in the SMC cluster NGC 121 based on WFPC2@HST archive and ACS proprietary observations (PI: J. Gallagher) of the cluster. Figure 21.2 shows the cluster CMD from the planetary camera of the WFPC2 with the candidate variable stars marked by different symbols. We have detected 27 candidate RR Lyrae stars and 20 Dwarf Cepheid (DC) candidates (SX Phoenicis and δ Scuti stars) in the central region of NGC 121 which is non resolved by ground-based data [3]. Our results confirm the "true" globular cluster nature of NGC 121, a cluster which is at the young end of the Galactic globulars age range [6].

The pioneering study by [2] showed that the GCs of M31 contain RR Lyrae star candidates, but the HST archive data were too few to derive periods and classify the clusters in Oosterhoff types. Seventy-eight orbits with WFPC2 were awarded in HST Cycle 15 (PI: G. Clementini) for studying the variable star population of six

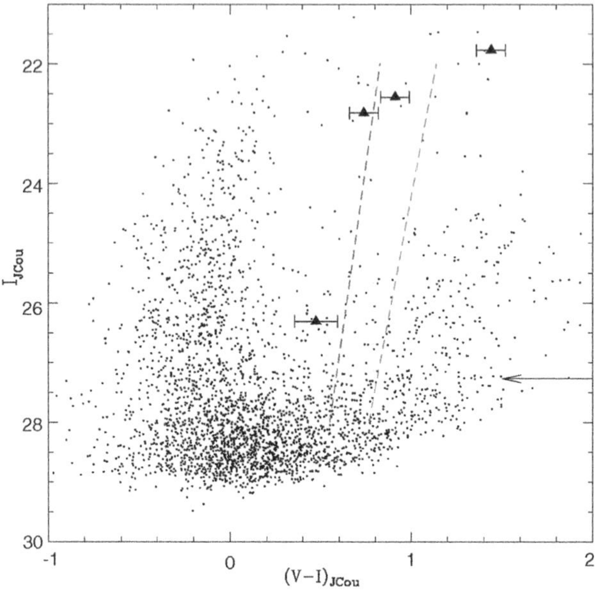

Fig. 21.3 (Color online) I, $V-I$ CMD of IZw 18 from ACS@HST observations, with the variable stars shown by *filled triangles*. The *arrow* locates the RGB Tip. *Dashed lines* show the boundaries of the theoretical instability strip of CCs at the metallicity of IZw 18 (from [12])

properly selected globular cluster of M31. Observations were completed in September 2007, data reduction is in progress. First preliminary results show light curves of excellent quality for a large number of variable stars [4].

21.3 IZw 18: The "Rosetta Stone" Without a Distance

IZw 18 is the most metal poor galaxy ever observed in the Local Universe. Due to the apparent lack of an old population it has long been regarded as a galaxy undergoing its first burst of star formation, and a local analogue of primordial galaxies in the distant Universe. We have discovered and obtained periods and light curves for Classical Cepheids (CCs) in IZw 18, by applying the Image Subtraction technique to proprietary ACS@HST time series-data (PI: A. Aloisi) of the galaxy [5]. Figure 21.3 shows the galaxy CMD with the variable stars marked by filled triangles. New theoretical models of CCs suited for the extremely low metallicity of this galaxy have been computed to interpret the properties of the variables discovered in IZw 18 [12]. Our theoretical and observational combined efforts have allowed to resolve the controversy on the nature of IZw 18. In fact, the direct determination of the galaxy's distance from its CCs allowed us to firmly locate the tip of a red giant branch (RGB Tip) population as old as at least 2 Gyr in IZw 18, thus demonstrating that the galaxy is older than previously thought [1].

References

1. Aloisi, A., et al., 2007. Astrophys. J. 667, L151.
2. Clementini, G., et al., 2001. Astrophys. J. 559, L109.
3. Clementini, G., et al., 2010. Astron. J., in preparation.
4. Clementini, G., et al., 2010. Astrophys. J., in preparation.
5. Fiorentino, G., et al., 2010. Astron. J., in preparation.
6. Glatt, K., et al., 2008. Astrophys. J. 136, 1703.
7. Greco, C., et al., 2007. Astrophys. J. 670, 332.
8. Greco, C., et al., 2007. In: Peletier, R.F., Vazdekis, A. (Eds.), IAU Symp., vol. 241, p. 343. Cambridge University Press, Cambridge.
9. Ibata, R.A., Gilmore, G., Irwin, M.J., 1994. Nature 370, 194.
10. Ibata, R., et al., 2001. Nature 412, 49.
11. Mackey, A.D., Gilmore, G., 2003. Mon. Not. R. Astron. Soc. 343, 747.
12. Marconi, M., et al., 2010. Astrophys. J., in preparation.
13. Oosterhoff, P.Th., 1939. Observatory 62, 104.

Chapter 22
Extremely Metal-Poor Star-Forming Dwarf Galaxies

Alessandra Aloisi

Abstract New deep HST/ACS photometry of the extremely metal-poor star-forming dwarf galaxy I Zw 18 is presented and compared to a similar study of its twin sister SBS 1415+437. Chemically-unevolved systems like I Zw 18 or SBS 1415+437 represent the closest analog to primordial galaxies in the early universe. It has also been argued that some of these objects (e.g., I Zw 18) may be genuine nearby 'young' galaxies. However, by revealing the existence of an evolved red stellar population in both I Zw 18 and SBS 1415+437, our HST/ACS observations do not support such an interpretation.

22.1 Introduction

In hierarchical formation scenarios, dwarf ($M < 10^9\ M_\odot$) galaxies are the first systems to collapse and start forming stars, supplying the building blocks for the formation of more massive systems. As remnants of this process, present-day dwarfs may have been sites of the earliest star-formation (SF) activity in the universe. However, the most metal-poor ($12+\log(O/H) < 7.6$, i.e. $Z < 1/20\ Z_\odot$) dwarf irregular (dIrr) and blue compact dwarf (BCD) galaxies have been repeatedly pointed out as candidate "primeval" galaxies in the nearby universe, with ages <100–500 Myr (e.g. [5]).

The only direct way to unambiguously infer the evolutionary status of a metal-poor dIrr/BCD is to resolve it into stars with deep HST observations, and study stellar features in the color–magnitude diagram (CMD). The brightest of these features that contains stars of significant age is the Red Giant Branch (RGB), formed by low-mass stars with ages ∼1–13 Gyr. In the last 15 years all metal-poor dIrrs in the Local Group and BCDs with $D < 15$ Mpc have been imaged with HST. An RGB has been detected in all those galaxies for which photometric data exist that go deep enough to reach the RGB tip (TRGB, brightest phase of the RGB) in the CMD.

The only possible exception so far is the BCD I Zw 18. With $12+\log(O/H) = 7.2$, i.e. $Z = 1/50\ Z_\odot$, I Zw 18 is the most metal-poor galaxy in the nearby universe.

A. Aloisi (✉)
Space Telescope Division of the European Space Agency, Space Telescope Science Institute, Baltimore, MD 21218, USA
e-mail: aloisi@stsci.edu

F.D. Macchetto (ed.), *The Impact of HST on European Astronomy*, Astrophysics and Space Science Proceedings,
DOI 10.1007/978-90-481-3400-7_22, © Springer Science+Business Media B.V. 2010

After other groups had already resolved its brightest individual stars, our group was the first to go deep enough to detect asymptotic giant branch (AGB) stars in HST/WFPC2 images with ages of at least several hundreds Myr and possibly up to a few Gyr [2, 8]. More recently, [6] presented new deep HST/ACS observations. Their I vs. V–I CMD shows no sign of an RGB, from which they concluded that the most evolved (AGB) stars are not older than 500 Myr. This result was subsequently challenged by [7] and our group [9] based on a better photometric analysis of the same data. This showed that many red sources do exist at the expected position of an RGB, and that their density drops exactly where a TRGB would be expected. However, small number statistics, large photometric errors, and incompleteness at the TRGB, did not allow a conclusive statement.

22.2 New HST/ACS Observations of I Zw 18

We were awarded 24 additional orbits with ACS over a three-month period starting in October 2005 (GO program 10586, PI Aloisi) to better understand the evolutionary status of I Zw 18. The observations were obtained in 12 different epochs in F606W and F814W to: (1) build a deeper CMD to search for RGB stars; (2) detect and characterize Cepheids at the lowest metallicity available in the local universe; and (3) use both Cepheids and a possible TRGB detection to determine an accurate distance to I Zw 18 (see [1]).

PSF-fitting photometry was performed on deep images that were obtained by combining the exposures in each filter with MultiDrizzle. After application of aperture corrections, the count rates were transformed to Johnson–Cousins V and I magnitudes. Values shown and discussed hereafter are corrected for $E(B-V) = 0.032$ mag of Galactic foreground extinction, but not for any extinction intrinsic to I Zw 18. The archival ACS data in F555W and F814W (GO program 9400, PI Thuan) were also re-processed in a similar manner. The two ACS datasets were then combined by demanding that stars should be detected in all the four deep images (V and I for both datasets).

22.3 Results and Interpretation

Figure 22.1a shows the resulting I vs. $V-I$ CMD of I Zw 18. The CMD shows faint red stars exactly at the position where an RGB would be expected. Figure 22.2 shows the luminosity function (LF) of the red stars. It shows a sharp drop towards brighter magnitudes, exactly as would be expected from a TRGB. The magnitude of the discontinuity, $I = 27.27 \pm 0.14$, implies a distance modulus $m - M = 31.30 \pm 0.17$, i.e., $D = 18.2 \pm 1.5$ Mpc. The TRGB distance is consistent with the distance as inferred from the analysis of the Cepheid variables identified by our program (Fig. 22.1, see also [1]). This agreement further supports our interpretation of the LF drop in Fig. 22.2 as a TRGB feature.

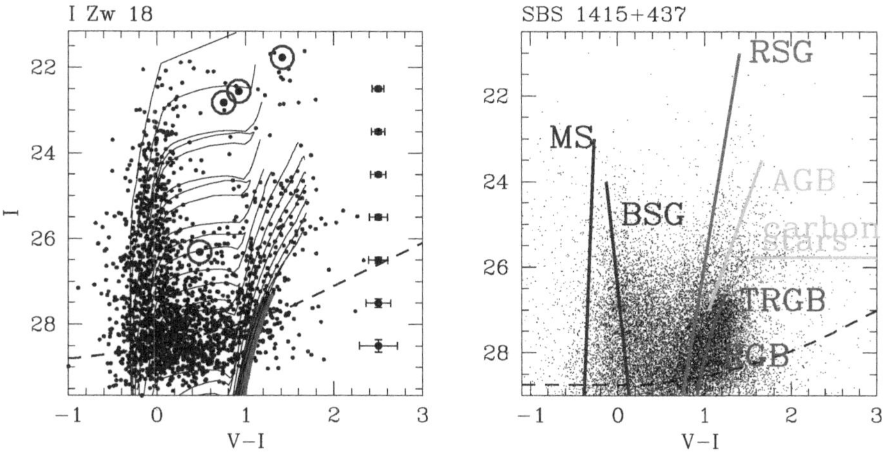

Fig. 22.1 (Color online) **(a)** HST/ACS CMD for I Zw 18 [1]. Median photometric errors at $V - I = 1$ (determined by comparison of measurements from GO-9400 and GO-10586) are also shown as function of I. Padua isochrones from 5.5 Myr to 10 Gyr are overlaid, with the RGB phase for isochrones from 1.7 to 10 Gyr colored *red*. The isochrones have metallicity $Z = 0.0004$ (as inferred from the HII regions of I Zw 18) and are shown for the distance $D = 18.2$ Mpc ($m - M = 31.30$). The CMD includes stars in both the main and secondary bodies of I Zw 18. *Blue open circles* highlight the four confirmed variables, which are plotted according to their intensity-averaged magnitudes. **(b)** HST/ACS CMD for SBS 1415+437 [3]. The main evolutionary sequences seen in the data are indicated in approximate sense as *colored straight lines*: main sequence (MS), blue supergiants (BSG), red supergiants (RSG), the red giant branch (RGB) with its tip (TRGB), the asymptotic giant branch (AGB), and carbon stars. Both CMDs are corrected for Galactic foreground extinction. *Dashed lines* are estimates of the 50% completeness level. The *vertical axes* of the *panels* are offset from each other by 0.61 mag, i.e. the difference in distance modulus between the galaxies (see Fig. 22.2). Some ~10 times more stars were detected in SBS 1415+437, owing to its smaller distance

The evidence for an RGB in I Zw 18 is further strengthened by comparison to another BCD, SBS 1415+437, observed by us with a similar HST/ACS set-up [3]. This galaxy is not quite as metal poor as I Zw 18 ($12 + \log(O/H) = 7.6$) and is somewhat nearer at $D \approx 13.6$ Mpc. But taking into account the differences in distance and completeness, the CMDs of these galaxies look very similar. Since SBS 1415+437 has an unmistakable RGB sequence, this suggests that such an RGB sequence exists in I Zw 18.

22.4 Conclusions

Our HST/ACS observations of I Zw 18 and SBS 1415+437 provide improved insight into the evolutionary status of metal-poor BCDs by indicating that underlying old (>1 Gyr) populations are present in even the most metal-poor systems. The coherent picture that emerges is that these galaxies did not form recently ($z < 0.1$) and

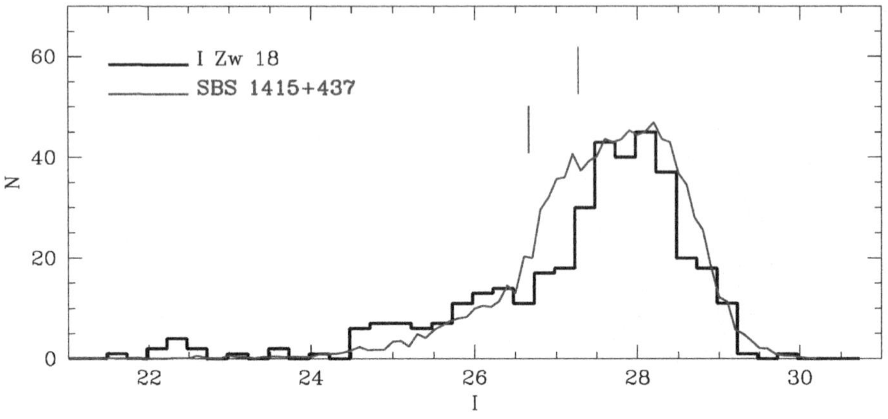

Fig. 22.2 (Color online) I-band LFs for stars with red colors in the range $V - I = 0.75$–1.5 mag, inferred from the CMDs in Fig. 22.1. Normalizations are arbitrary. *Vertical marks* indicate the positions of the TRGB, as determined using a Savitzky–Golay filtering technique developed by one of us [4]. At these magnitudes there is a steep LF drop towards brighter magnitudes, due to the end of the RGB sequence. By contrast, the LF drop towards fainter magnitudes at $I > 28$ mag is due to incompleteness in both cases. Apart from a shift $\Delta(m - M) \approx 0.61$, these metal-poor BCD galaxies have very similar LFs

may well be as old as the first systems that collapsed in the early universe. Deeper studies (well below the TRGB) will be needed to pinpoint the exact onset of the star formation in these extreme objects.

Acknowledgements The author would like to thank all her collaborators that made these projects possible, in particular F. Annibali, G. Clementini, R. Contreras, G. Fiorentino, C. Leitherer, J. Mack, M. Marconi, I. Musella, A. Saha, M. Sirianni, M. Tosi, and R. van der Marel.

References

1. Aloisi, A., et al., 2007. Astrophys. J. 667, L151.
2. Aloisi, A., Tosi, M., Greggio, L., 1999. Astron. J. 118, 302.
3. Aloisi, A., van der Marel, R.P., Mack, J., Leitherer, C., Sirianni, M., Tosi, M., 2005. Astrophys. J. 631, L45.
4. Cioni, M.-R.L., van der Marel, R.P., Loup, C., Habing, H.J., 2000. Astron. Astrophys. 359, 601.
5. Izotov, Y.I., Thuan, T.X., 1999. Astrophys. J. 511, 639.
6. Izotov, Y.I., Thuan, T.X., 2004. Astrophys. J. 616, 768.
7. Momany, Y., et al., 2005. Astron. Astrophys. 439, 111.
8. Östlin, G., 2000. Astrophys. J. 535, L99.
9. Tosi, M., Aloisi, A., Mack, J., Maio, M., 2007, in: Combes, F., Palous, J. (Eds.), Galaxy Evolution Across the Hubble Time, IAU Symp., vol. 235, p. 65. Cambridge University Press, Cambridge.

Chapter 23
Resolved Stellar Populations in Nearby Galaxy Halos

Marina Rejkuba

Abstract The results of the three different studies of resolved stellar populations in galaxies across the Hubble sequence based on HST ACS and WFPC2 data are summarized in this contribution. (1) The ACS observations of the stellar halo of the gE galaxy NGC 5128 revealed the core helium-burning stars located in the red clump. Combining the metallicity distribution measured from the upper red giant branch (RGB) and the luminosities of the red clump and asymptotic giant branch (AGB) bump stars the constraints on the star formation history are given. (2) The WFPC2 observations supplemented by the deep near-IR observations with ISAAC on the ESO VLT have been used to derive the ratios of the intermediate-age to old stars in two early-type dwarf companions of NGC 5128. Their intermediate-age star fractions are lower than in outlying dSph satellites of the Milky Way. (3) The ACS observations of the color distribution of the RGB stars in the halo of the edge-on spiral galaxy NGC 891 gives the possibility to derive the metallicity distribution. The average halo metallicity at $\sim$10 kpc above the galaxy disk is $[Fe/H] = -0.9$ dex, significantly higher than that of the Milky Way halo.

23.1 Introduction

The resolved stellar populations of nearby galaxy halos contain the important historical record with the oldest and most metal-poor components of galaxies. Therefore they provide clues to the understanding of how galaxies have assembled their mass and constrain the early phases of galaxy formation. I summarize the main results of three studies of resolved stellar populations in different types of galaxies outside the Local Group: in gE/S0 galaxy NGC 5128, in its dSph/dE companions and in the edge-on spiral galaxy NGC 891. More details can be found in the original papers [7, 12, 13].

23.2 The Remote Metal-Rich Halo of NGC 5128

At 3.8 Mpc NGC 5128 (= Centaurus A) is the nearest easily observable gE/S0 galaxy [1, 11]. Its structure is that of a giant elliptical galaxy crossed by a warped

M. Rejkuba (✉)
ESO, Karl-Schwarzschild-Strasse 2, 85748 Garching b. München, Germany
e-mail: mrejkuba@eso.org

F.D. Macchetto (ed.), *The Impact of HST on European Astronomy*, Astrophysics and Space Science Proceedings,
DOI 10.1007/978-90-481-3400-7_23, © Springer Science+Business Media B.V. 2010

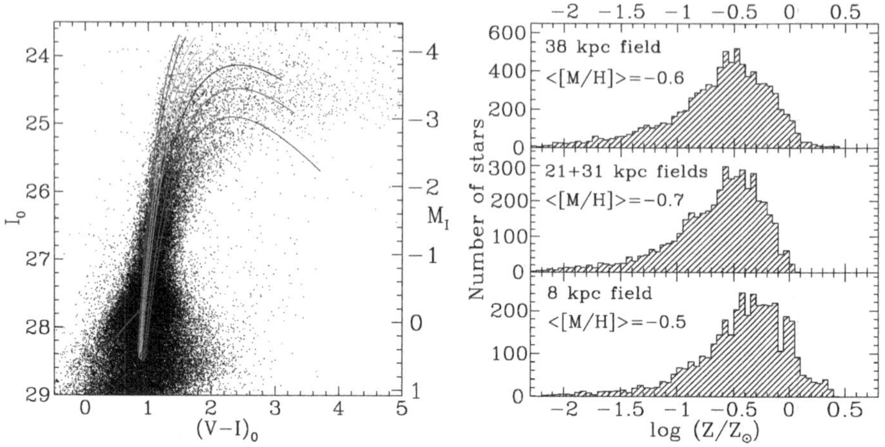

Fig. 23.1 (Color online) *Left*: VI CMD of the stars in the ACS field of NGC 5128. The RGB tip is detected at $I\,(\mathrm{RGBT}) = 24.05$, the AGB bump is at $I\,(\mathrm{AGBb}) = 26.77$ and the mean magnitude of the red clump is $I\,(\mathrm{RC}) = 27.873$. The *overplotted lines* are the empirical fits to the MW globular cluster RGBs from [14]. *Right*: The MDFs of three fields in NGC 5128. The average metallicity for each field is shown

dust lane in the centre. Numerous HST orbits have been dedicated to the observations of the resolved stellar populations in NGC 5128, mostly using the WFPC2 camera [1–4, 9, 15]. The first view of the resolved stellar populations in this galaxy halo in the near-IR was presented by [8] based on the NICMOS data. The deepest optical observations were obtained with the latest camera on board of HST, the Advanced Camera for Surveys (ACS) [12]. Here I summarize the main results of this last study.

The ACS observations of a field located ~38 kpc south of the galaxy center were taken in Cycle 11. They consist of 24 full orbit exposures evenly split between F606W (V) and F814W (I) bands. The PSF fitting photometry on pipeline processed images was performed with DAOPHOT suite of programmes. The resulting color–magnitude diagram (CMD) of 77,810 stars (Fig. 23.1) is dominated by an old stellar population of RGB stars. The RGBT is detected at $I\,(\mathrm{RGBT}) = 24.05 \pm 0.05$ for stars bluer than $(V - I) \leq 1.8$. It yields the distance modulus in agreement with previous RGBT measurements [3, 15]. Due to a very wide metallicity of the RGB stars the I magnitude becomes fainter as $V - I$ gets redder or the stars more metal-rich. The slope of the I vs. $V - I$ relation for the metal-rich stars is well reproduced by the Teramo stellar evolution models [10].

The metallicity distribution function (MDF) is shown in Fig. 23.1. The slightly lower average metallicity of the 21+31 kpc fields is due to photometric incompleteness at the metal-rich end for these observations [2, 12]. Figure 23.1 shows that up to ~7 R_e away from the center of galaxy there is no appreciable metallicity gradient in the halo. Very recently [5] have shown that in order to find the metal-poor halos in gE galaxies one should probably look beyond 10–12 R_e.

Comparing the luminosity of the observed asymptotic giant branch (AGB) bump and red clump (RC) stars, assuming the measured MDF, with the stellar evolutionary models it is possible to give the luminosity weighted mean age of the stars in the halo. The resulting average luminosity weighted age of the halo stars is $8^{+3}_{-3.5}$ Gyr.

23.3 Dwarf Spheroidal Galaxies in Cen A Group

Dwarf spheroidal galaxies in Cen A group AM 1339-445 and AM 1343-452 have been observed with the WFPC2 camera in F606W and F814W bands [6]. The data have been retrieved from the HST science archive, and after the pipeline processing the PSF photometry was done with the HSTphot package (see [13] for details). The optical CMDs of these two galaxies show old RGB stars. The tip of the RGB is clearly detected and together with the $(V - I)_{0,3.5}$, the dereddened color of the RGB at $M_I = -3.5$, is used to derive distance moduli to the two galaxies. They are $(m - M)_0(\text{AM 1339-445}) = 27.24 \pm 0.20$, and $(m - M)_0(\text{AM 1343-452}) = 27.86 \pm 0.20$. The mean metallicities are determined from the mean color of the RGBs: $\langle[\text{Fe/H}]\rangle = -1.4 \pm 0.2$ for AM 1339-445 and $\langle[\text{Fe/H}]\rangle = -1.6 \pm 0.2$ for AM 1343-452.

Above the tip of the RGB a small population of bright stars is detected. Some are expected to belong to the foreground Milky Way (MW) population, but some could be metal-poor upper-AGB stars, present in an intermediate-age population. In order to put the constraints on the number and maximum luminosity of the AGB population in these galaxies the optical WFPC2 photometry was combined with the near-IR J and Ks magnitudes obtained from the deep ISAAC images. The upper-AGB star candidates were selected according to the following selection criteria: in optical $1.6 \leq (V - I)_0 \leq 3.3$ and $-4.3 \geq M_I \geq -5.5$, while in near-IR the stars had to be redder than most of the MW stars, i.e. $(J - K)_0 > 0.90$, and brighter than the RGB tip $K_0 < 21.4$ (AM 1339-445) or $K_0 < 21.7$ (AM 1343-452). This resulted in 11 upper-AGB candidates in AM 1339-445 and nine in AM 1343-452. Two additional AGB candidates were found in AM 1343-452 based on the presence of variability in Ks observations.

Using an empirical AGB tip luminosity vs. age relation (Fig. 23.2), the brightness of the most luminous AGB candidates indicate that the last significant star formation episode was ~4 Gyr ago in AM 1343-452 and ~6.5 Gyr ago in AM 1339-445. Their relative numbers with respect to the number of RGBT stars gives a fraction of intermediate-age stellar populations in both galaxies of ~15%. This is significantly less than what is found in outlying MW satellites, indicating possibly the role of the harsher environment of the Cen A group on the evolution of these dwarf galaxies.

23.4 The Milky Way Analogue NGC 891

The archival ACS data from Cycle 11 have been used to construct deep CMDs, reaching ~2 mag fainter than the tip of the RGB, in three different fields in the edge-on spiral galaxy NGC 891 [7]. The tip of the RGB brightness measurement yields

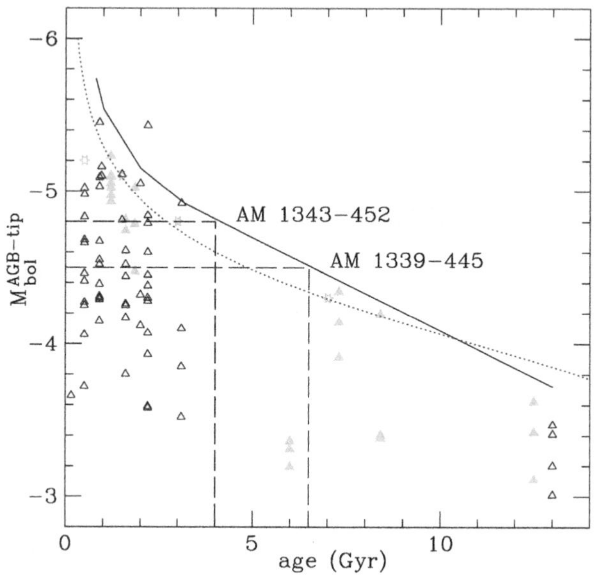

Fig. 23.2 (Color online) The empirical AGB tip luminosity–age relation constructed based on the LMC and SMC cluster data (*blue and cyan triangles*). See [13] for details

the distance modulus of $(m - M)_0 = 29.94 \pm 0.04$(random) $\pm$ 0.16(systematic). The observed fields span a range of distances parallel to the disk from ~ -1.5 to 17 kpc, while the perpendicular distance extends up to ~12 kpc above the disk. The MDF of the stars located at 10 ± 1.5 kpc distance from the disk probes the halo of the galaxy. The average metallicity is $[Fe/H] \approx -0.9$ dex, significantly higher than that in the Milky Way halo. This result has implications for the formation and extent of the metal-poor halos in the spiral galaxies.

References

1. Ferrarese, L., et al., 2007. Astrophys. J. 654, 186.
2. Harris, G.L.H., Harris, W.E., 2000. Astron. J. 120, 2423.
3. Harris, G.L.H., Harris, W.E., Poole, G.B., 1999. Astron. J. 117, 855.
4. Harris, W.E., Harris, G.L.H., 2002. Astron. J. 123, 3108.
5. Harris, W.E., et al., 2007. Astrophys. J. 666, 903.
6. Karachentsev, I.D., et al., 2002. Astron. Astrophys. 385, 21.
7. Mouhcine, M., Rejkuba, M., Ibata, R., 2007. Mon. Not. R. Astron. Soc. 381, 873.
8. Marleau, F.R., et al., 2000. Astron. J. 120, 1779.
9. Mould, J.R., et al., 2000. Astrophys. J. 536, 266.
10. Pietrinferni, A., et al., 2004. Astrophys. J. 612, 168.
11. Rejkuba, M., 2004. Astron. Astrophys. 413, 903.
12. Rejkuba, M., et al., 2005. Astrophys. J. 631, 262.
13. Rejkuba, M., et al., 2006. Astron. Astrophys. 448, 983.
14. Saviane, I., et al., 2000. Astron. Astrophys. 355, 966.
15. Soria, R., et al., 1996. Astrophys. J. 465, 79.

Chapter 24
The Kinematics of Core and Cusp Galaxies: Comparing HST Imaging and Integral-Field Observations

J. Falcón-Barroso, R. Bacon, M. Cappellari, R.L. Davies, P.T. de Zeeuw, E. Emsellem, D. Krajnović, H. Kuntschner, R.M. McDermid, R.F. Peletier, M. Sarzi, and G. van de Ven

Abstract In this proceeding we look at the relationship between the photometric nuclear properties of early-type galaxies from Hubble Space Telescope imaging and their overall kinematics as observed with the SAURON integral-field spectrograph. We compare the inner slope of their photometric profiles and the Slow/Fast rotator classes, defined by the amplitude of a newly defined λ_R parameter, to show that slow rotators tend to be more massive systems and display shallower inner profiles and fast rotators steeper ones. It is important to remark, however, that there is not a one-to-one relationship between the two photometric and kinematic groups.

J. Falcón-Barroso (✉)
European Space and Technology Centre, Keplerlaan 1, 2200 AG Noordwijk, The Netherlands
e-mail: jfalcon@rssd.esa.int

R. Bacon · E. Emsellem
Université de Lyon 1, CRAL, Observatoire de Lyon, 9 av. Charles André, 69230 Saint-Genis Laval, France

M. Cappellari · R.L. Davies · D. Krajnović
Sub-Department of Astrophysics, University of Oxford, Denys Wilkinson Building, Keble Road, Oxford OX1 3RH, UK

P.T. de Zeeuw · R.M. McDermid
Sterrewacht Leiden, Universiteit Leiden, Postbus 9513, 2300 RA, Leiden, The Netherlands

H. Kuntschner
Space Telescope European Coordinating Facility, European Southern Observatory, Karl-Schwarzschild-Str. 2, 85748 Garching, Germany

R.F. Peletier
Kapteyn Astronomical Institute, University of Groningen, 9700 AV Groningen, The Netherlands

M. Sarzi
Centre for Astrophysics Research, University of Hertfordshire, Hatfield, Herts AL10 9AB, UK

G. van de Ven
Institute for Advanced Study, Einstein Drive, Princeton, NJ 08540, USA

F.D. Macchetto (ed.), *The Impact of HST on European Astronomy*, Astrophysics and Space Science Proceedings,
DOI 10.1007/978-90-481-3400-7_24, © Springer Science+Business Media B.V. 2010

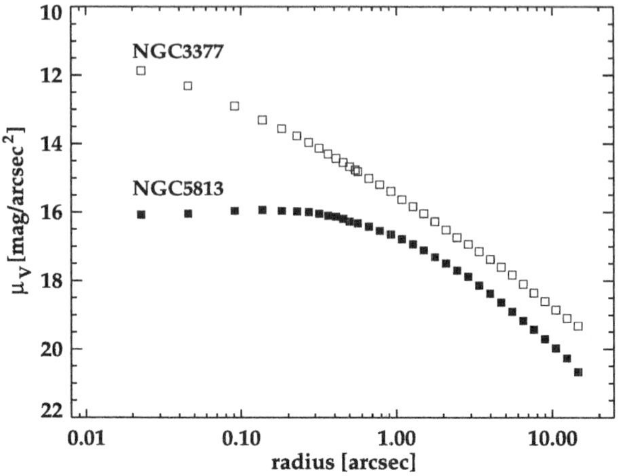

Fig. 24.1 An illustration of a core and a cuspy nuclear profiles of two galaxies in the SAURON sample from HST V-band observations. Data from [12]

24.1 Introduction

The study of the galactic nuclei in early-type galaxies has played a fundamental role in our understanding of how galaxies form and evolve. The arrival of high resolution instrumentation with the Hubble Space Telescope (HST) in the early 1990s opened a new window into the analysis of the nuclear properties of these systems. One of the major discoveries by HST in this respect is the *apparent* dichotomy in the nuclear structural properties of these galaxies (see Fig. 24.1). First highlighted by [7] and soon after extended by [11], the core/cusp properties of early-type galaxies have been, and still are, under severe scrutiny by many groups around the world (e.g. [10, 14, 15, 17]). Recently the existence of a dichotomy has been questioned by [8] from ACS observations as part of the ACS Virgo Cluster Survey [2] (see also A. Jordán contribution in these proceedings — Chap. 18). The topic, however, is far from settled as illustrated by the numerous recent papers debating the existence of the two classes (e.g. see [9, 13] and references therein for an in-depth discussion on the subject). Independently of the debate, it is unquestionable that HST is the only current facility in the world that can allow us to revisit this issue.

Almost in parallel to the progress made by HST, ground-based integral-field spectroscopy slowly started to emerge allowing us to look at galaxies in a different new way. With the first observations in 1999, the SAURON project [3] has been one of the pioneers in the exploitation of this technology to study the kinematical properties of galaxies. In this contribution we made use of one of the latest results from our survey to connect the nuclear photometric properties of early-type galaxies with their overall level of rotation. A more detailed analysis of the results presented here is the matter of study in [5].

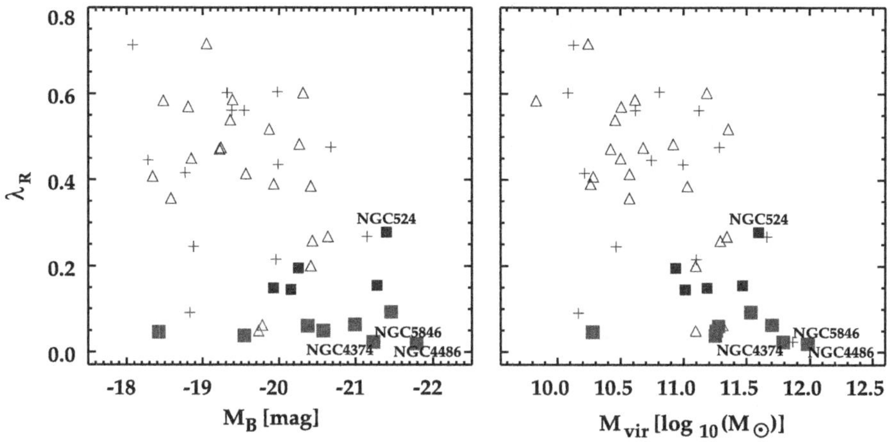

Fig. 24.2 (Color online) λ_R versus absolute magnitude M_B (*left panel*) and virial mass M_{vir} (*right panel*) for the 48 E and S0 of the SAURON sample. In both *panels*, symbols correspond to the inner slope classification [6, 12, 14, 15] with power-laws as *open triangles*, cores as *filled squares*, and *crosses* indicating galaxies for which there is no published classification. Slow rotators are coloured in *red* and fast rotators in *blue*

24.2 Core and Cusp Galaxies in the SAURON Sample

There are 33 galaxies in the SAURON sample of 48 elliptical and lenticular galaxies with known values for their inner profile slope, which separates them in "core" and "power-law" galaxies [6, 12, 14, 15]. From the point of view of their stellar kinematics (see [4]) early-type galaxies appear in two broad flavours, depending on whether they exhibit clear large-scale rotation or not. We measure the level of rotation via a new parameter (λ_R) and use it as a basis for a new kinematic classification that separates early-type galaxies into slow and fast rotators in [5]. We have defined the new quantity λ_R:

$$\lambda_R \equiv \frac{\langle R|V|\rangle}{\langle R\sqrt{V^2+\sigma^2}\rangle},$$

that measures the amount of specific (projected) angular momentum from the velocity maps. The parameter has been defined such that is insensitive to small features in the maps, and therefore provides a robust measurement of the global rotation. As we go from galaxies with low to high λ_R values, the overall velocity amplitude naturally tends to increase. More importantly, there seems to be a change in the observed stellar velocity structures.

In the left panel of Fig. 24.2, we show the distribution of λ_R as a function of absolute magnitude M_B. The three slowest rotators (NGC 4486, NGC 4374, NGC 5846) are among the brightest galaxies in our sample with $M_B < -21$ mag. Other slow rotators tend to be bright but are spread over a wide range of absolute magnitude. Most fast rotators are fainter than $M_B > -20.5$ mag. In terms of core/cusp distribution, we find that most slow rotators are core galaxies while most fast rotators display power-law profiles. Interestingly there are no core galaxies with $\lambda_R > 0.3$, although the inclined core galaxy NGC 524 would probably have a very high λ_R value if seen edge-on. The general behavior that core galaxies have lower λ_R than cusp galaxies is expected, since both classifications show trends with total luminosity, with brighter members tending to be core galaxies, and lower luminosity ones having power-law profiles [6]. Indeed, all galaxies with $\lambda_R > 0.3$ have $M_B > -20.7$. It is important to notice though that there is not a one-to-one correspondance between the different photometric and kinematic groups, since we find both "power-laws" in slow rotators and "cores" in fast rotator. One can appreciate in the figure that there is a domain in luminosity and also in mass where both cusp and core galaxies coexist.

In the right panel of Fig. 24.2, we show the same trends but now as a function of the total mass of the galaxies. We have derived the mass values assuming it approximates to the virial mass derived from the best-fitting M_{vir}–σ relation presented in [1]. The figure displays a trend with λ_R such that smaller values are found in more massive galaxies. The three slowest rotators are in the high range of M_{vir} with values above $10^{11.5}$ $M_{\odot}$. There is a clear overlap in mass between fast and slow rotators for M_{vir} between 10^{11} and $10^{11.5}$ $M_{\odot}$. However, all slow rotators, have $M_{vir} > 10^{11}$ $M_{\odot}$, whereas most fast rotators have $M_{vir} < 10^{11}$ $M_{\odot}$, lower masses being reached as the value of λ_R increases.

The lack of one-to-one relationship between the photometric and kinematic classes is probably highlighting the complex merging histories these galaxies go through in their evolution. Based on the studies showing the incidence of gas in our sample [5, 16], it is likely that dissipation is one of the dominant factors in the way galaxies in the different classes evolve. In a forthcoming paper we will combine HST and ground-based imaging with our integral-field observations to extend the analysis presented in this contribution, and to investigate in more detail the links between core/cups and slow/fast rotators in our sample of galaxies.

References

1. Cappellari, M., et al., 2006. Mon. Not. R. Astron. Soc. 366, 1126.
2. Côté, P., et al., 2004. Astrophys. J. Suppl. Ser. 153, 223.
3. de Zeeuw, P.T., et al., 2002. Mon. Not. R. Astron. Soc. 329, 513.
4. Emsellem, E., et al., 2004. Mon. Not. R. Astron. Soc. 352, 721.
5. Emsellem, E., et al., 2007. Mon. Not. R. Astron. Soc. 379, 401.
6. Faber, S.M., et al., 1997. Astron. J. 114, 1771.
7. Ferrarese, L., van den Bosch, F.C., Ford, H.C., Jaffe, W., O'Connell, R.W., 1994. Astron. J. 108, 1598.
8. Ferrarese, L., et al., 2006. Astrophys. J. Suppl. Ser. 164, 334.
9. Ferrarese, L., et al., 2006. In: Karas, V., Matt, G. (Eds.), Black Holes: From Stars to Galaxies — Across the Range of Masses, Proceedings IAU Symposium, vol. 238. astro-ph/0609762.
10. Graham, A.W., Erwin, P., Trujillo, I., Asensio Ramos, A., 2003. Astron. J. 125, 2951.
11. Lauer, T.R., et al., 1995. Astron. J. 110, 2622.
12. Lauer, T.R., et al., 2005. Astron. J. 129, 2138.
13. Lauer, T.R., et al., 2007. Astrophys. J. 664, 226.

14. Ravindranath, S., Ho, L.C., Peng, C.Y., Filippenko, A.V., Sargent, W.L.W., 2001. Astron. J. 122, 653.
15. Rest, A., van den Bosch, F.C., Jaffe, W., Tran, H., Tsvetanov, Z., Ford, H.C., Davies, J., Schafer, J., 2001. Astron. J. 121, 2431.
16. Sarzi, M., et al., 2006. Mon. Not. R. Astron. Soc. 366, 1151.
17. Trujillo, I., Erwin, P., Asensio Ramos, A., Graham, A.W., 2004. Astron. J. 127, 1917.

Chapter 25
Resolving Extragalactic Star Clusters with HST/ACS

Søren S. Larsen

Abstract With HST, colour–magnitude diagrams (CMDs) can be obtained for young star clusters well beyond the Local Group. Such data can help constrain cluster ages and metallicities, and also provide a reference against which intermediate- and high mass stellar models can be compared. Here, CMDs are presented for two massive ($>10^5$ $M_\odot$) clusters and compared with Padua and Geneva isochrones. The problem of the ratio of blue to red supergiants is also addressed.

Star clusters remain the best approximation provided by Nature to "simple stellar populations" and have a long history as important test labs for models of stellar evolution. The rich old globular clusters (GCs) in the Milky Way have played a vital role for testing and calibrating models for low-mass stars [11]. However, observational tests of intermediate- and high mass stars remain more scarce. The problem is largely one of statistics: for any realistic stellar initial mass function (IMF), massive stars constitute only a small fraction of the total number of stars in a cluster, and have the shortest life times. Where evolved stars are concerned, this difficulty becomes especially acute: a 10^5 $M_\odot$ cluster with an age of 10^7 years and a Kroupa IMF [5] is only expected to contain about 25 red supergiants, while a 10 Gyr old GC of the same mass contains $\sim$1,000 post-main sequence stars. Clearly, evolved massive stars are rare in typical open clusters, and even within the entire Local Group the number of potential targets is limited if one wishes to obtain useful constraints on quantities such as the ratio of blue to red supergiants.

Fortunately, several galaxies within distances of a few Mpc contain significant populations of young "massive" clusters (YMCs) with masses in excess of 10^5 $M_\odot$ [7]. In such clusters, the late stages of stellar evolution start to be reasonably well sampled even for young ($\sim 10^7$–10^8 years) ages. With HST, colour–magnitude diagrams (CMDs) are within reach for some of these clusters, although care has to be taken in order to deal with the severe crowding. In the following, early results for two illustrative cases are presented.

S.S. Larsen (✉)
Astronomical Institute, Utrecht University, Princetonplein 5, 3584 CC, Utrecht, The Netherlands
e-mail: larsen@astro.uu.nl

F.D. Macchetto (ed.), *The Impact of HST on European Astronomy*, Astrophysics and Space Science Proceedings,
DOI 10.1007/978-90-481-3400-7_25, © Springer Science+Business Media B.V. 2010

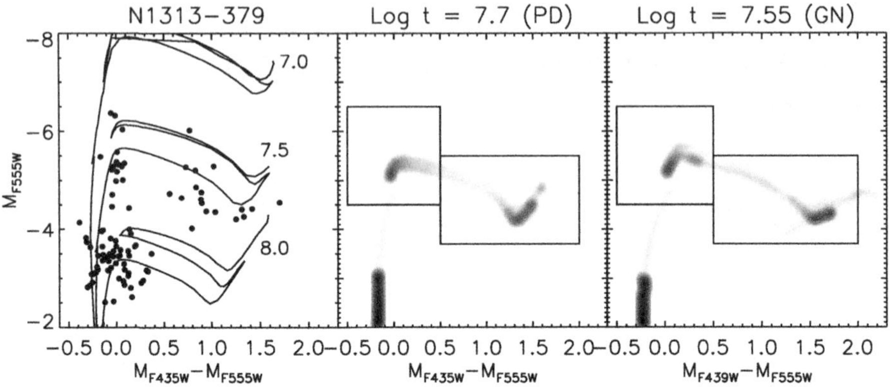

Fig. 25.1 *Left*: observed colour–magnitude diagram for a massive star cluster in NGC 1313. *Centre* and *right*: Hess diagrams showing the density of stars predicted by isochrones from the Padua (*centre*) and Geneva groups (*right*). *Boxes* indicate regions where red and blue supergiants are counted, as explained in the text

25.1 Case 1: A 35–50 Myr Old Cluster in NGC 1313

The Magellanic-type galaxy NGC 1313 ($D \sim 4.1$ Mpc) is known from ground-based imaging to host several YMCs [7]. The western part of the galaxy appears to have experienced a recent burst of star formation during which also a particularly massive star cluster was formed (#379 in the list of [7]). The cluster has an age of ~50 Myr (see below), consistent with the peak of field star formation activity in the region [8]. The integrated magnitude $M_V = -10.9$ corresponds to a mass of 175,000 $M_\odot$ for a Chabrier IMF [1, 2].

The cluster is well resolved on our ACS/WFC images (Progr. ID 9774; P.I. S.S. Larsen), and the CMD is shown in Fig. 25.1 (left). Several distinct features are seen: The main sequence turn-off (MSTO) is at $m_{F435} - m_{F555W} \approx 0$ and $M_V \approx -3.5$. At brighter absolute magnitudes ($M_V \sim -5.5$) are the blue core He burning stars ("blue supergiants", BSGs). The red core He burning stars ("red supergiants", RSGs) are seen at $m_{F435} - m_{F555W} > 0.5$ and $M_V \sim -4.5$. Also shown are Padua isochrones [4] for $Z = 0.008$ and $\log t = 7.0$, 7.5 and 8.0. The cluster appears slightly older than $\log t = 7.5$.

Isochrones do not fully illustrate the distribution of stars in the CMD. In particular, the gap between the MSTO and the BSGs is only revealed when the *density* of stars in the CMD is shown. The centre panel shows the synthetic Hess diagram for a $\log t = 7.7$ Padua isochrone, while the right-hand panel shows a $\log t = 7.55$ Geneva isochrone. These ages correspond to MSTO masses of 7.4 and 8.1 $M_\odot$, respectively. Both sets of models reproduce the overall distribution of stars in the CMD fairly well, albeit for somewhat different ages. The RSGs may be slightly too red (cool) in both cases. The apparent difference in the model colours of the RSGs is mostly due to the fact that WFPC2 F439W–F555W colours are shown for the Geneva isochrones, while the proper ACS/WFC F435W–F555W colours are used for the Padua isochrones. In reality, the difference in $T_{\rm eff}$ between the Padua and Geneva RSGs is small, with the Geneva RSGs being some 150 K cooler.

One long-standing problem concerns the relative fraction of RSGs and BSGs [3, 6]. Models tend to predict a decreasing BSG/RSG ratio with increasing metallicity, while observations show the opposite. Fortunately, a comparison with the data is possible without an exact calibration between fundamental stellar properties (L, $T_{\rm eff}$) and colours. The boxes drawn in Fig. 25.1 indicate regions of the CMD where we count RSGs and BSGs. In the observed CMD we find 16 BSGs and 17 RSGs, corresponding to a ratio of BSG/RSG = 0.94 ± 0.33. Within the error, this is consistent with the model predictions: BSG/RSG = 0.73 and 0.74 for the Padua and Geneva models.

To summarize, the agreement between models and observations seems rather satisfactory for this cluster. This is true both for the colours and luminosities of various types of stars and the BSG/RSG ratio. Since the stars in this cluster only barely qualify for the label "massive", this was perhaps to be expected. It should be noted, however, that the CMD-based age could be either 50 or 35 Myr, depending on the choice of models. An important corollary is that absolute ages derived from *integrated* colours (or spectra) are likely to be uncertain by at least the same amount.

25.2 Case 2: Cluster NGC 1569-B

The dwarf irregular galaxy NGC 1569 is well known for hosting two very compact, massive young clusters. Although it is about a factor of two closer than NGC 1313, observations of the clusters in NGC 1569 are challenging due to both their more compact structure, the general degree of crowding, and the significant Galactic foreground reddening ($b = 11°$). Nevertheless, both clusters A and B are resolved into individual stars in ACS/HRC data (Progr. ID. 9300; P.I. H. Ford); here we concentrate on cluster B for which we also have Keck/NIRSPEC near-infrared spectroscopy [9]. The reddening-corrected absolute V magnitude is $M_V = -12.2$ [10], corresponding to a mass of about 280,000 $M_\odot$ for an age of 15 Myr.

The CMD for NGC 1569-B is shown in Fig. 25.2. The RSGs can again be clearly discerned, but the identification of BSGs is less obvious and the MSTO is too faint to be detected. The best fitting Padua models have an age of $\log t = 7.25$ while the best fitting Geneva models are slightly younger; $\log t = 7.15$. The corresponding MSTO masses are 13 and 14 $M_\odot$, respectively. We have again assumed $Z = 0.008$ models, which give the best fits, although this may not exactly match the actual metallicity of NGC 1569. However, neither set of models provides a very good match to the observations for any metallicity. Even for fairly generous limits in the selection of BSGs, the observed BSG/RSG ratio of $28/60 = 0.47 \pm 0.11$ is well below the model predictions (BSG/RSG = 1.28 and 0.87 for the Padua and Geneva models). The CMD also shows a small number of very bright ($M_V < -8$), blue stars which are not predicted by any model for this age. These are somewhat redder than the fainter blue stars, and thus unlikely to be simple blends. One possibility is that these stars have somewhat younger ages than the majority of stars in cluster B.

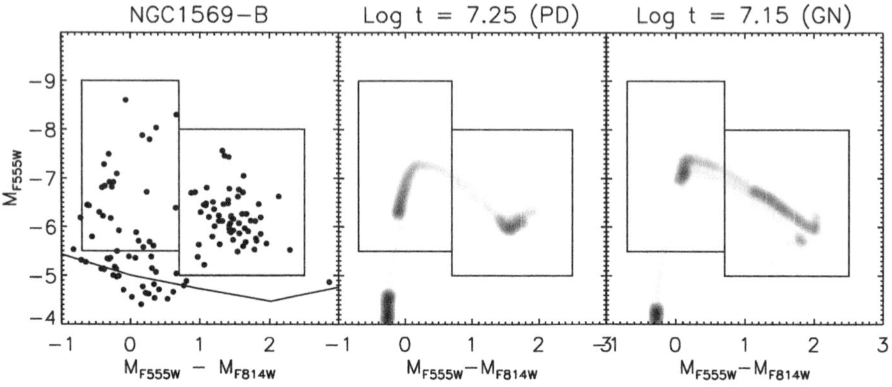

Fig. 25.2 As Fig. 25.1, but for NGC 1569-B. The 75% completeness limit is indicated in the *left-hand panel*

25.3 Concluding Remarks

The two cases discussed here demonstrate that useful information can be extracted from the CMDs of star clusters well beyond the Local Group. Young, massive ($>10^5$ $M_\odot$) clusters offer significant samples of stars with similar age and chemical composition, offering a potentially powerful way to test models for massive stars. Most current alternatives either suffer from small number statistics (e.g. open clusters in the Milky Way) or from difficulties disentangling the contributions from stars of different ages, hence masses (e.g. resolved stellar populations in Local Group galaxies).

References

1. Bruzual, G., Charlot, S., 2003. Mon. Not. R. Astron. Soc. 344, 1000.
2. Chabrier, G., 2003. Publ. Astron. Soc. Pac. 115, 763.
3. Eggenberger, P., et al., 2002. Astron. Astrophys. 386, 576.
4. Girardi, L., et al., 2002. Astron. Astrophys. 391, 195.
5. Kroupa, P., 2002. Science 295, 82.
6. Langer, N., Maeder, A., 1995. Astron. Astrophys. 295, 685.
7. Larsen, S.S., Richtler, T., 1999. Astron. Astrophys. 345, 59.
8. Larsen, S.S., et al., 2007. In: Vazdekis, A., Peletier, R. (Eds.), IAU Symp., vol. 241, p. 435. Cambridge University Press, Cambridge.
9. Larsen, S.S., Origlia, L., et al., 2008. Mon. Not. R. Astron. Soc. 838, 263.
10. Origlia, L., et al., 2001. Astron. J. 122, 815.
11. Renzini, A., Fusi Pecci, F., 1988. Annu. Rev. Astron. Astrophys. 26, 199.

Chapter 26
Tracing Galaxy Evolution in Clusters and Groups at $z \approx 1$

Simona Mei and the ACS IDT Team

Abstract We present recent results on a comparative analysis of the evolution of the Color–Magnitude Relation (CMR) at $z \approx 1$ on a sample of eight clusters from the ACS (Advanced Camera for Surveys) Intermediate Cluster Survey. Our clusters have redshifts between $z = 0.8$ and ≈ 1.3, and we also studied five galaxy groups at $z = 1.26$. We find that the early-type CMR (red sequence) is already on place at these redshifts and that CMR parameters are constant over a large range of redshifts, with no significant evolution up to redshift $z \approx 1.3$.

26.1 Introduction

We present a comparative analysis of the color–magnitude relation (hereafter CMR, see e.g. [1–6, 9, 10, 15–23, 25, 26, 29] from our Advanced Camera for Surveys (ACS; [13]) Intermediate Redshift Cluster Survey [14] sample. The sample consists of eight galaxy clusters with redshift between 0.8 and ~ 1.3 whose properties are shown in Table 26.1.

26.2 CMR of the ACS Intermediate Cluster Sample

CMR were calculated for each cluster in a series of dedicated papers, [4, 18, 22, 23], from now on the CMR paper series. The main aim in the CMR paper series was to constrain galaxy ages and to study variations in CMR parameters as a function of galaxy morphology and structural properties (e.g. effective radii, ellipticities, etc.). We show the mean luminosity weighted ages derived for the elliptical population in Table 26.1, using stellar population models from [7] (hereafter BC03). We found that the elliptical population has ages ranging between 2.5 and 3.5 Gyr depending on the cluster redshift, with an average formation redshift $z_f > 2$.

S. Mei (✉)
University of Paris 7 Denis Diderot, 75205 Paris Cedex 13, France
e-mail: simona.mei@obspm.fr

S. Mei
GEPI, Observatoire de Paris, Section de Meudon, 5 Place J. Janssen, 92195 Meudon Cedex, France

F.D. Macchetto (ed.), *The Impact of HST on European Astronomy*, Astrophysics and Space Science Proceedings,
DOI 10.1007/978-90-481-3400-7_26, © Springer Science+Business Media B.V. 2010

Table 26.1 ACS Intermediate Redshift Cluster Survey sample

Cluster	z	v (km/s)	$L_{bol}^{X_a}$ ($10^{44}h_{70}^{-2}$ erg/sec)	R_{200}^{b} (Mpc)	$M_{tot}^{X_a}$ (10^{14} $M_{\odot}$)	Age (Gyr)
MS 1054-0321	0.831	1156 ± 82	28.48 ± 2.96	1.8	21.27 ± 3.96	3.5
RXJ 0152.7-1357 N	0.834	888^{+152}_{-75}	10.67 ± 0.67	1.4	2.5 ± 0.9	3.5
RXJ 0152.7-1357 S	0.830	301^{+122}_{-107}	7.73 ± 0.4	0.5	1.1 ± 0.4	3.5
CL1604+4304	0.897	989^{+98}_{-76}	2.0 ± 0.3	1.5		3.5
CL1604+4321	0.924	649^{+59}_{-46}	<0.7	0.95		3.5
RDCS J0910+5422	1.106	675 ± 190	2.83 ± 0.35	0.89	4.91 ± 2.93	3.1
RXJ1252.9-292	1.237	747^{+74}_{-84}	5.99 ± 1.1	0.94	1.59 ± 0.35	2.7
RX J0849+4452	1.261	740^{+113}_{-134}	2.83 ± 0.17	0.90	2.85 ± 1.48	2.5
RX J0848+4453	1.270	650 ± 170	1.04 ± 0.73	0.78	1.37 ± 0.98	2.5

[a]Derived within a overdensity $\Delta_z = 500$ relative to an Einstein–de Sitter universe

[b] R_{200} is the virial radius (the radius at which the cluster mean density is 200 times the critical density)

In Fig. 26.1 we show the color–magnitude relations for our sample. The early-type red sequence is well defined and tight out to redshifts $z \approx 1.3$. Elliptical and early-type galaxies lie on similar CMRs. We observe the presence of bright blue late-type galaxies that start appearing at higher redshifts and in less massive clusters, and that are not observed in local samples.

When analyzing CMR scatters, we notice two trends. Firstly, for bright galaxies ($M_B < -21$ mag), the CMR scatter of the elliptical populations are smaller than those of the S0 population, while they become similar at fainter magnitudes. Secondly, scatters in the central regions of the cluster (e.g. we show scatters within one and one half the cluster virial radius) are smaller. If we interpret these results as due to a difference in galaxy age, bright elliptical galaxies are on average older (by ~0.5 Gyr from a single burst solar metallicity BC03 stellar population model) than S0 galaxies, and galaxies closest to the cluster core are on average older than galaxies closest to the virial radius. At faint magnitudes ($M_B > -21$ mag) all populations present similar scatters/ages.

CMR slope and scatter have constant behavior as a function of redshift. In Fig. 26.2 we show elliptical galaxy CMR slope and scatter as a function of redshift for our sample and compare to local cluster samples. CMR parameters are shown to be constant on a large range of redshifts, showing no significant evolution up to redshift $z \approx 1.3$.

A preliminary analysis of five galaxy groups at $z = 1.26$ by [24] shows larger scatters in these less massive structures.

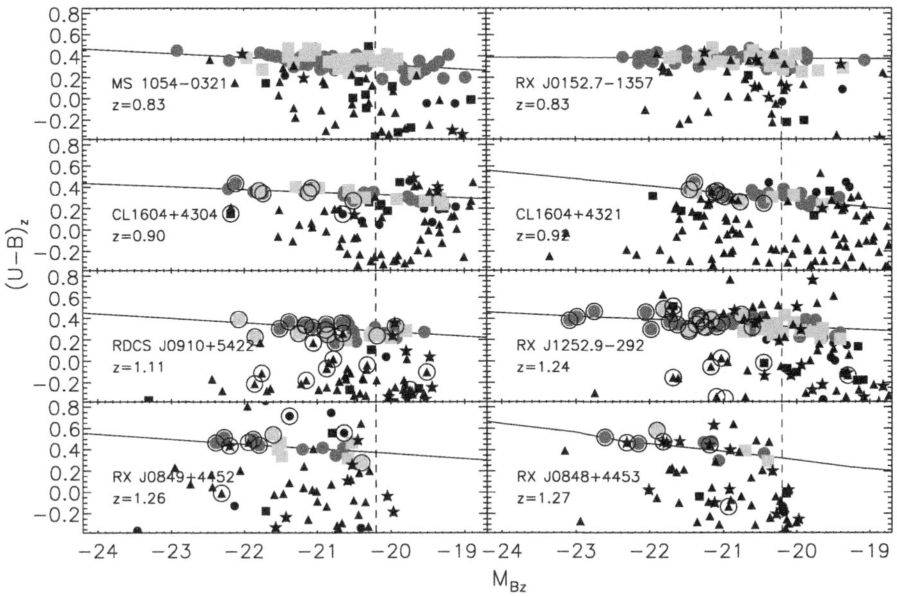

Fig. 26.1 (Color online) Color–magnitude relations in the rest-frame $(U - B)_z$ color vs the absolute rest frame B magnitude M_B for the eight clusters of the ACS Intermediate Cluster Survey. Elliptical galaxies are shown as *circles*, S0 as *squares*, S0/a as *stars*, spirals as *triangles*. The *continuous line* shows the CMR relations for the elliptical sample calculated within the virial radius, to the same limit in rest-frame absolute magnitude $M_B = -20.2$ mag (shown by the *horizontal dashed line*). *Larger red circles* and *larger yellow squares* show elliptical and S0 galaxies, respectively, contained within three times the scatter around the CMR. Galaxies plotted in the two most massive clusters, RX J0152.7-1357 and MS 1054-0321 are all spectroscopically confirmed. The *rings around symbols* denote spectroscopically confirmed members in the other clusters [11, 12, 27, 28, 30]

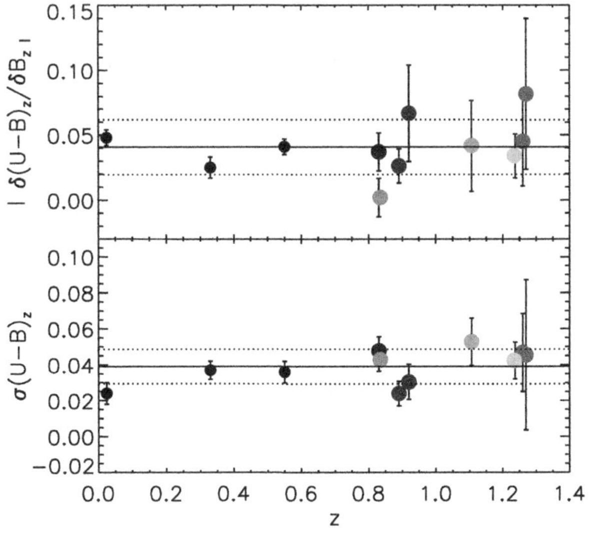

Fig. 26.2 (Color online) CMR evolution for the elliptical galaxy sample as compared to local cluster CMR parameters. CMR absolute slope $|\frac{\delta(U-B)_z}{\delta B_z}|$ and scatter $\sigma(U - B)_z$ for ellipticals as a function of redshift are shown by *circles* from [6] results for the Coma and Virgo clusters; [29] results for MS 1054-03; [8] results for a sample of nearby clusters of galaxies (from *left* to *right*, in order of increasing redshift). The *continuous line* shows the average parameter value and the *dotted line* the 1 σ range. The *colored larger circles* are our results

References

1. Baldry, I., et al., 2004. Astrophys. J. 600, 681.
2. Bell, E.F., et al., 2004. Astrophys. J. 608, 883.
3. Bernardi, M., et al.., 2005. Astron. J. 129, 61.
4. Blakeslee, J.P., Franx, M., Postman, M., et al., 2003. Astrophys. J. Lett. 596, 143.
5. Blakeslee, J.P., et al., 2006. Astrophys. J. 644, 30.
6. Bower, R.G., Lucey, J.R., Ellis, R.S., 1992. Mon. Not. R. Astron. Soc. 254, 589.
7. Bruzual, A.G., Charlot, S., 2003. Mon. Not. R. Astron. Soc. 344, 1000 (BC03).
8. Ellis, R.S., et al., 1997. Astrophys. J. 483, 582.
9. Cucciati, O., et al., 2006. Astron. Astrophys. 458, 39.
10. De Lucia, G., et al., 2004. Astrophys. J. Lett. 610, 77.
11. Demarco, R., et al., 2005. Astron. Astrophys. 432, 381.
12. Demarco, R., et al., 2007. Astrophys. J. 663, 164.
13. Ford, H.C., et al., 2002. Proc. SPIE 4854, 81.
14. Ford, H.C., et al., 2004. astro-ph/0408165.
15. Gallazzi, A., et al., 2006. Mon. Not. R. Astron. Soc. 370, 1106.
16. Hogg, D.W., et al., 2004. Astrophys. J. Lett. 601, 29.
17. Holden, B.P., et al., 2004. Astron. J. 127, 248.
18. Homeier, N.L., et al., 2006. Astrophys. J. 647, 256.
19. Lidman, C., et al., 2004. Astron. Astrophys. 416, 829.
20. López-Cruz, O., Barkhouse, W.A., Yee, H.K.C., 2004. 614, 679.
21. McIntosh, D.H., Zabludoff, A.I., Rix, H.-W., Caldwell, N., 2005. Astrophys. J. 619, 193.
22. Mei, S., et al., 2006. Astrophys. J. 639, 81.
23. Mei, S., et al., 2006. Astrophys. J. 644, 759.
24. Nakata, F., et al., 2005. Mon. Not. R. Astron. Soc. 357, 1357.
25. Stanford, S.A., et al., 2006. Astrophys. J. 646, L13.
26. Tanaka, M., et al., 2005. Mon. Not. R. Astron. Soc. 362, 268.
27. Tran, K.H., et al., 2005. Astrophys. J. 627, L25.
28. Tran, K.H., et al., 2007. Astrophys. J. 661, 750.
29. van Dokkum, P.G., et al., 1998. Astrophys. J. 500, 714.
30. van Dokkum, P., et al., 2000. Astrophys. J. 541, 95.

Chapter 27
The Dawn of Galaxies

Piero Madau

Abstract The development of primordial inhomogeneities into the non-linear regime and the formation of the first astrophysical objects within dark matter halos mark the transition from a simple, neutral, cooling universe — described by just a few parameters — to a messy ionized one — the realm of radiative, hydrodynamic, and star formation processes. The *Wilkinson Microwave Anisotropy Probe* (*WMAP*) polarization data show that this transition must have begun quite early, and that the universe was fully reionized some 350 million years after the Big Bang. It is a young generation of extremely metal-poor massive stars and/or 'seed' accreting black holes in subgalactic halos that may have generated the ultraviolet radiation and mechanical energy that reheated and reionized most of the hydrogen in the cosmos. The detailed thermal, ionization, and chemical enrichment history of the universe during the crucial formative stages around redshift 10 depends on the power-spectrum of density fluctuations on small scales, the stellar initial mass function and star formation efficiency, a complex network of poorly understood 'feedback' mechanisms, and remains one of the crucial missing links in galaxy formation and evolution studies.

27.1 Preamble

Hydrogen in the universe recombined about half a million years after the Big Bang, and cooled down to a temperature of a few kelvins until the first non-linearities developed, and evolved into stars, galaxies, and black holes that lit up the universe again. In currently popular cold dark matter flat cosmologies (ΛCDM), some time beyond a redshift of 10 the gas within halos with virial temperatures $T_{\rm vir} \gtrsim 10^4$ K — or, equivalently, with masses $M \gtrsim 10^8[(1+z)/10]^{-3/2}\,{\rm M}_\odot$ — cooled rapidly due to the excitation of hydrogen Lyα and fragmented. Massive stars formed with some initial mass function, synthesized heavy elements, and exploded as Type II supernovae after a few $\times 10^7$ yr, enriching the surrounding medium: these subgalactic stellar systems, aided perhaps by an early population of accreting black holes in their nuclei, generated the ultraviolet radiation and mechanical energy that contributed to the reheating and reionization of the cosmos. It is widely believed that collisional excitation of molecular hydrogen may have allowed gas in even

P. Madau
Department of Astronomy and Astrophysics, University of California, Santa Cruz, CA 95064, USA

F.D. Macchetto (ed.), *The Impact of HST on European Astronomy*, Astrophysics and Space Science Proceedings,
DOI 10.1007/978-90-481-3400-7_27, © Springer Science+Business Media B.V. 2010

smaller systems — virial temperatures of a thousand K, corresponding to masses around $5 \times 10^5[(1+z)/10]^{-3/2}\ M_\odot$ — to cool and form stars at even earlier times [1, 6, 49]. Throughout the epoch of structure formation, the all-pervading intergalactic medium (IGM), which contains most of the ordinary baryonic material left over from the Big Bang, becomes clumpy under the influence of gravity, and acts as a source for the gas that gets accreted, cools, and forms stars within subgalactic fragments, and as a sink for the metal enriched material, energy, and radiation which they eject. The well-established existence of heavy elements like carbon and silicon in the Lyα forest clouds at $z = 2$–6 [42, 46] may be indirect evidence for such an early episode of pregalactic star formation. The recently released five-year *WMAP* data require the universe to be fully reionized by redshift 11 ± 1.4 [15], another indication that significant star-formation activity started at very early cosmic times.

The last decade has witnessed great advances in our understanding of the high redshift universe, thanks to breakthroughs achieved with satellites, 8–10 m class telescopes, and cosmic microwave background (CMB) experiments. Large surveys such as the *Sloan Digital Sky Survey* (SDSS), together with the use of novel instruments and observational techniques have led to the discovery of galaxies and quasars at redshifts in excess of 6. At the time of writing, nine quasars have already been found with $z > 6$ [17], and one actively star-forming has been spectroscopically confirmed at $z = 6.96$ [25]. These sources probe an epoch when the universe was <7% of its current age. *Keck* and *VLT* observations of redshifted H I Lyα ('forest') absorption have been shown to be sensitive probe of the distribution of gaseous matter in the universe (e.g. [41]). Gamma-ray bursts have recently displayed their potential to replace quasars as the preferred probe of early star formation and chemical enrichment: GRB050904, the most distant event known to date, is at $z = 6.39$ [23]. The underlying goal of all these efforts is to understand the growth of cosmic structures, the properties of galaxies and their evolution, and ultimately to map the transition from the cosmic "dark age" to a ionized universe populated with luminous sources.

Progress has been equally significant on the theoretical side. The key idea of currently popular cosmological scenarios, that primordial density fluctuations grow by gravitational instability driven by collisionless CDM, has been elaborated upon and explored in detail through large-scale numerical simulations on supercomputers, leading to a hierarchical ('bottom-up') scenario of structure formation. In this model, the first objects to form are on subgalactic scales, and merge to make progressively bigger structures ('hierarchical clustering'). Ordinary matter in the universe follows the dynamics dictated by the dark matter until radiative, hydrodynamic, and star formation processes take over. According to these calculations, a truly inter- and proto-galactic medium (the main repository of baryons at high redshift) collapses under the influence of dark matter gravity into flattened and filamentary structures, which are seen in absorption against background QSOs. Gas condensation in the first baryonic objects is possible through the formation of H_2 molecules, which cool via roto-vibrational transitions down to temperatures of a few hundred kelvins. In the absence of a UV photodissociating flux and of ionizing X-ray radiation, three-dimensional simulations of early structure formation show that the fraction of cold,

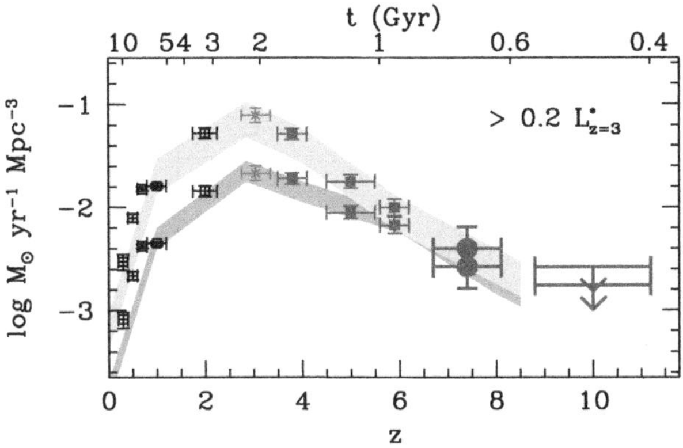

Fig. 27.1 (Color online) Estimated star formation rate (SFR) density as a function of redshift (integrated down to $0.2\ L^*_{z=3}$). The *lower set of points* give the SFR density without a correction for dust extinction, and the *upper set of points* give the SFR density with such a correction. This is also indicated with the *shaded blue and red regions*, respectively, where the width of these regions show the approximate uncertainties. (From [4])

dense gas available for star formation and accretion onto seed black holes exceeds 20% for halos more massive than $10^6\ M_\odot$ already at redshifts 20 [30, 32].

In spite of some significant achievements in our understanding of the formation of cosmic structures, there are still many challenges facing hierarchical clustering theories. While quite successful in matching the observed large-scale density distribution (like, e.g., the properties of galaxy clusters, galaxy clustering, and the statistics of the Lyα forest), CDM simulations appear to produce halos that are too centrally concentrated compared to the mass distribution inferred from the rotation curves of (dark matter-dominated) dwarf galaxies, and to predict too many dark matter subhalos compared to the number of dwarf satellites observed within the Local Group [28, 36, 39]. Another perceived difficulty (arguably connected with the "missing satellites problem", see e.g. [7]) is our inability to predict when and how the universe was reheated and reionized, i.e. to understand the initial conditions of the galaxy formation process and the basic building blocks of today's massive baryonic structures. We know that at least some galaxies and quasars had already formed when the universe was less than 10^9 yr old, and have made great progress in mapping the cosmic star formation history of bright galaxies up to redshift 7 or so [4, 33] (see Fig. 27.1). But when did the first luminous systems form, was star formation efficient in objects below the atomic cooling mass, and what was the impact of the first stars on the surrounding intergalactic gas? The crucial processes of star formation and "feedback" (e.g. the effect of the energy input from the earliest generations of sources on later ones) in the nuclei of galaxies are still poorly understood. Accreting black holes can release large amounts of energy to their surroundings, and may play a role in regulating the thermodynamics of the interstellar, intracluster, and intergalactic medium (e.g. [14]). The detailed astrophysics of these processes is, however, unclear. Although we may have a sketchy history of the production of the chemical elements in the universe, we know little about how and where exactly they were produced and how they are distributed in the IGM and in the intracluster medium. Finally, where are the first stars and their remnants now, and why do the hundreds of early-forming, massive satellites predicted to survive today in the Milky Way halo remain dark?

Fig. 27.2 (Color online) Evolution of the radiation (*long-dashed line*, labeled *CMB*) and gas (*solid line*, labeled *IGM*) temperatures after recombination. The universe is assumed to be reionized by ultraviolet radiation at $z \sim 10$. The *short-dashed line* is the extrapolated gas temperature in the absence of any reheating mechanism

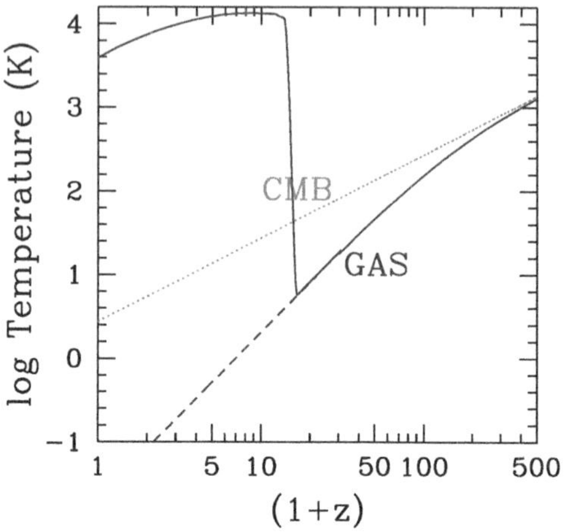

27.2 The Dark Age

The universe became optically thin to Thomson scattering at redshift 1,100, and entered a 'dark age'. At this epoch the electron fraction dropped below 15%, and the primordial radiation cooled below 3,000 K, shifting first into the infrared and then into the radio. We understand the micro-physics of the post-recombination universe well. The fractional ionization froze out to the value $\sim 3 \times 10^{-4}$: these residual electrons were enough to keep the matter in thermal equilibrium with the radiation via Compton scattering until a thermalization redshift of $z_t \sim 150$, i.e. well after the universe became transparent. Thereafter, the matter temperature decreased as $(1+z)^2$ due to adiabatic expansion (Fig. 27.2) until primordial inhomogeneities in the density field evolved into the non-linear regime.

The minimum mass scale for the gravitational aggregation of cold dark matter particles is negligibly small. One of the most popular CDM candidates is the neutralino: in neutralino CDM, collisional damping and free streaming smear out all power of primordial density inhomogeneities only below Earth-mass scales. Baryons, however, respond to pressure gradients and do not fall into dark matter clumps below the cosmological Jeans mass (in linear theory this is the minimum mass-scale of a perturbation where gravity overcomes pressure),

$$M_J = 4\pi \frac{\rho}{3} \left(\frac{5\pi k_B T_e}{12 G \rho m_p \mu} \right)^{3/2} \approx 9 \times 10^4 \ \mathrm{M}_\odot \, (a T_e/\mu)^{3/2}. \qquad (27.1)$$

Here $a = (1+z)^{-1}$ is the scale factor, ρ the total mass density including dark matter, μ the mean molecular weight, and T_e the gas temperature. In the post-recombination universe, the baryon-electron gas remained thermally coupled to the CMB, $T_e \propto a^{-1}$, and the Jeans mass was independent of redshift and comparable to the mass of globular clusters, $M_J \approx 10^6 \ \mathrm{M}_\odot$. At $z < z_t$, the temperature of the baryons dropped as $T_e \propto a^{-2}$, and the Jeans mass decreased with time, $M_J \propto a^{-3/2}$. This trend was reversed by the reheating of the IGM. The energy released by the

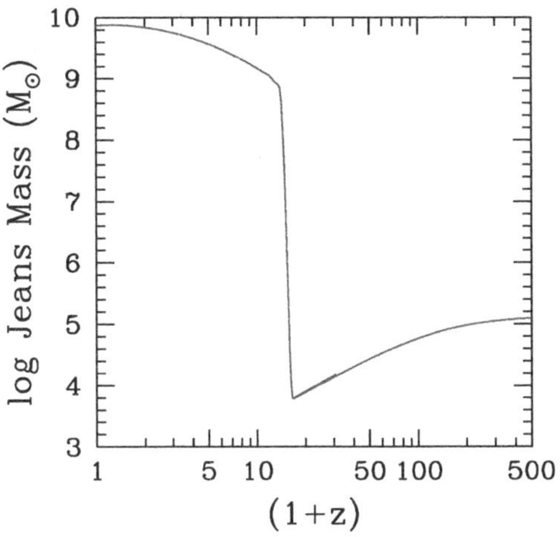

Fig. 27.3 (Color online) Cosmological (gas + dark matter) Jeans mass for a universe reionized by ultraviolet radiation

first collapsed objects drove the Jeans mass up to galaxy scales: baryonic density perturbations stopped growing as their mass dropped below the new Jeans mass. Photo-ionization by the ultraviolet radiation from the first stars and quasars heated the IGM to temperatures of $\approx 10^4$ K (corresponding to a Jeans mass $M_J \lesssim 10^{10}$ $M_\odot$ at $z \simeq 11$), suppressing gas infall into low mass halos and preventing new (dwarf) galaxies from forming (see Fig. 27.3).

27.3 The Emergence of Cosmic Structure

As mentioned above, some shortcomings on galactic and sub-galactic scales of the currently favored model of hierarchical galaxy formation in a universe dominated by CDM have appeared in the last decade. The significance of these discrepancies is still debated, and 'gastrophysical' solutions involving feedback mechanisms may offer a possible way out. Other models have attempted to solve the apparent small-scale problems of CDM at a more fundamental level, i.e. by reducing small-scale power. Although the 'standard' ΛCDM model for structure formation assumes a scale-invariant initial power spectrum of density fluctuations, $P(k) \propto k^n$ with $n = 1$, there is strong evidence in the *WMAP* data for a departure from scale invariance, with a best fit value $n = 0.963^{+0.014}_{-0.015}$ [15]. The one and three-year *WMAP* data showed some preference for a scale-dependent index, $dn/d\ln k < 0$, i.e. for a model in which the spectral index varies as a function of wavenumber k [47]. The five-year *WMAP* data do not significantly prefer such a 'running index'. Models with either $n < 1$ or $dn/d\ln k < 0$ predict a significantly lower amplitude of fluctuations on small scales than standard ΛCDM. The suppression of small-scale power has the advantage of reducing the amount of substructure in galactic halos and makes small halos form later (when the universe was less dense) hence less concentrated [52]. But it makes early reionization a challenge.

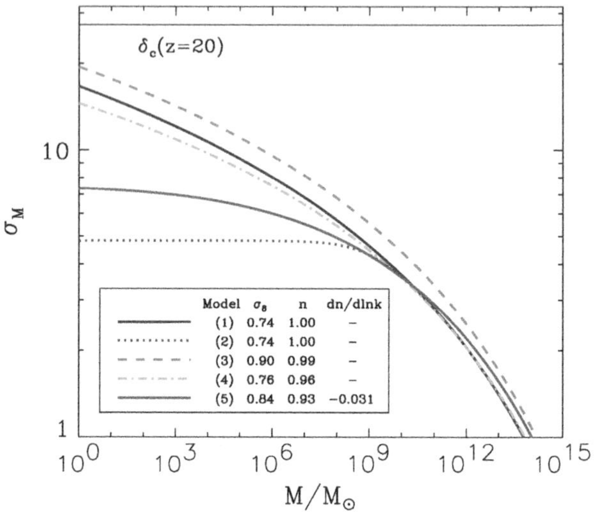

Fig. 27.4 (Color online) The variance of the matter-density field vs. mass M, for different power spectra. All models assume a 'concordance' cosmology with parameters $(\Omega_M, \Omega_\Lambda, \Omega_b, h) = (0.29, 0.71, 0.045, 0.7)$. *Solid curve*: standard ΛCDM with no tilt, cluster normalized. *Dotted curve*: ΛWDM with a particle mass $m_X = 2$ keV, cluster normalized, no tilt. *Dashed curve*: tilted *WMAP* model, *WMAP* data only. *Dash-dotted curve*: tilted *WMAP* model, including 2dFGRS and Lyman-α data. *Dash-triple dotted curve*: running spectral index *WMAP* model, including 2dFGRS and Lyman-α data. Here n refers to the spectral index at $k = 0.05$ Mpc^{-1}. The *horizontal line at the top* of the figure shows the value of the extrapolated collapse overdensity $\delta_c(z)$ at $z = 20$

Figure 27.4 shows the linearly extrapolated (to $z = 0$) variance of the mass-density field for different power spectra. In the CDM paradigm, structure formation proceeds 'bottom-up', i.e., the smallest objects collapse first, and subsequently merge together to form larger objects. It then follows that the loss of small-scale power modifies structure formation most severely at the highest redshifts, significantly reducing the number of self-gravitating objects then. This, of course, will make it more difficult to reionize the universe early enough. It has been argued, for example, that one popular modification of the CDM paradigm, warm dark matter (WDM), has so little structure at high redshift that it is unable to explain the *WMAP* observations of an early epoch of reionization [47]. The running-index model may suffer from a similar problem. A look at Fig. 27.4 shows that 10^6 M$_\odot$ halos will collapse at $z = 20$ from 2.9σ fluctuations in a tilted ΛCDM model with $n = 0.99$ and $\sigma_8 = 0.9$, from 4.6σ fluctuations in a running-index model, and from 5.7σ fluctuations in a WDM cosmology. The problem is that scenarios with increasingly rarer halos at early times require even more extreme assumptions (i.e. higher star formation efficiencies and UV photon production rates) in order to be able to reionize the universe suitably early [8, 9, 22, 45, 51].

The study of the non-linear regime for the baryons is far more complicated than that of the dark matter because of the need to take into account pressure gradients and radiative processes. As a dark matter halo grows and virializes above the cosmological Jeans mass through merging and accretion, baryonic material will

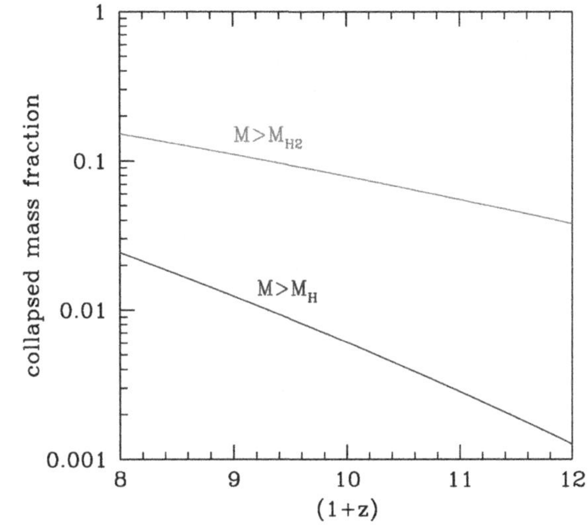

Fig. 27.5 (Color online) *Solid lines*: Total mass fraction in all collapsed dark matter halos above the molecular cooling and the atomic cooling masses, M_{H_2} and M_{H}, as a function of redshift, for standard ΛCDM cosmology

be shock heated to the effective virial temperature of the host and compressed to the same fractional overdensity as the dark matter. The subsequent behavior of gas in a dark matter halo depends on the efficiency with which it can cool. It is useful here to identify two mass scales for the host halos: (1) a *molecular cooling mass* M_{H_2} above which gas can cool via roto-vibrational levels of H_2 and contract, $M_{\mathrm{H}_2} \approx 10^5[(1+z)/10]^{-3/2}$ $\mathrm{M}_\odot$ (virial temperature above 200 K); and (2) *an atomic cooling mass* M_{H} above which gas can cool efficiently and fragment via excitation of hydrogen Lyα, $M_{\mathrm{H}} \approx 10^8[(1+z)/10]^{-3/2}$ $\mathrm{M}_\odot$ (virial temperature above 10^4 K). Figure 27.5 shows the fraction of the total mass in the universe that is in collapsed dark matter halos with masses greater than M_{H_2} and M_{H} at different epochs.

27.4 The Epoch of Reionization

Since hierarchical clustering theories provide a well-defined framework in which the history of baryonic material can be tracked through cosmic time, probing the reionization epoch may then help constrain competing models for the formation of cosmic structures. Quite apart from uncertainties in the primordial power spectrum on small scales, however, it is the astrophysics of baryons that makes us unable to predict when reionization actually occurred. Consider the following illustrative example:

Hydrogen photo-ionization requires more than one photon above 13.6 eV per hydrogen atom: of order $t/\bar{t}_{\mathrm{rec}} \sim 10$ (where $\bar{t}_{\mathrm{rec}}$ is the volume-averaged hydrogen recombination timescale) extra photons appear to be needed to keep the gas in overdense regions and filaments ionized against radiative recombinations [21, 35]. A 'typical' stellar population produces during its lifetime about 4,000 Lyman continuum (ionizing) photons per stellar proton. A fraction $f \sim 0.25\%$ of cosmic baryons must then condense into stars to supply the requisite ultraviolet flux. This estimate assumes a standard (Salpeter) initial mass function (IMF), which determines the relative abundances of hot, high mass stars versus cold, low mass ones.

The very first generation of stars ('Population III') must have formed, however, out of unmagnetized metal-free gas: numerical simulations of the fragmentation of pure H and He molecular clouds [1, 6] have shown that these characteristics likely led to a 'top-heavy' IMF biased towards very massive stars (VMSs, i.e. stars a few hundred times more massive than the Sun), quite different from the present-day Galactic case. Metal-free VMSs emit about 10^5 Lyman continuum photons per stellar baryon [5], approximately 25 times more than a standard stellar population. A corresponding smaller fraction of cosmic baryons would have to collapse then into VMSs to reionize the universe, $f \sim 10^{-4}$. There are of course further complications. Since, at zero metallicity, mass loss through radiatively-driven stellar winds is expected to be negligible [29], Population III stars may actually die losing only a small fraction of their mass. If they retain their large mass until death, VMSs with masses $140 \lesssim m \lesssim 260\ \mathrm{M}_\odot$ will encounter the electron-positron pair instability and disappear in a giant nuclear-powered explosion [19], leaving no compact remnants and polluting the universe with the first heavy elements. In still heavier stars, however, oxygen and silicon burning is unable to drive an explosion, and complete collapse to a black hole will occur instead [3]. Thin disk accretion onto a Schwarzschild black hole releases about 50 MeV per baryon. The conversion of a trace amount of the total baryonic mass into early black holes, $f \sim 3 \times 10^{-6}$, would then suffice to reionize the universe.

Even if the IMF at early times were known, we still would remain uncertain about the fraction of cold gas that gets retained in protogalaxies after the formation of the first stars (this quantity affects the global efficiency of star formation at these epochs) and whether — in addition to ultraviolet radiation — an early input of mechanical energy may also play a role in determining the thermal and ionization state of the IGM on large scales. The same massive stars that emit ultraviolet light also explode as supernovae (SNe), returning most of the metals to the interstellar medium of pregalactic systems and injecting about 10^{51} ergs per event in kinetic energy. A complex network of feedback mechanisms is likely at work in these systems, as the gas in shallow potential is more easily blown away [11], thereby quenching star formation. Furthermore, as the blast waves produced by supernova explosions — and possibly also by winds from 'miniquasars' — sweep the surrounding intergalactic gas, they may inhibit the formation of nearby low-mass galaxies, and drive vast portions of the IGM to a significantly higher temperature than expected from photo-ionization, so as to 'choke off' the collapse of further galaxy-scale systems. Note that this type of global feedback is fundamentally different from the 'in situ' heat deposition commonly adopted in galaxy formation models, in which hot gas is produced by supernovae within the parent galaxy.

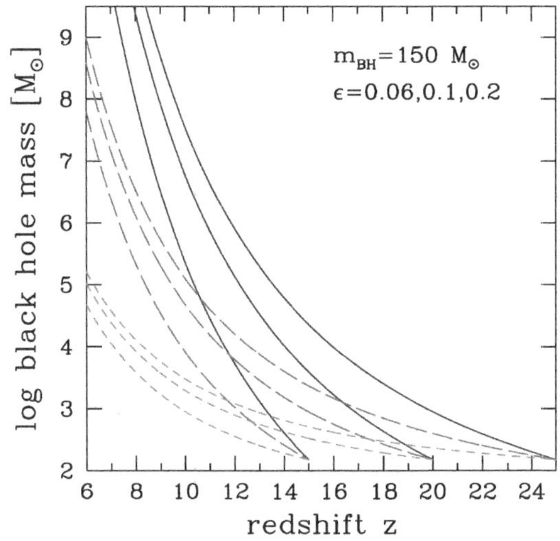

Fig. 27.6 (Color online) Growth of MBHs from early epochs down to $z = 6$, the redshift of the most distant SDSS quasars. The three sets of curves assume Eddington-limited accretion with radiative efficiency $\epsilon = 0.06$ (*solid lines*), 0.1 (*long-dashed lines*), and 0.2 (*short-dashed lines*). Gas accretion starts at $z = 15$, 20, 25 onto a seed black of mass $m_{BH} = 150\ M_\odot$

27.5 Massive Black Holes and Galaxy Formation

The strong link observed between the masses of supermassive black holes (MBHs) at the center of most galaxies and the gravitational potential wells that host them suggests a fundamental mechanism for assembling black holes and forming spheroids in galaxy halos. The m_{BH}–σ relation [18, 20] implies a rough proportionality between MBH mass and the mass of the baryonic component of the bulge. It is not yet understood whether this relation was set in primordial structures, and consequently how it is maintained throughout cosmic time with such a small dispersion, or indeed which physical processes established such a correlation in the first place (e.g. [44]).

In CDM-dominated cosmologies, galaxy halos experience multiple mergers during their lifetime, with those between comparable-mass systems ("major mergers") expected to result in the formation of elliptical galaxies [24]. Simple models in which MBHs are also assumed to grow during major mergers and to be present in every galaxy at any redshift — while only a fraction of them is "active" at any given time — have been shown to explain many aspects of the observed evolution of quasars [27]. The coevolution of MBHs and their host galaxies in hierarchical structure formation scenarios gives origin to a number of important questions, most notably:

- Did the first MBHs form in subgalactic units far up in the merger hierarchy, well before the bulk of the stars observed today? The seeds of the $z \sim 6$ quasars discovered in the *Sloan Digital Sky Survey* (see Fig. 27.6) had to appear at very high redshift, $z \gtrsim 10$, if they are accreting no faster than the Eddington rate. In hierarchical cosmologies, the ubiquity of MBHs in nearby luminous galaxies can arise even if only a small fraction of halos harbor MBHs at very high redshift [37].
- How massive were the initial seeds, and is there a population of relic pregalactic MBHs lurking in present-day galaxy halos? A clue to these questions may

lie in the numerous population of ultraluminous off-nuclear ("non-AGN") X-ray sources that have been detected in nearby galaxies. Assuming isotropic emission, the inferred masses of these "ULXs" may suggest intermediate-mass black holes with masses $\gtrsim$ a few hundred $M_{\odot}$ [26].
- Do MBH binaries form and coalescence in large numbers? If MBHs were common in the past (as implied by the notion that many distant galaxies harbor active nuclei for a short period of their life), and if their host galaxies undergo multiple mergers, then MBH binaries will inevitably form in large numbers during cosmic history. MBH pairs that are able to coalesce in less than a Hubble time will give origin to the loudest gravitational wave events in the universe [43].
- If was first proposed by [16] that the heating of the surrounding stars by a decaying MBH pair would create a low-density stellar core out of a preexisting cuspy (e.g. $\rho_* \propto r^{-2}$) stellar density profile. If stellar dynamical processes can efficiently drive wide MBH binaries to the gravitational wave (GW) emission stage, what is the cumulative dynamical effect of multiple black hole mergers on galaxy stellar cusps?
- Active galactic nuclei powered by supermassive holes keep the universe ionized at $z \lesssim 4$, structure the intergalactic medium (IGM), and probably regulate star formation in their host galaxies [14]. Intermediate-mass holes accreting gas from the surrounding medium may shine as "miniquasars" at redshifts as high as $z \sim 20$. What is the thermodynamic effect of miniquasars on the IGM at early times?

27.6 Concluding Remarks

This dialog should make it clear that, despite much recent progress in our understanding of the formation of early cosmic structure and the high-redshift universe, the astrophysics of first light remains one of the missing links in galaxy formation and evolution studies. We are left very uncertain about the whole era from 10^8 to 10^9 yr — the epoch of the first galaxies, stars, supernovae, and massive black holes. Some of the issues discussed above are likely to remain a topic of lively controversy until the launch of the *James Webb Space Telescope* (*JWST*), ideally suited to image the earliest generation of stars in the universe. If the first massive black holes form in pregalactic systems at very high redshifts, they will be incorporated through a series of mergers into larger and larger halos, sink to the center owing to dynamical friction, accrete a fraction of the gas in the merger remnant to become supermassive, and form binary systems [50]. Their coalescence would be signaled by the emission of low-frequency gravitational waves detectable by the planned *Laser Interferometer Space Antenna* (*LISA*).

An alternative way to probe the end of the dark age and discriminate between different reionization histories is through 21 cm tomography [34]. Prior to the epoch of full reionization, 21 cm spectral features will display angular structure as well as structure in redshift space due to inhomogeneities in the gas density field, hydrogen ionized fraction, and spin temperature. Radio maps will show a patchwork (both in angle and in frequency) of emission signals from H I zones modulated by H II regions

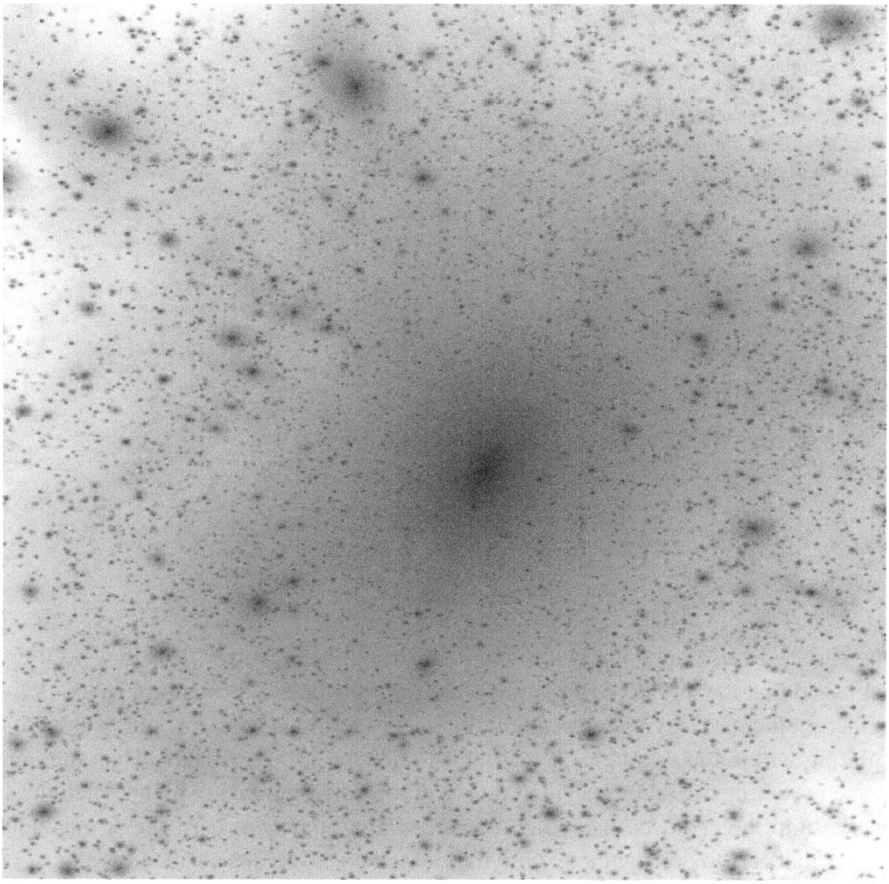

Fig. 27.7 (Color online) Projected dark matter density-squared map of the Via Lactea halo at the present epoch. The simulation follows the growth of a Milky Way-size halo in a ΛCDM Universe from redshift 50 to the present. It was performed with the parallel treecode PKDGRAV [48] and samples the galaxy-forming region with 234 million particles of mass 2.1×10^4 $M_\odot$. The image covers an area of 800×600 kpc, and the projection goes through a 600 kpc-deep cuboid containing a total of 110 million particles. The logarithmic color scale covers 20 decades in density-square. Note the wealth of resolved substructure

where no signal is detectable against the CMB [9]. The search at 21 cm for the epoch of first light, while remaining an extremely challenging project, remains a tantalizing possibility within range of the next generation of radio arrays.

While the above cosmological puzzles can be tackled directly by studying distant structures, it has recently become clear that many of today's "observables" within the Milky Way and nearby galaxies relate to events occurring at very high redshifts, during and soon after the epoch of reionization (see e.g. [40] and references therein). In this sense, galaxies in the Local Group can provide a crucial diagnostic link to the physical processes that govern structure formation and evolution in

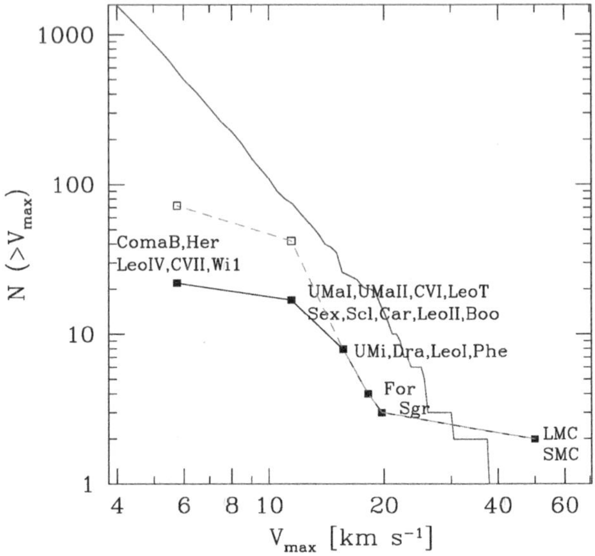

Fig. 27.8 (Color online) Cumulative number of Via Lactea subhalos (*solid curve*) as well as all Milky Way satellite galaxies within 420 kpc (*filled squares*), as a function of circular velocity. The data points assume a maximum circular velocity of $V_{\max} = \sqrt{3}\sigma$, where σ is the measured stellar velocity dispersion. The *short-dashed curve* connecting the *empty squares* shows the expected abundance of luminous satellites after correcting for the sky coverage of the SDSS. (From [36])

the early universe, an approach termed "near-field cosmology". It is now well established that the hierarchical mergers that form the halos surrounding galaxies are rather inefficient, leaving substantial amounts of stripped halo cores or "subhalos" orbiting within these systems. Small halos collapse at high redshift when the universe is very dense, so their central densities are correspondingly high. When these merge into larger hosts, their high densities allow them to resist the strong tidal forces that acts to destroy them. Gravitational interactions appear to unbind most of the mass associated with the merged progenitors, but a significant fraction of these small halos survives as distinct substructure. The "Via Lactea Project", a suite of the most detailed N-body simulations of Milky Way CDM substructure to date [12, 13, 31, 36], has shown that, in the standard CDM paradigm, galaxy halos should be filled with tens of thousands subhalos that appear to have no optically luminous counterpart (see Fig. 27.7).

As shown in Fig. 27.8, such finding appears to exacerbate the so-called "missing satellite problem", the large mismatch between the twenty or so dwarf satellite galaxies observed around the Milky Way and the predicted large number of massive CDM subhalos. Even if most dark matter satellites have no optically luminous counterparts, the substructure population may be detectable via flux ratio anomalies in strong gravitational lenses [38], or possibly via γ-rays from dark matter annihilation in their cores (e.g. [2, 10]). We are coming into a new era of galaxy formation and evolution studies, in which fossil signatures accessible today within nearby galaxy halos will allow us to probe back to early epochs, and in which the basic building blocks of galaxies will become recognizable in the near-field.

References

1. Abel, T., Bryan, G.L., Norman, M.L., 2000. Astrophys. J. 540, 39.
2. Bergstrom, L., Edsjo, J., Gondolo, P., Ullio, P., 1999. Phys. Rev. D 59, 043506.
3. Bond, J.R., Arnett, W.D., Carr, B.J., 1984. Astrophys. J. 280, 825.
4. Bouwens, R.J., Illingworth, G.D., Franx, M., Ford, H., 2008. Astrophys. J. 686, 230.
5. Bromm, V., Kudritzki, R.P., Loeb, A., 2001. Astrophys. J. 552, 464.
6. Bromm, V., Coppi, P.S., Larson, R.B., 2002. Astrophys. J. 564, 23.
7. Bullock, J.S., Kravtsov, A.V., Weinberg, D.H., 2000. Astrophys. J. 539, 517.
8. Cen, R., 2003. Astrophys. J. 591, L5.
9. Ciardi, B., Ferrara, A., White, S.D.M., 2003. Mon. Not. R. Astron. Soc. 344, L7.
10. Colafrancesco, S., Profumo, S., Ullio, P., 2006. Astron. Astrophys. 455, 21.
11. Dekel, A., Silk, J., 1986. Astrophys. J. 303, 39.
12. Diemand, J., Kuhlen, M., Madau, P., 2007. Astrophys. J. 657, 262.
13. Diemand, J., Kuhlen, M., Madau, P., 2007. Astrophys. J. 667, 859.
14. Di Matteo, T., Springel, V., Hernquist, L., 2005. Nature 433, 604.
15. Dunkley, J., et al.., 2009. Astrophys. J. Suppl. Ser. 180, 306.
16. Ebisuzaki, T., Makino, J., Okumura, S.K., 1991. Nature 354, 212.
17. Fan, X., 2006. New Astron. Rev. 60, 665.
18. Ferrarese, L., Merritt, D., 2000. Astrophys. J. 539, L9.
19. Fryer, C.L., Woosley, S.E., Heger, A., 2001. Astrophys. J. 550, 372.
20. Gebhardt, K., et al., 2000. Astrophys. J. 543, L5.
21. Gnedin, N.Y., 2000. Astrophys. J. 542, 535.
22. Haiman, Z., Holder, G.P., 2003. Astrophys. J. 595, 1.
23. Haislip, J., et al., 2006. Nature 440, 181.
24. Hernquist, L., 1992. Astrophys. J. 400, 460.
25. Iye, M., et al., 2006. Nature 443, 186.
26. Kaaret, P., et al., 2001. Mon. Not. R. Astron. Soc. 321, L29.
27. Kauffmann, G., Haehnelt, M.G., 2000. Mon. Not. R. Astron. Soc. 311, 576.
28. Klypin, A., Kravtsov, A.V., Valenzuela, O., Prada, F., 1999. Astrophys. J. 522, 82.
29. Kudritzki, R.P., 2000. In: Weiss, A., Abel, T., Hill, V. (Eds.), The First Stars, p. 127. Springer, Heidelberg.
30. Kuhlen, M., Madau, P., 2005. Mon. Not. R. Astron. Soc. 363, 1069.
31. Kuhlen, M., Diemand, J., Madau, P., 2007. Astrophys. J. 671, 1135.
32. Machacek, M.M., Bryan, G.L., Abel, T., 2003. Mon. Not. R. Astron. Soc. 338, 273.
33. Madau, P., et al., 1996. Mon. Not. R. Astron. Soc. 283, 1388.
34. Madau, P., Meiksin, A., Rees, M.J., 1997. Astrophys. J. 475, 429.
35. Madau, P., Haardt, F., Rees, M.J., 1999. Astrophys. J. 514, 648.
36. Madau, P., Diemand, J., Kuhlen, M., 2008. Astrophys. J., 679, 1260.
37. Menou, K., Haiman, Z., Narayanan, V.K., 2001. Astrophys. J. 558, 535.
38. Metcalf, R.B., Madau, P., 2001. Astrophys. J. 563, 9.
39. Moore, B., Ghigna, S., Governato, F., Lake, G., Quinn, T., Stadel, J., Tozzi, P., 1999. Astrophys. J. 524, L19.
40. Moore, B., Diemand, J., Madau, P., Zemp, M., Stadel, J., 2006. Mon. Not. R. Astron. Soc. 368, 563.
41. Rauch, M., 1998. Annu. Rev. Astron. Astrophys. 36, 267.
42. Ryan-Weber, E.V., Pettini, M., Madau, P., 2006. Astrophys. J. 371, L78.
43. Sesana, A., Haardt, F., Madau, P., Volonteri, M., 2004. Astrophys. J. 611, 623.
44. Silk, J., Rees, M.J., 1998. Astron. Astrophys. 331, L1.
45. Somerville, R.S., Bullock, J.S., Livio, M., 2003. Astrophys. J. 593, 616.
46. Songaila, A., 2001. Astrophys. J. 561, 153.
47. Spergel, D.N., et al., 2003. Astrophys. J. Suppl. Ser. 148, 175.
48. Stadel, J., 2001. PhD thesis, U. Washington.
49. Tegmark, M., Silk, J., Rees, M.J., Blanchard, A., Abel, T., Palla, F., 1997. Astrophys. J. 474, 1.

50. Volonteri, M., Haardt, F., Madau, P., 2003. Astrophys. J. 582, 559.
51. Wyithe, J.S.B., Loeb, A., 2003. Astrophys. J. 588, L69.
52. Zentner, A.R., Bullock, J.S., 2002. Phys. Rev. D 66(4), 043003.

Chapter 28
A Simple Physical Model for Young Galaxies in the Early Universe

G. De Zotti, J. Mao, A. Lapi, G.L. Granato, and L. Danese

Abstract We present a simple physical model featuring an evolutionary link between Lyα emitting (LAE), Lyman break (LBG), sub-millimeter bright (SMB), and passively evolving galaxies, in the framework of the approach by Granato et al. (Astrophys. J. 600:580, 2004), according to which the evolution of massive galaxies is tied to the growth of the supermassive black holes at their centers. A key role in determining the properties of high-z galaxies is played by dust absorption. We model its evolution adopting a power-law relationship between UV extinction, star formation rate and metallicity, whose parameters are determined from a fit to the luminosity-reddening correlation for Lyman Break Galaxies, found by Shapley et al. (Astrophys. J. 562:95, 2001). The model yields quantitative relations between luminosities and basic physical quantities such as stellar and halo masses, star formation rates, galactic ages, metal abundances and dust extinction. It reproduces the luminosity functions (LFs) of LBGs and of LAEs at different redshifts, in the framework of the scenario that was previously shown to account for the wealth of observational data on SMB and passively evolving galaxies, as well as on high redshift quasars.

28.1 Introduction

Different techniques and selection criteria have produced samples of high-z galaxies, dubbed Lyman Break Galaxies (LBGs), Lyα Emitters (LAEs), Sub-Mm Bright

G. De Zotti (✉) · G.L. Granato
INAF, Osservatorio Astronomico di Padova, Vicolo dell'Osservatorio 5, 35122 Padova, Italy
e-mail: gianfranco.dezotti@oapd.inaf.it

G.L. Granato
e-mail: gianluigi.granato@oapd.inaf.it

G. De Zotti · J. Mao · A. Lapi · G.L. Granato · L. Danese
Astrophysics Sector, SISSA/ISAS, Via Beirut 2-4, 34014 Trieste, Italy

J. Mao
e-mail: mao@sissa.it

A. Lapi
e-mail: lapi@sissa.it

L. Danese
e-mail: danese@sissa.it

F.D. Macchetto (ed.), *The Impact of HST on European Astronomy*, Astrophysics and Space Science Proceedings,
DOI 10.1007/978-90-481-3400-7_28, © Springer Science+Business Media B.V. 2010

Galaxies (SMBGs). Substantial numbers of passively evolving high-z galaxies have also been found. In this paper we present a simple physical model aiming at integrating within a unified scheme these galaxy populations, interpreted as corresponding to different phases of the galaxy evolution. The model extends to earlier evolutionary phases the approach by [3], which relates the star formation rate (SFR) to the distribution of gas and dark matter (DM) in proto-galaxies, taking into account the effects of the energy fed back to the intra-galactic gas by supernova explosions and by accretion onto the nuclear, supermassive black hole (BH). These energy feedbacks actually *reverse* the formation sequence of the stellar component of galaxies compared to that of DM halos: the star formation and the buildup of central BHs are completed more rapidly in the more massive haloes, thus accounting for the phenomenon now commonly referred to as *downsizing*. A summary of earlier results obtained with this model can be found in [7]. Here we focus on the interpretation of the population properties of LBGs and LAEs. More details can be found in [8]. Throughout the paper we adopt a flat cosmology with matter density $\Omega_M = 0.3$, Hubble constant $H_0 = 70\,\mathrm{km\,s^{-1}\,Mpc^{-1}}$, and normalization of the mass variance $\sigma_8 = 0.8$.

28.2 The UV Luminosity Function of LBGs

The model of [3] model predicts the evolution with galaxy age of the SFR, of the metal abundance Z, of the stellar mass, and of the mass stored in the central supermassive BH, for any given galaxy halo mass and virialization redshift. Once the star formation and the chemical evolution history of a galactic halo of given mass and virialization redshift have been computed, the SED as function of time from extreme UV to radio frequencies is estimated through the GRASIL code [13]. Coupling these results with the halo formation rate, we can obtain the luminosity functions (LFs) of galaxies as a function of cosmic time.

An application of the model to the earliest evolutionary phases, when metals were first produced, requires a specific treatment of dust extinction, which plays an important role in the derivation of high-z galaxy properties, since even small amounts of dust strongly affect the rest-frame UV emission. Theoretically it is expected (see e.g. [5]) that the dust attenuation correlates with the gas metallicity, Z, and with the SFR, $\dot{M}_\star$. Reference [8] have shown that the relation

$$A_{1,350} \approx 0.35 \left(\frac{\dot{M}_\star}{\mathrm{M}_\odot\,\mathrm{yr}^{-1}} \right)^{0.45} \left(\frac{Z}{Z_\odot} \right)^{0.8}, \qquad (28.1)$$

where $A_{1,350}$ is the attenuation at 1,350 Å, provides a very good fit of the luminosity-reddening relation found by [12] for LBGs. This equation implies that extinction is small for LBGs hosted by low-mass halos, whose SFRs and metallicities are relatively low, while it quickly increases with time for higher masses, so that more massive galaxies are UV-bright for shorter times. The more massive galaxies spend approximately 90% of their burst time in an interstellar medium optically thick to their UV emission.

Adopting the above attenuation law [8] obtained a very good agreement with observational estimates of the LFs at different redshifts *without resorting to any tunable parameter*: all parameters of the model were fixed to the values previously determined from fits to other data sets.

28.3 The Luminosity Function of LAEs

The Lyα line typically arises from ionizing photons absorbed by nearby hydrogen gas, about 2/3 of which are converted into Lyα photons [9]. The interstellar dust attenuates the intrinsic Lyα luminosity by a factor $\exp(-\tau_{\mathrm{Ly}\alpha})$, where $\tau_{\mathrm{Ly}\alpha}$ is the optical depth at the Lyα wavelength (1,216 Å). We compute it from Eq. (28.1), extrapolated with the extinction curvei of [2]. Furthermore, only a fraction f_{IGM} of Lyα photons survives the passage through the IGM. Therefore the luminosity we see is

$$L^{\mathrm{obs}}_{\mathrm{Ly}\alpha} \approx 5.6 \times 10^{42} \left(\frac{\dot{M}_\star}{\mathrm{M}_\odot\,\mathrm{yr}^{-1}} \right) (1 - f_{\mathrm{HI}}) e^{-\tau_{\mathrm{Ly}\alpha}} f_{\mathrm{IGM}}\ \mathrm{erg\,s^{-1}}, \tag{28.2}$$

for the IMF of [11] adopted here; for a Salpeter IMF the normalization is a factor of about 1.6 lower. The fraction f_{HI} of ionizing photons surviving HI absorption can be written as $f_{\mathrm{HI}} = (R_{\mathrm{esc}}/R_{\mathrm{int}}) \exp(\tau_{912} - \tau_{1,350})$, where $R_{\mathrm{int}} \equiv L^{\mathrm{int}}_{912}/L^{\mathrm{int}}_{1,350}$ is given by the model and $(\tau_{912} - \tau_{1,350})$ is obtained from Eq. (28.1) plus the extinction curve of [2], leaving R_{esc} as the only tunable parameter. A good fit to the observed LAE LFs is obtained with $R_{\mathrm{esc}} = 0.15$ [8].

A clear-cut prediction of our model is that the Lyα luminosity of the most massive galaxies decreases abruptly after few $\times 10^7$ yr, implying that, with rare exceptions, bright LBGs should exhibit relatively low Lyα luminosity. A deficiency of objects with large Lyα equivalent width among bright LBGs has recently been reported by [1].

28.4 Conclusions

We have extended to the earliest evolutionary phases of high redshift galaxies the physical model worked out by [3] which proved to be capable of accounting for the wealth of data on the heavily obscured active star formation phase probed by (sub-)mm surveys, on the subsequent passive evolution of spheroidal galaxies, and on the cosmic epoch dependent optical and X-ray AGN luminosity functions [7].

According to this model, the LAE phase is confined to the youngest galaxy ages (up to few $\times 10^7$ yr), except for the low mass galaxies ($M_H \lesssim 10^{11}\ \mathrm{M}_\odot$), for which the fraction of Lyα photons escaping the galaxy keeps high for much longer times.

However, galaxies with $M_H \ll 10^{11}$ $M_\odot$ are too weak to be represented in the available LFs. This conclusion is at variance with that of [12], who argued that LBGs with Lyα in emission have ages larger than few $\times 10^8$ yr while younger system are dustier and have Lyα in absorption. On the other hand, the analysis of $z \sim 4$ LBGs observed in the GOODS-S survey [10] indicates that those with Lyα emission are on average much younger, less massive and less dusty, consistent with expectations from the present model.

At the highest z most of the star formation is associated to the LBG phase, but about 50% of galaxies with the highest star formation rates are predicted to be dust-obscured (and therefore sub-mm bright) even at $z = 10$. This is because, in the present framework, the more massive galaxies have higher SFRs, yielding higher metallicities and higher dust extinctions. As z decreases an increasing fraction of the cosmic SFR occurs in very dusty galaxies.

Another implication of the larger extinction for more massive objects is that the UV LF evolves between $z \approx 6$ and $z \approx 3$ much less than the halo mass function. This is due to the fact that the higher dust extinction mitigates the fast increase with decreasing redshift of the massive halo density. The effect of increasing opacity with increasing mass is even stronger for LAEs and is responsible for the weaker evolution of the LAE LF, compared to the LBG's, between $z = 6$ and $z = 3$. The indication of a weak evolution of the LAE LF between $z = 6.6$ and $z = 3.4$ is one of the key results of the search for LAEs at $z \simeq 6.6$ by [14].

In the LBG phase the nuclear black holes are still very far from their final masses (proportional to the halo masses) and only those associated to halos at the upper limit of the range represented in the available samples (a few percent of the total) can reach Eddington limited accretion rates large enough to yield $L_X \gtrsim 10^{42}$ erg s^{-1}. In the later sub-mm bright phase, the central BHs have grown much bigger (although they are still well below the final mass), and the halo mass distribution is shifted to higher values. As a consequence, the model predicts [4] a much higher fraction of X-ray bright sub-mm sources, compared to LBGs, as observed [6].

Acknowledgements This work is partially supported by ASI, INAF and MIUR grants.

References

1. Ando, M., et al., 2006. Astrophys. J. 645, L9.
2. Calzetti, D., et al., 2000. Astrophys. J. 533, 682.
3. Granato, G.L., et al., 2004. Astrophys. J. 600, 580.
4. Granato, G.L., et al., 2006. Mon. Not. R. Astron. Soc. 368, L72.
5. Jonsson, P., Cox, T.J., Primack, J.R., Somerville, R.S., 2006. Astrophys. J. 637, 255.
6. Laird, E.S., Nandra, K., Hobbs, A., Steidel, C.C., 2006. Mon. Not. R. Astron. Soc. 373, 217.
7. Lapi, A., et al., 2006. Astrophys. J. 650, 42.
8. Mao, J., et al., 2007. Astrophys. J. 667, 655.
9. Osterbrock, D.E., 1989. Astrophysics of Gaseous Nebulae and Active Galactic Nuclei. Univ. Science Books, Mill Valley.
10. Pentericci, L., et al., 2007. Astron. Astrophys. 471, 433.
11. Romano, D., et al., 2002. Mon. Not. R. Astron. Soc. 334, 444.

12. Shapley, A.E., et al., 2001. Astrophys. J. 562, 95.
13. Silva, L., Granato, G.L., Bressan, A., Danese, L., 1998. Astrophys. J. 509, 103.
14. Taniguchi, Y., et al., 2005. Publ. Astron. Soc. Jpn. 57, 165.

Chapter 29
Diffuse Ionized Gas Halos Seen with HST

Ralf-Jürgen Dettmar, Jörn Rossa, Michael Dahlem, and Roeland van der Marel

Abstract In order to better characterize the morphology and to determine physical parameters of possible outflows from star-forming galactic disks we have observed four edge-on galaxies with HST/ACS in Hα. Here we present first results for two of these objects, namely NGC 4364 and NGC 7090.

Over the last few years studies of the various phases of the interstellar medium (ISM) in the halos of edge-on galaxies have come to the conclusion that the presence of halo gas is generally well correlated with the star-formation rate in the underlying disk. This holds for the X-ray emitting hot ISM (HIM) (e.g., [11–13]), the diffuse ionized gas (DIG or warm ionized medium WIM) [5, 8, 9], as well as for the cosmic ray component (e.g., [2]). The required transport of gas from the disk into the halo is well modeled in the case of concentrated nuclear star-bursts [14]. Less well understood is this phenomenon of disk–halo interaction in the case of more normal disk galaxies with star-forming regions distributed all over the disk. Early models of the disk–halo interface used blow-outs in so-called "chimneys" to estimate relevant physical parameters [6]. However, even in the new high-resolution HI surveys of the Milky Way Galaxy the signature of the predicted chimney structures is found in only a few cases [4], orders of magnitudes less frequent than predicted.

A few studies of nearby galaxies with sufficiently high angular resolution using HST or Chandra (e.g., [10, 15]) also show a much more complex morphology of the DIG and HIM in the disk–halo interface. This can be understood in the framework of more realistic hydrodynamical models of the ISM in which clustered supernova

R.-J. Dettmar (✉)
Astronomical Institute, Ruhr-University Bochum, Bochum, Germany
e-mail: dettmar@astro.rub.de

J. Rossa
Department of Astronomy, University of Florida, Gainesville, FL, USA
e-mail: jrossa@astro.ufl.edu

M. Dahlem
Australia Telescope National Facility, Narrabri, NSW, Australia
e-mail: Michael.Dahlem@csiro.au

R. van der Marel
Space Telescope Science Institute, Baltimore, MD, USA
e-mail: marel@stsci.edu

F.D. Macchetto (ed.), *The Impact of HST on European Astronomy*, Astrophysics and Space Science Proceedings,
DOI 10.1007/978-90-481-3400-7_29, © Springer Science+Business Media B.V. 2010

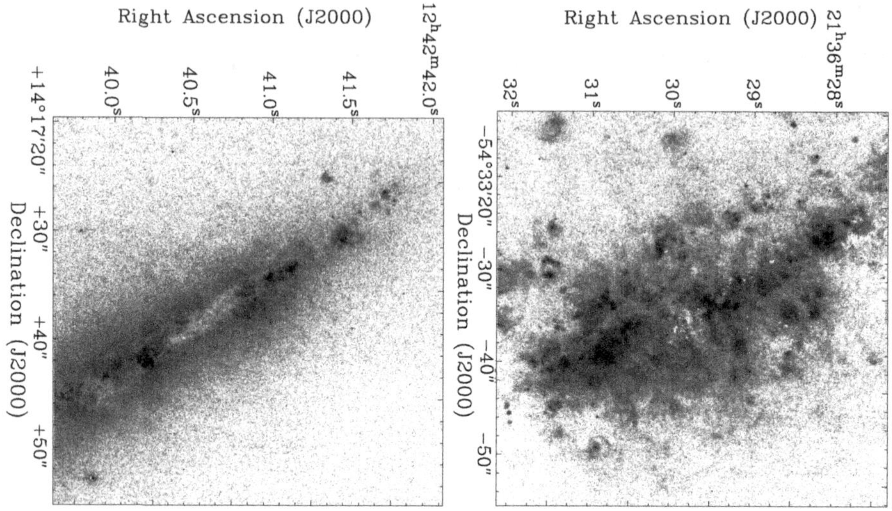

Fig. 29.1 Continuum subtracted Hα images obtained with HST/ACS of NGC 4634 (*left*) and NGC 7090 (*right*). In order to allow for sufficient resolution the fields shown cover only a representative part of the ACS field of view. The total integration time in the F658N filter is 6,880 sec for NGC 4634 and 7,469 sec for NGC 7090

explosions occur in an already highly structured medium (e.g., [3]) rather than in an hydrostatic equilibrium situation for the gas.

In order to better characterize the morphology and to determine physical parameters of possible outflows from star-forming galactic disks we have observed four edge-on galaxies with HST/ACS in Hα.

Here we present first results for two of these objects, namely NGC 4364 and NGC 7090 — see Fig. 29.1. They are chosen to display very different properties with regard to the morphology of the disk–halo interface. NGC 4634 displays a very smoothly distributed, homogeneous thick layer of H^+. In contrast, the warm ionized medium in NGC 7090 is much more confined into bubbles, shells, and filaments on smaller scales. This obvious difference cannot be explained by the difference in distances (D = 19 Mpc for NGC 4634 and D = 11 Mpc for NGC 7090) and rather has to be attributed to very different physical conditions in the ISM. The variations in the small scale structure of the ISM are most likely due to differences in the ambient interstellar pressure, influencing the super-bubble evolution as described by [7] or [1].

Acknowledgements RJD is supported at Ruhr-University Bochum by DLR through 50 OR 0503.

References

1. Clarke, C., Oey, M.S., 2002. Mon. Not. R. Astron. Soc. 337, 1299.

2. Dahlem, M., Lisenfeld, U., Rossa, J., 2006. Astron. Astrophys. 457, 121.
3. de Avillez, M.A., Breitschwerdt, D., 2004. Astron. Astrophys. 425, 899.
4. McClure-Griffiths, N.M., et al., 2006. Astrophys. J. 638, 196.
5. Miller, S.T., Veilleux, S., 2003. Astrophys. J. Suppl. Ser. 148, 383.
6. Norman, C.A., Ikeuchi, S., 1989. Astrophys. J. 345, 372.
7. Oey, M.S., Garcia-Seguro, G., 2004. Astrophys. J. 613, 302.
8. Rossa, J., Dettmar, R.-J., 2003. Astron. Astrophys. 406, 493.
9. Rossa, J., Dettmar, R.-J., 2003. Astron. Astrophys. 406, 505.
10. Rossa, J., et al., 2004. Astron. J. 128, 674.
11. Strickland, D.K., et al., 2004. Astrophys. J. Suppl. Ser. 151, 193.
12. Tüllmann, R., et al., 2006. Astron. Astrophys. 448, 43.
13. Tüllmann, R., et al., 2006. Astron. Astrophys. 457, 779.
14. Veilleux, S., Cecil, G., Bland-Hawthorn, J., 2005. Annu. Rev. Astron. Astrophys. 43, 769.
15. Wang, Q.D., et al., 2001. Astrophys. J. 555, L99.

Part III
Deep Fields, AGN, Black Holes and Radio Galaxies

Chapter 30
The Host Galaxy Properties of Powerful Radio Sources Across Cosmic Time

Robert A.E. Fosbury

Abstract The powerful extragalactic radio sources are believed to mark the most massive galaxies at every epoch. Using a combination of sensitive observational techniques covering different regions of the electromagnetic spectrum, both the stellar and interstellar properties of these host galaxies are now being investigated in detail. This talk reports some results from two major observing programmes. Rest-frame UV-optical spectroscopy and polarimetry of a sample of radio galaxies at $z \sim 2.5$ using the VLT and the Keck telescopes has provided high quality measurements of the Lyα to Hα region from which the spatial, kinematic and chemical properties of the ISM are being investigated. A comprehensive Spitzer survey of 69 radio galaxies from $1 \leq z \leq 5.2$, supported by archival HST imaging and other observations, is being used to construct SED covering a wide wavelength range which includes the rest-frame optical and NIR spectrum. This enables reliable estimates to be made of the galaxy stellar masses from the rest-frame H-band stellar emission. The derived luminosities imply stellar masses of 10^{11}–10^{12} $M_{\odot}$ even at the highest redshifts. The rather complete SED that we can compile for some of the sources allows us to distinguish between the bolometric emissions from the stars and the AGN.

30.1 Introduction

A radio galaxy was the first extragalactic source to be identified with an optical counterpart in 1949. The identification of Centaurus A and Virgo A were soon followed by Cygnus A in 1951, but the notion of sources being bright in the radio yet faint in the optical was initially difficult to accept. The realization that the radio emission resulted from the synchrotron process, driven by relativistic electrons in the presence of a magnetic field, removed this objection. The identification of radio galaxies was followed by the discovery of quasars in the early 1960s.

It is now appreciated that both of these classes of object are massive elliptical 'host' galaxies containing an active nucleus powered by accretion onto a — probably spinning — supermassive black hole (SMBH). The desire to get the clearest

R.A.E. Fosbury (✉)
ST-ECF, Garching bei München, Germany
e-mail: rfosbury@eso.org

F.D. Macchetto (ed.), *The Impact of HST on European Astronomy*,
Astrophysics and Space Science Proceedings,
DOI 10.1007/978-90-481-3400-7_30, © Springer Science+Business Media B.V. 2010

possible view of the host galaxy leads to the study of the radio galaxies in preference to the quasars since the now widely accepted orientation-based unification scheme reassures us that they are intrinsically similar objects with the radio galaxies offering the advantage of an inbuilt coronograph.

Given their ease of identification, the radio galaxies have played a pioneering role in the exploration of the high redshift Universe. They were the first galaxies to be found above redshifts 1, 2, 3, and 4. They are now understood, basically from their position in the K-band Hubble diagram, to be the most massive galaxies at every epoch [12] and they also appear to mark the positions of the first protoclusters [15].

In developing an understanding of the relationships that exist between the SMBH and their host galaxies, such as the $M_{\rm BH}$–$\sigma_{\rm bulge}$ relation [8, 10], the juxtaposition of powerful sources of radiation derived from nucleosynthesis in stars and from gravitational collapse in AGN leads to an intense interest in the nature of the feedback mechanisms that may be operating in such objects.

This talk is based on the results of two substantial observing programmes that were designed to investigate the nature of the interaction between the AGN and the host galaxy and to measure the properties of the latter in ways that not readily available in non-active systems. It is an example of the way that the accumulated imaging data with HST over the years has formed the foundation for major programmes of spectroscopy with the largest ground-based telescopes and, subsequently, the extension into the infrared using observations from Spitzer.

30.2 The Programmes

The first of the programmes, known as $z2p5$, was designed to elucidate the sources of line and continuum radiation at rest-frame wavelengths between Lyα and Hα. Using visible (3,900–9,000 Å) spectropolarimetry at $R \sim 600$ with LRISp on the Keck II telescope and J, H and K-band spectroscopy $R \sim 500$ with ISAAC on the VLT, excellent quality spectra were obtained for a partially overlapping sample of powerful radio galaxies selected from the Ultra-Steep Spectrum (USS) survey [13]. With one exception, the galaxies have redshifts between 2 and 3, allowing the measurement of most of the major emission lines in the UV/optical spectrum.

The results from the programme of spectropolarimetry are presented in [2] and [16] (see Fig. 30.1). Further analyses of the emission line spectra have been carried out by: [4–6, 17–20]. The first results from the ISAAC spectroscopy will appear in [7] (see Fig. 30.2).

The second of the programmes, the Spitzer High redshift Radio Galaxy Survey (SHzRGS), was designed to carry out a comprehensive MIR survey of ~70 high redshift radio galaxies at $1 \leq z \leq 5.2$. By tracking with increasing redshift the ~1.5 μm H^- peak of the evolved stellar population SED to provide a rest-frame H-band Hubble diagram, reliable stellar mass estimates can be made. For these powerful radio sources, however, the contributions from the obscured AGN must be identified and removed, requiring observations longward of the redshifted H^- peak in order to characterise the hot dust re-emission from material close

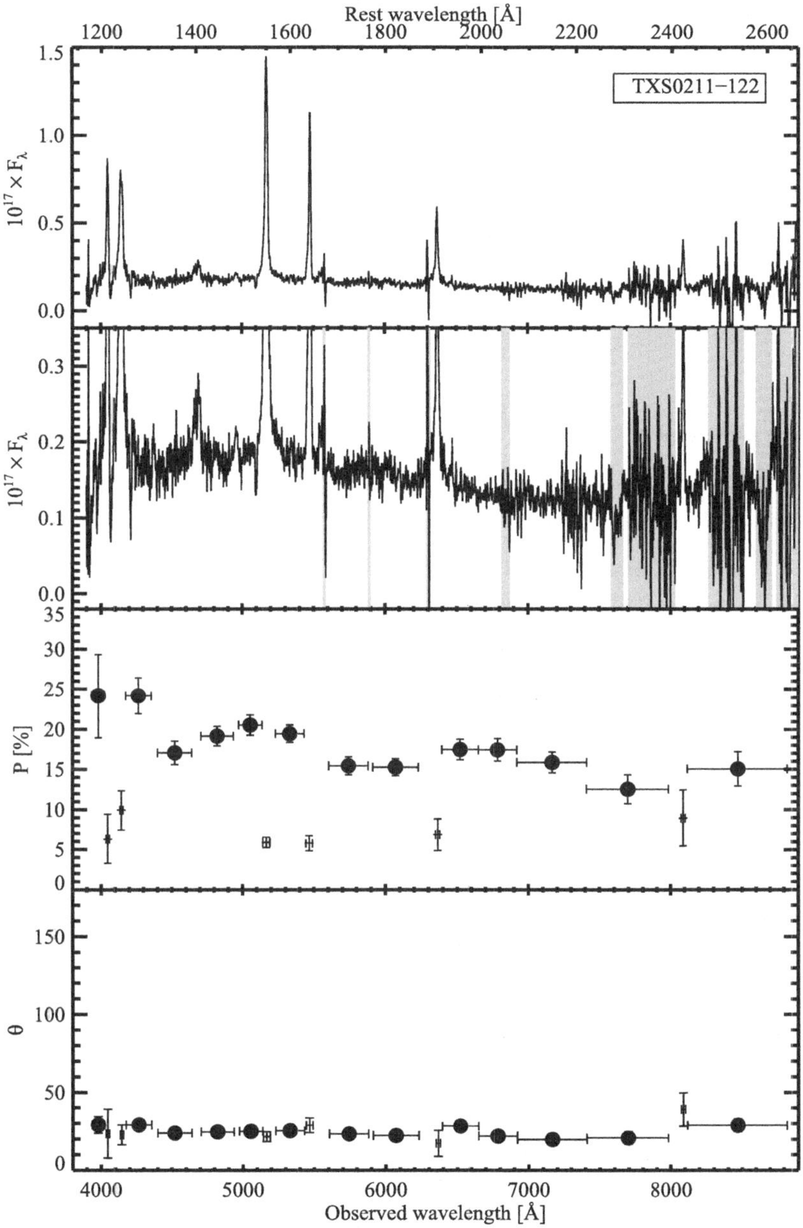

Fig. 30.1 Flux spectrum (*top two panels*, units of 10^{-17} erg s^{-1} cm^{-2} Å^{-1}), fractional linear polarization (*third panel*) and position angle — in degrees North through East — of the electric vector (*bottom panel*) for the most highly polarized of the $z2p5$ sample, TXS 0211-122 ($z = 2.340$). *Filled circles* and *crosses* respectively indicate continuum and narrow emission lines (including their underlying continuum). *Shaded regions* indicate strong night sky emission

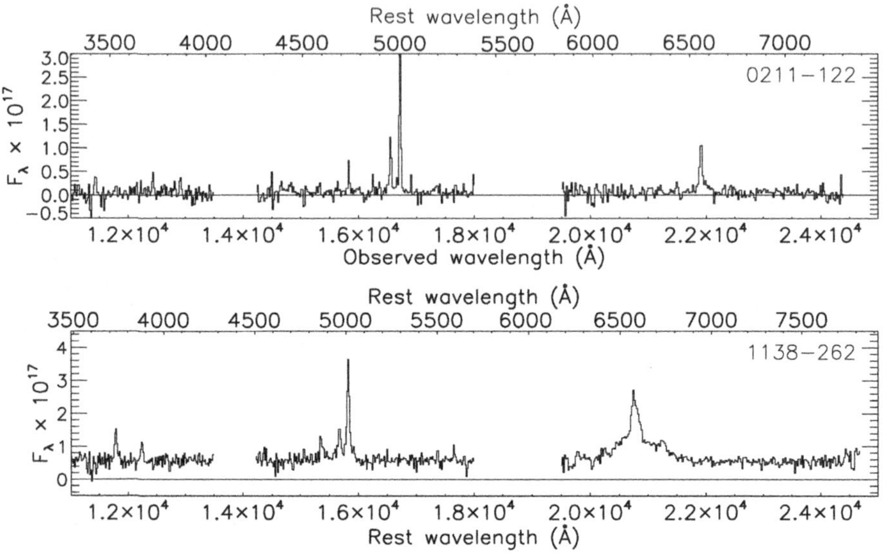

Fig. 30.2 ISAAC J, H and K-band spectra of two of the $z2p5$ sources, TXS 0211-122 and MRC 1138-262 ($z = 2.156$). The units are 10^{-17} erg s^{-1} cm^{-2} Å^{-1}

to the AGN. The sample was chosen to span a substantial range in radio power ($26 \leq \log(L_{3\mathrm{GHz}}\ \mathrm{WHz}^{-1}) \leq 28.5$) as well as redshift. A description of the programme and the first derivations of stellar masses using this method are presented in [14].

In addition to the datum provided by the rest-frame H-band luminosity, additional photometric and polarimetric data are being obtained in order to build a comprehensive library of SED measurements from the X-ray to the radio bands. With the polarimetry being used to estimate the scattered AGN contribution to the UV and optical emissions, the first global attempts are being made to separate the bolometric outputs of the AGN and the stellar population. The imaging data from HST, Spitzer, XMM-Newton, Chandra and ground-based telescopes also provide a powerful resource for studying the (proto-)cluster environments of these sources.

30.3 Summary of Results

30.3.1 z2p5

With the spectrograph slit aligned with the radio axis and typical exposure times of 20 ks, the Keck LRISp observations provide high s/n, spatially resolved, line and continuum measurements of 9 high power radio galaxies. The continuum polarization, in a line-free region from ∼1,250–1,400 Å, ranges from ≤1% to 20% in these sources, indicating that the scattered AGN can be a major, or even the dominant, contributor to the rest-frame UV continuum. The orientation of the electric vector is more closely perpendicular to the UV extension seen in the HST images that it is to the radio source axis.

It was shown that a clumpy, dusty obscuring region would naturally, by a process of finding a path of maximum escape efficiency, result in grey scattering. The energy absorbed by the illuminated scattering region was unlikely, however, to provide the powerful FIR re-emission seen in some of these sources with sub-mm instruments, suggesting that, as well as the AGN, these objects were the sites of intense star formation. While most of this star formation activity is heavily obscured, there must be sufficient UV escaping to dilute the scattered light in the lower polarization sources. Further evidence for the star formation is contained in the great strength of the Lyα emission line seen in some of the higher redshift objects [20], which cannot readily be explained as resulting from AGN photoionization processes alone. Also, the sources with larger Lyα/C IV ratios tend to have the lowest polarization suggesting a link between the starlight-diluted continuum and the excess Lyα emission.

The long-slit spectra reveal very extended emission line halos with sizes sometimes exceeding 100 kpc. This extended emission divides into kinematically perturbed (FWHM $\geq$ 1,000 km s^{-1}) regions associated with the radio jets and quiescent halos with FWHM $\leq$ 250 km s^{-1}. An analysis in combination with the NIR (rest-frame optical) spectroscopy shows that the ionized ISM in these galaxies has a metallicity that is close to Solar with little variation from source to source, a conclusion similar to that reached for low redshift radio galaxies by [11].

30.3.2 SHzRGS

Using IRAC (3.6–8.0 μm), IRS (16 μm) and MIPS (24–160 μm; a subset of 26 of the original 69 sources) imaging, the rest-frame optical to infrared spectral energy distributions have been decomposed into stellar, AGN, and dust components and the contribution of the host galaxy stellar emission to the rest-frame H-band has been derived (see Fig. 30.3). It is found that the fraction of emitted light at rest-frame H-band from stars is >60% for ~75% of the high redshift radio galaxies. The resultant stellar luminosities imply stellar masses of $10^{11-11.5}$ $M_{\odot}$ even at the highest redshifts.

30.4 Conclusions

The most powerful extragalactic radio sources are hosted by the most massive galaxies at every epoch. The obscured AGN provide copious ionizing radiation that creates huge narrow emission line regions, sometimes in excess of 100 kpc across. At high redshifts, they are also the sites of vigorous star formation which contributes,

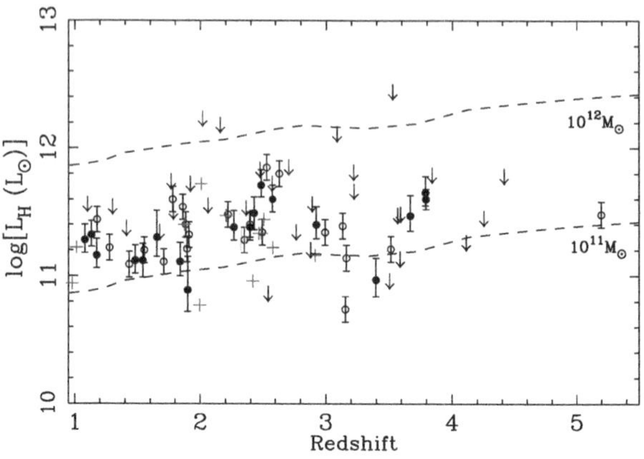

Fig. 30.3 (Color online) Rest-frame H-band stellar luminosity versus redshift for the SHzRGS sample, derived from the best-fit models to the multi-band photometry. *Solid circles* indicate those luminosities derived from HzRGs with MIPS detections, whilst *open circles* are luminosities derived from HzRGs without MIPS detections. Those radio galaxies where only a maximum fit of the stellar SED was possible have upper limits, indicated by *arrows*. The *dashed lines* represent the luminosities of elliptical galaxies with $z_{\rm form} = 10$ taken from the `PÉGASE.2` models [3] and normalized to 10^{11} and 10^{12} $M_\odot$. *Crosses* mark the stellar luminosity of sub-millimetre galaxies from [1], re-derived in the same fashion as the SHzRGS sample

via re-emission from cold dust, the bulk of the FIR/sub-mm emission from these luminous objects. This star formation is also revealed by an excess of Lyα emission, over and above that produced by the AGN radiation field. Studies of rest-frame UV-optical emission lines show that the gas metallicity varies little from source to source and is close to Solar in value, reminiscent of the low redshift sources studied two decades ago.

Polarimetry allows the unambiguous detection of indirect light from the obscured AGN and demonstrates that this radiation can dominate the UV continuum by effectively grey scattering from a clumpy, dusty medium.

MIR photometry with Spitzer allows reliable measurements, that can be corrected for AGN contamination, of the evolved stellar emission that peaks in the rest-frame H-band. By fitting appropriate stellar populations, stellar masses of $10^{11-11.5}$ $M_\odot$ are derived, subject to the uncertainties in the modelling of the thermally-pulsating AGB phase of stellar evolution at relatively young ages [9].

Acknowledgements The work described in this talk was carried out by two teams. The $z2p5$ programme included: Sperello di Serego Alighieri (Arcetri), Andrea Cimatti (Bologna), Montse Villar-Martín (Granada), Joël Vernet (ESO), Marshall Cohen (Caltech), Bob Goodrich (Caltech), Andrew Humphrey (UNAM) and Ilse van Bemmel (Leiden). The SHzRGS programme included: Daniel Stern — PI and Peter Eisenhardt (JPL); Carlos De Breuck — Co-PI, Joël Vernet (ESO), Alessandro Rettura (JHU); Andrew Zirm (JHU); Brigitte Rocca-Volmerange (IAP); Nick Seymour, Mark Lacy and Harry Teplitz (SSC); Arjun Dey and Mark Dickinson (NOAO); Wil van Breugel and Adam Stanford (UCLLNL); George Miley and Huub Röttgering (Leiden); Partick McCarthy (OCIW) and David Alexander (Durham).

References

1. Borys, C., et al., 2005. Astrophys. J. 635, 853.
2. Cimatti, A., di Serego Alighieri, S., Vernet, J., Cohen, M., Fosbury, R.A.E., 1998. Astron. Astrophys. 499, L21.
3. Fioc, M., Rocca-Volmerange, B., 1997. Astron. Astrophys. 326, 950.
4. Fosbury, R.A.E., et al., 2003. New Astron. Rev. 47, 299.
5. Humphrey, A., Villar-Martín, M., Fosbury, R., Vernet, J., di Serego Alighieri, S., 2006. Mon. Not. R. Astron. Soc. 369, 1103.
6. Humphrey, A., et al., 2007. Mon. Not. R. Astron. Soc. 375, 705.
7. Humphrey, A., et al., 2007. Mon. Not. R. Astron. Soc. 382, 1729.
8. Magorrian, J., et al., 1998. Astron. J. 115, 2285.
9. Maraston, C., 2005. Mon. Not. R. Astron. Soc. 362, 799.
10. Richstone, D., et al., 1998. Nature 395, 14.
11. Robinson, A., Binette, L., Fosbury, R.A.E., Tadhunter, C.N., 1987. Mon. Not. R. Astron. Soc. 227, 97.
12. Rocca-Volmerange, B., Le Borgne, D., De Breuck, C., Fioc, M., Moy, E., 2004. Astron. Astrophys. 415, 931.
13. Roettgering, H.J.A., Miley, G.K., Chambers, K.C., Macchetto, F., 1995. Astron. Astrophys. Suppl. Ser. 114, 51.
14. Seymour, N., et al., 2007. Astrophys. J. Suppl. Ser. 171, 353.
15. Venemans, B.P., et al., 2007. Astron. Astrophys. 461, 823.
16. Vernet, J., et al., 2001. Astron. Astrophys. 366, 7.
17. Villar-Martín, M., et al., 2002. Mon. Not. R. Astron. Soc. 336, 436.
18. Villar-Martín, M., et al., 2003. New Astron. Rev. 47, 291.
19. Villar-Martín, M., et al., 2003. Mon. Not. R. Astron. Soc. 346, 273.
20. Villar-Martín, M., et al., 2007. Mon. Not. R. Astron. Soc. 375, 1299.

Chapter 31
A New View of the Origin of the Radio-Quiet/Radio-Loud AGN Dichotomy?

Alessandro Capetti and Barbara Balmaverde

Abstract Using HST and Chandra observations of nearby early-type galaxies we re-explored the classical issue of the connection between the properties of AGN and the characteristics of their hosts. We have found evidence that the radio-loud/radio-quiet dichotomy is directly connected to the structure of the inner regions of their host galaxies in the following sense: (1) Radio-loud AGN are associated with galaxies with shallow cusp in their light profiles; (2) Radio-quiet AGN are only hosted by galaxies with steep cusps.

Since the brightness profile is determined by the galaxy's evolution, through its merger history, our results suggest that the same process sets the AGN flavour. This provides us with a novel tool to explore the co-evolution of galaxies and supermassive black holes, and it opens a new path to understand the origin of the radio-loud/radio-quiet AGN dichotomy.

31.1 Introduction

It is becoming increasingly clear that not only most massive galaxies host a supermassive black hole (SMBH) but also that SMBH and host galaxies follow a common evolutionary path, as suggested by the tight relationships between the SMBH mass and galaxies properties (e.g. [9, 10]). In the co-evolution of the SMBH/galaxy system, nuclear activity plays a major role, generating a feed-back acting on the host galaxy (e.g. [6]). Thus AGN represent our best tool to investigate the growth of SMBH but also to explore galaxies evolution.

But, despite this breakthrough in our understanding of the SMBH/galaxy system, we still lack a clear picture of the connection between the properties of AGN and of their host galaxies. One of the crucial issues is to explain the origin of the AGN radio loudness dichotomy (e.g. [12]); this is important from the point of view of AGN physics, but it is also related the feedback process, as it corresponds to different modes of energy transfer from the AGN into the ambient medium.

A. Capetti (✉) · B. Balmaverde
INAF, Osservatorio Astronomico di Torino, Strada Osservatorio 20, 10025 Pino Torinese, Italy
e-mail: capetti@oato.inaf.it

B. Balmaverde
e-mail: balmaverde@oato.inaf.it

F.D. Macchetto (ed.), *The Impact of HST on European Astronomy*, Astrophysics and Space Science Proceedings,
DOI 10.1007/978-90-481-3400-7_31, © Springer Science+Business Media B.V. 2010

Previous studies have not provided clear-cut answers, since e.g. spiral galaxies only harbor radio-quiet (RQ) AGN, but early-type galaxies (ETG) can host both radio-loud (RL) and RQ AGN. Similarly, RL AGN are generally associated with the most massive SMBH, but the RL and RQ mass distributions are broad and overlap considerably (e.g. [7]).

31.2 Results from HST and Chandra Data

We re-explored the connection between the AGN properties in nearby ETG and the characteristics of their hosts. ETG appear to be the critical class of objects, as they host AGN of both classes of radio-loudness. Starting from an initial sample of 332 galaxies, we selected 116 AGN candidates (requiring a minimum radio flux of $\sim$1 mJy). In [2] we analyzed the 65 objects with available HST images, distinguishing them on the basis of the nuclear slope of their brightness profiles following the modeling scheme proposed by [13]. A classification into core and power-law galaxies was possible for 51 objects. We explored the robustness of this result against a different choice of the analytic form for the brightness profiles, using a Sérsic law. Reassuringly, we found a complete correspondence between galaxies reproduced by core-Sérsic models [11] with core-Nuker galaxies. Thus, the separation of ETG into two classes, with and without a shallow core is independent of the modeling strategy (at least for galaxies brighter than $\sim M^* + 2$).

We used HST and Chandra data to isolate the nuclear emission of these galaxies. Core galaxies invariably host a RL nucleus [1] with a median radio-loudness of $\mathrm{Log\,R} = 3.6$. Power-law galaxies show a substantial excess of optical and X-ray emission with respect to core galaxies at the same level of radio luminosity [3]. Conversely, their radio-loudness parameter, $\mathrm{Log\,R} \sim 1.6$ is similar to that measured e.g. in Seyfert galaxies ($\mathrm{Log\,R} \sim 1.9$, see [15]).

RL and RQ nuclei cannot be distinguished on the basis of host's luminosity or black-hole mass; they differ only statistically (see Fig. 31.1). Only the brightness profiles provide a full separation between the two classes (see Fig. 31.2).

We extended our analysis to a sample of nearby ($V_{rec} < 7{,}000$ km/s) Seyfert galaxies hosted by ETG [4]. Their brightness profiles can be modeled with a Nuker law with a steep nuclear cusp ($\gamma \sim 0.5$–1.1), as expected for these radio-quiet AGN based on our previous findings. This extends the validity of the connection between brightness profile and radio-loudness to AGN of a far higher luminosity (see Fig. 31.3).

31.3 Conclusions

Taking together these results with those of [5] (where we showed that low-luminosity radio-galaxies are hosted by core galaxies), we have covered the different manifestations of nuclear activity in the local Universe. We consistently recovered

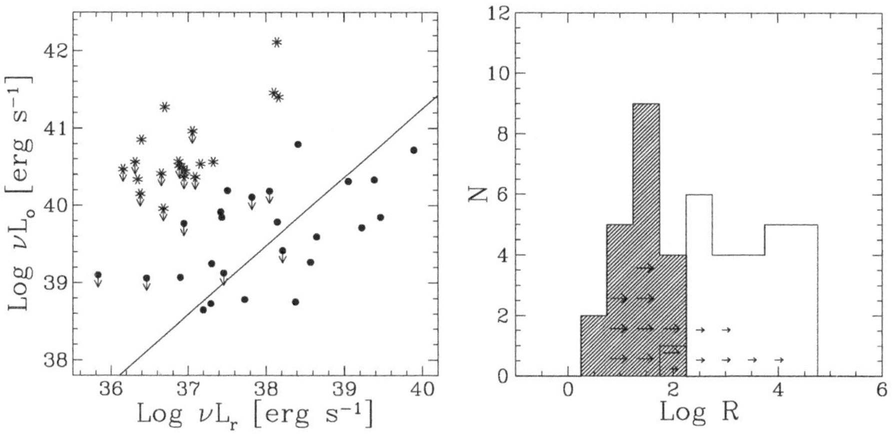

Fig. 31.1 *Left panel*: radio vs. optical nuclear luminosity for power-law (*stars*) and core galaxies (*filled circles*). The *solid line* represents the correlation between the nuclear luminosities for CoreG and the 3C/FR I sample derived by [1]. *Right panel*: radio-loudness parameter $R = L_{5\,GHz}/L_B$ for PlawG (*filled histogram*) and CoreG (*empty histogram*)

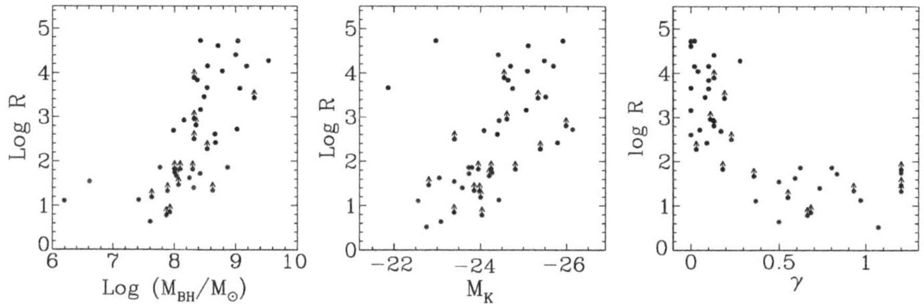

Fig. 31.2 Radio-loudness parameter R vs. (*left*) estimated black-hole mass and (*center*) galaxies absolute magnitude. No clear separation between RL and RQ nuclei can be obtained based on these two quantities. (*Right*) radio-loudness vs. γ, the logarithmic slope of the brightness profile. Only the brightness profiles provide a full separation between the two classes

the association of radio-loud AGN with core galaxies and of radio-quiet AGN with non-core galaxies.

Since the brightness profile is determined by the galaxy's evolution through its merger history (e.g. [8]), our results suggest that the same process sets the AGN flavor. In this scenario, the black holes hosted by the merging galaxies rapidly sink toward the center of the newly formed object, setting its nuclear configuration, described by e.g. the total mass, spin, mass ratio, or separation of the SMBHs. These parameters are most likely at the origin of the different levels of the AGN radio-loudness. For example, it has been proposed that a core galaxy is the result of (at least) one major merger and that the core formation is related to the dynamical effects of the binary black holes on the stellar component (e.g. [14]). From the AGN

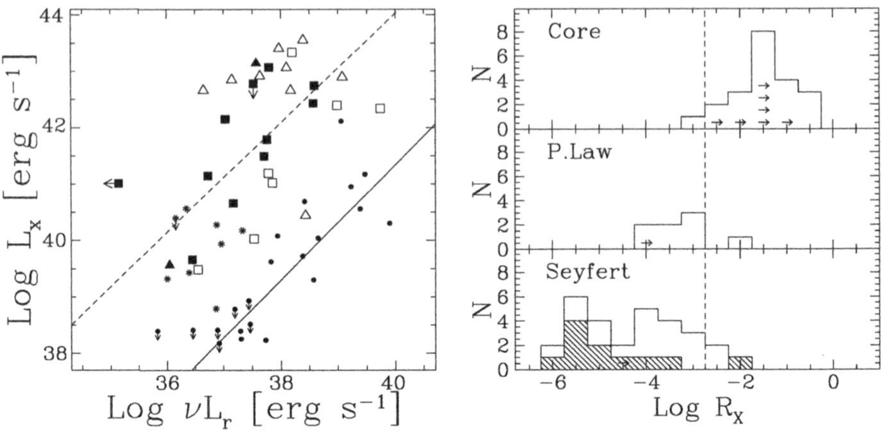

Fig. 31.3 *Left panel*: Radio vs. X-ray nuclear luminosity for Seyfert galaxies (*triangles* are Sy 1, *squares* are Sy 2), power-law (*stars*) and core galaxies (*filled circles*). The *solid line* represents the correlation found for CoreG and the 3C/FR I. The *dashed line* represents the relation derived by [15] from their sample of Seyfert galaxies. *Right panel*: radio-loudness parameter $R_X = (\nu L_r/L_X)$, for Seyferts (*bottom*), power-law (*middle*), core galaxies (*top*)

point of view, [16] suggested that a RL source can form only after the coalescence of two SMBH of similar (high) mass, forming a highly spinning nuclear object, from which the energy necessary to launch a relativistic jet can be extracted. In this situation (the merging of two large galaxies of similar mass), the expected outcome is a massive core galaxy in line with our results. The connection of the radio-loudness with the host's brightness profile might open a new path toward understanding the origin of the RL/RQ AGN dichotomy, and it provides us with a further tool for exploring the co-evolution of galaxies and supermassive black holes.

References

1. Balmaverde, B., Capetti, A., 2006. Astron. Astrophys. 447, 97.
2. Capetti, A., Balmaverde, B., 2005. Astron. Astrophys. 440, 73.
3. Capetti, A., Balmaverde, B., 2006. Astron. Astrophys. 453, 27.
4. Capetti, A., Balmaverde, B., 2007. Astron. Astrophys. 469, 75.
5. de Ruiter, H., et al., 2005. Astron. Astrophys. 439, 487.
6. Di Matteo, T., Springel, V., Hernquist, L., 2005. Nature 433, 604.
7. Dunlop, J., et al., 2003. Mon. Not. R. Astron. Soc. 340, 1095.
8. Faber, S., et al., 1997. Astron. J. 114, 1771.
9. Ferrarese, L., Merritt, D., 2000. Astrophys. J. 539, L9.
10. Gebhardt, K., et al., 2000. Astrophys. J. 539, L13.
11. Graham, A., et al., 2003. Astron. J. 125, 2951.
12. Kellermann, K., et al., 1994. Astron. J. 108, 1163.
13. Lauer, T., et al., 1995. Astron. J. 110, 2622.
14. Milosavljević, M., et al., 2002. Mon. Not. R. Astron. Soc. 331, L51.
15. Panessa, F., et al., 2007. Astron. Astrophys. 467, 519.
16. Wilson, A., Colbert, E., 1995. Astrophys. J. 438, 62.

Chapter 32
The Bright and the Dark Side of Malin 1

Renzo Sancisi and Filippo Fraternali

Abstract Malin 1 has long been considered a prototype giant, dark matter dominated Low Surface Brightness galaxy. Two recent studies, one based on a re-analysis of VLA HI observations and the other on an archival Hubble I-band image, throw a new light on this enigmatic galaxy and on its dark/luminous matter properties.

32.1 Introduction

Malin 1 is a highly unusual disk galaxy characterized by an enormous, H I rich and extremely low surface brightness disk [2, 3]. Recent, deep R-band data by [4] show an exponential disk extending out to 124 kpc $h^{-1}_{[75]}$ (scale length 53 kpc). This corresponds to the H I extent. There is also a prominent *bulge*-like component. According to [3] the rotation curve has the slowly rising shape typical of the less luminous, "dark matter dominated" LSB galaxies. This slow rise of the rotation curve in the presence of a luminous central component is in marked contrast with the rule that there is a close correlation between the distribution of light and the shape of the rotation curve [6]. However, [3] do point out that their rotation curve is very uncertain because of the low resolution of the observations, the low signal/noise ratio and the strong warping. This has induced us to carry out a re-analysis of the H I data with special attention for the beam-smearing effects and to make a new comparison with the luminosity profile. The results of our analysis agree with those of a recent HST optical study of the bright central component [1]: Malin 1 is a normal, early-type galaxy surrounded by a huge, low-surface-brightness outer disk.

R. Sancisi (✉)
INAF, Observatory of Bologna, Bologna, Italy
e-mail: renzo.sancisi@oabo.inaf.it

R. Sancisi
Kapteyn Astronomical Institute, Groningen, Netherlands

F. Fraternali
Department of Astronomy, University of Bologna, Bologna, Italy
e-mail: filippo.fraternali@unibo.it

F.D. Macchetto (ed.), *The Impact of HST on European Astronomy*, Astrophysics and Space Science Proceedings,
DOI 10.1007/978-90-481-3400-7_32, © Springer Science+Business Media B.V. 2010

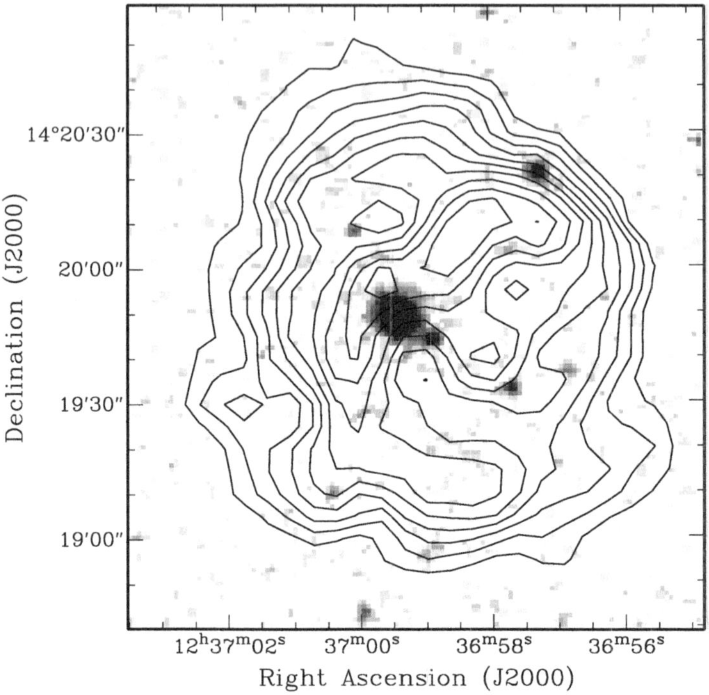

Fig. 32.1 Total H I map of Malin 1 overlaid on the optical DSS image. *Contours* run from 7.7 to 49.1×10^{19} atom cm^{-2} with an increment of 4.6×10^{19} atom cm^{-2}. The HPBW is 21″ (∼29 kpc)

32.2 New Rotation Curve and Comparison with the Luminosity Profile

The H I data cube obtained with the VLA by [3] has been re-analyzed. The large extent of the H I disk is shown in Fig. 32.1 superposed on the optical (DSS) image. The H I radius corresponds approximately to that of the extended, faint optical disk. A new velocity field has been derived. The 21-cm line profiles are strongly affected by beam smearing and are very asymmetric. Instead of the intensity-weighted mean velocities used by Pickering et al., which suffer heavily from beam smearing, we have taken the velocities at the profile peaks, close to the high rotation velocity side. Subsequently, the rotation curve has been derived from the velocity field following well-known standard procedures. This rotation curve has been used to construct model data cubes to verify its correctness. The new and the old [3] rotation curves are shown in Fig. 32.2 (bottom). Amplitude and flat outer part are the same. In the inner parts the new curve rises much more steeply and reaches higher values inside 20 arcsec (∼30 kpc) in correspondence of the central concentration in the luminosity profile (Fig. 32.2, top). Figure 32.3 shows the "maximum disk" decomposition, with isothermal halo and H I disk. The R-band profile [4] has been used. The maximum disk M/L ratio is 5.2. This is in the range of the values found for luminous early-type galaxies [5].

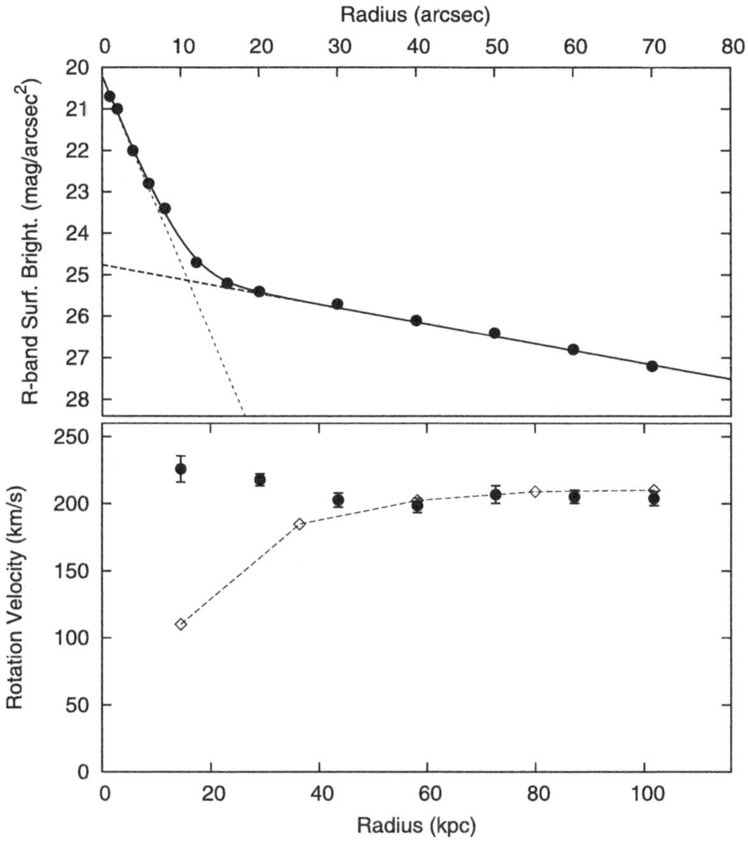

Fig. 32.2 Mass follows light. *Upper panel*: R-band luminosity profile derived from [4] and fitted with two exponential disks. *Lower panel*: new rotation curve of Malin 1 derived as described in the text (*filled circles*). The *open diamonds* show the rotation curve derived by [3]

Fig. 32.3 New rotation curve (*squares*) of Malin 1 and "maximum disk" mass decomposition. The contributions from the stellar (*thick dash*) and gaseous (*short dash*) disks and the DM isothermal halo (*dot-dash*) are shown. The *thick line* shows the sum of the three

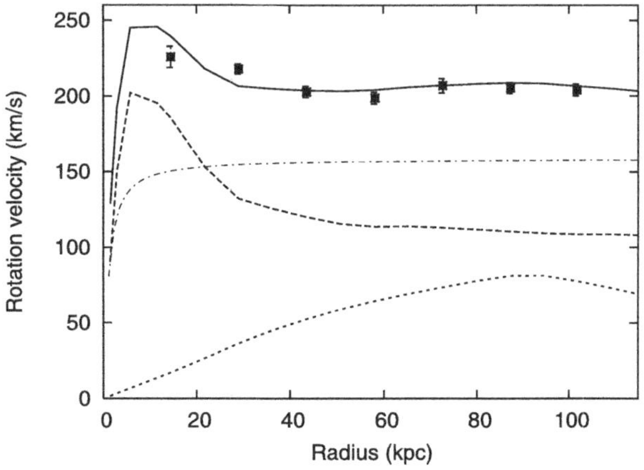

32.3 Conclusions

The present study is based on a re-analysis of existing H I observations of Malin 1 obtained with the VLA. A new rotation curve has been derived. This rotation curve shows a close correlation with the luminosity distribution, in line with the "rule" suggested by [6]. Also in this galaxy the mass seems to follow the light. The rotation curve has the shape (steep inner rise) typical of high surface brightness (HSB) galaxies. The classical disk–halo decomposition of the rotation curve has shown that a maximum disk solution is possible. Clearly, in its inner luminous part, Malin 1 has the characteristics of an early-type HSB galaxy.

Reference [1] has recently published a study of Malin 1 based on archival Hubble I-band data. He has examined the structure and the properties of the inner bright parts and has concluded that Malin 1 has a normal stellar disk and that, out to a radius of ~10 kpc, its structure is that of a typical SB0/a galaxy.

The new H I analysis and the optical study throw new light on Malin 1 and on its dark/luminous matter properties. Both point at the same conclusion: Malin 1 is a normal, luminous early-type galaxy. The enigma of the huge (120 kpc), low-surface-brightness stellar and H I disk surrounding the bright inner parts remains. In view of the large-scale symmetry and regularity and of the large orbital period (~3.5 billion years) in the outer parts, it seems unlikely that the formation of this extended structure is due to recent accretion and mergers.

Acknowledgements We thank Tim Pickering and his co-authors for kindly making their H I data cube available to us.

References

1. Barth, A.J., 2007. Astron. J. 133, 1085.
2. Bothun, G.D., Impey, C.D., Malin, D.F., Mould, J.R., 1987. Astron. J. 94, 23.
3. Pickering, T.E., Impey, C.D., van Gorkom, J.H., Bothun, G.D., 1997. Astron. J. 114, 1858.
4. Moore, L., Parker, Q.A., 2006. Publ. Astron. Soc. Pac. 23, 165.
5. Noordermeer, E., van der Hulst, J.M., Sancisi, R., Swaters, R.S., van Albada, T.S., 2007. Mon. Not. R. Astron. Soc. 376, 1513.
6. Sancisi, R., 2004. IAUS 220, 233S.

Chapter 33
Cluster Lensing with Hubble

Jean-Paul Kneib

Abstract I review here some of the most important gravitational lensing discoveries obtained with *Hubble* observations of cluster of galaxies. We can identify three categories of results: (i) Dark matter mapping: the measurement of the mass distribution of clusters conducted either on individual clusters, or in a statistical way, looking at a cluster sample aiming to understand cluster physics; (ii) Gravitational Telescope: the study of distant galaxies that can be conducted in great details thanks to the gravitational magnification of massive clusters; (iii) Cosmography: as lensing is sensitive to cosmological distances, one can expect to put constraints on the geometrical parameters that defines the Universe geometry. Although, *Hubble* was not build to explore gravitational lensing in clusters, *Hubble* unique capabilities in terms of spatial resolution and sensitivity as well its field of view that matches more or less the cluster core has permitted many discoveries. *Hubble* observations will certainly continue to produce more important results in cluster lensing until the next generation of space optical/near-infrared observatories are taking data.

33.1 Introduction

Gravitational Lensing in cluster of galaxies was first observed on ground based telescope at the end of the 1980s (e.g. [1]) by the detection of large arcs near cluster cores. The launch of *Hubble* just followed this important discovery. However, it is only after the first servicing mission that *Hubble* was recognised as a useful and unique observatory to study cluster lenses. This was captioned by the exceptional WFPC2 image of Abell 2218 released to the public on April 5, 1995 by W. Couch and collaborators. In this image a plethora of arcs and arclets could be readily identified demonstrating the power of lensing.

With post-COSTAR images, *Hubble* has provided unique view of multiply imaged galaxies: allowing better identification, detecting fainter images, revealing internal morphologies, all these leading to better understanding of the mass distribution in cluster cores.

The first in-depth analysis of the Abell 2218 lensing data [2] revealed five multiply images systems. For three of them, deep spectroscopy on the *William Herschel*

J.-P. Kneib (✉)
Laboratoire d'Astrophysique de Marseille, CNRS–Université de Provence, Technopole de Marseille-Etoile, 38, rue Frédéric Joliot-Curie, 13388 Marseille cedex 13, France
e-mail: jean-paul.kneib@oamp.fr

F.D. Macchetto (ed.), *The Impact of HST on European Astronomy*, Astrophysics and Space Science Proceedings,
DOI 10.1007/978-90-481-3400-7_33, © Springer Science+Business Media B.V. 2010

telescope determined their redshift allowing a detailed mass reconstruction of the cluster mass distribution [3]. In particular, this observation found that: about 10% of the total mass is located in cluster galaxies, and that Dark Matter was primarily distributed on scales larger than galaxy scale.

One of the multiple images was identified spectroscopically at redshift $z = 2.515$ [3] and revealed the first detailed morphology of a Lyman-break galaxy, hence paving the way of using cluster as gravitational telescope.

In this short review, I will stress some of the most striking results obtained with *Hubble* using cluster lenses as a tool to probe: the Dark Matter distribution in clusters, distant galaxies using the natural magnification factor of cluster lenses, and finally cluster lensing cosmography. To conclude, I will envision what could be achieved with *Hubble* before it will retire.

33.2 Dark Matter Distribution

Following the first lensing mass map of Abell 2218 ($z = 0.175$) obtained with *Hubble* [2], other clusters have been studied using either strong or weak lensing techniques or both. The first weak lensing study was conducted on the optically selected cluster Cl0939+4713 ($z = 0.41$) revealing a complex mass distribution [4].

Using WFPC2, ten X-ray luminous clusters (with $L_X > 8 10^{44}$ erg/s) at $z \sim 0.2$ have been observed with *Hubble*, together with extensive follow-up using ground-based telescope (to measure spectroscopic redshift) and deep *Chandra* X-ray observation. Using strong and weak lensing as measured by these observations, we have explored for the first the relation between the dark matter mass and the X-ray luminosity and temperature [5] as well as putting constraints on the σ_8 cosmological parameter [6].

The very limited field of view of a single WFPC2 pointing is however limiting the measurement of the cluster total mass and the determination of the mass profile on large scales. To work around this limitation, either wider-field ground based data are necessary or multi-pointing mosaic with *Hubble*.

The mass model of Abell 2218 demonstrated the need to put in galaxy mass contribution in order to reach a good fit with a limited number of free parameters. Building on this idea a weak and strong lensing method was developed to probe the mass distribution of cluster galaxies [7] and shortly after was applied on a deep six pointing WFPC2-mosaic of the cluster AC114 ($z = 0.31$) [8]. The favored model indicated that S0 galaxy halos are less massive than Elliptical galaxy halos.

On even larger scale, reaching the virial radius and beyond, *Hubble* has compete with wide-field ground-based telescope in measuring the cluster dark matter profile. The first completed analysis was conducted on the cluster Cl0024+1654 ($z = 0.395$) [9]. Aiming to measure the mass profile of this cluster beyond the virial radius, an original observing strategy was used. 38 WFPC2 pointing were sparsely distributed over a ~5 Mpc radius making the widest coverage of a cluster. Using a careful analysis that combined the weak lensing data of both STIS and WFPC2 cameras a detailed mass map and profile was reconstructed [10]. An excellent agreement was found between the K-band light distribution and the total mass distribution on scale larger than 100 kpc. Indeed a constant Mass-to-light ratio could be measured over most of the radial profile except in the very central region.

Furthermore at large radii, the dark matter profile was found to be steeper than an isothermal sphere, suggesting that the NFW profile was a good representation of cluster mass profile on large scale. Surprisingly however, the concentration parameter was found to be much larger than found in numerical simulations. Different reasons could explain this result. In particular, the fact that the cluster has just suffered a major merger along the line of sight as found by dynamical analysis [11] may have increased temporarily the cluster concentration, an other likely possibility is the uncertainty in the redshift distribution of faint galaxies used in reconstructing the weak lensing mass profile which could also have artificially increased the concentration. Other WFPC2-mosaic were obtained in particular one on Abell 2218 and one on Abell 1689 ($z = 0.183$).

The installation on *Hubble* of the Advanced Camera for Surveys (ACS) has been a major improvement in efficiency for lensing studies. The wider field, and improved sensitivity compared to WFPC2 has brought new fascinating discoveries. One of the major discovery was the deep multi-colour image of the massive cluster Abell 1689 that display more than hundred images part of more than 30 multiple-image systems and has allowed detailed photometric redshift measurement of the faintest multiple images [12].

This stunning image has motivated new developments in cluster strong lensing mass reconstruction techniques [13–15] and resulted in various attempts in constraining Abell 1689 mass distribution [16–20]. Although there is an excellent agreement in the strong lensing modeling in terms of the overall mass distribution (a two DM clump is required) as well as the mass enclosed in the Einstein radius and the best fitted NFW parameters, the agreement on larger scales combining both the ACS strong lensing results and wide-field ground based weak lensing measurement is still a matter of debate [17, 21], specially regarding the value of the concentration parameter.

Other ACS observation have allowed to measure the detailed mass distribution of clusters. For example the ACS confirmation of the large discrepancy in the dark matter map and the X-ray gas map of the Bullet Cluster (aka 1E0657-56, $z = 0.296$) which is probably the best evidence of the existence of collision-less Dark Matter [22]. The detection of a possible Dark Matter ring in the cluster Cl0024+1654 [23] which may be due to the line of sight collision of two clusters [11] is a potentially very interesting result. The revisited Abell 2218 lensing mass models unambiguously confirms the need of cluster galaxy halos to produce a good fit as well as confirming the importance of the second clump and argues that Abell 2218 has just suffered a merger [24]. The discovery of a the fifth de-magnified image of the largest separation multiple quasar SDSSJ1004+4112 produced by a massive cluster at $z = 0.68$ argues for a shallower (than NFW) mass distribution for Dark Matter in this particular cluster center [25].

Finally, the recent detailed mass reconstruction of the magnificent Abell 1703 ($z = 0.28$) based on the identification of 42 images part of 16 multiple-image systems is favoring a NFW-like profile in the very central part of the cluster [26] (see Fig. 33.1).

ACS SNAPshot data have been gathered on different cluster samples particularly the MACS cluster sample at $z > 0.3$ [27] and the LOCUSS cluster sample at $0.15 < z < 0.3$. For both observing programs about 50% of the massive cluster display stunning strong lensing configurations arguing for a large number of massive clusters that can in the long term (providing deeper multi-colour data, and spectroscopic measurement of some of the arcs) lead to similar detailed mass model of these massive clusters as well as provide cosmological constraint through statistical approaches in a similar way as [6].

The ACS camera has also delivered wide field mosaic images of blank fields. The largest ACS contiguous mosaiced-image corresponds to the COSMOS field covering 1.65 square degrees [28, 29]. The complementary multi-wavelength data coming from all major observatories has provided a wealth of information on galaxy evolution and the large scale structures [30]. The weak lensing analysis of the ACS COSMOS data [31, 32] have lead to the first tridimensional Dark Matter mass map [33]. Furthermore, direct measurement of individual clusters down to a few 10^{13} solar masses, to lower-mass X-ray selected clusters through the "stacking" of the lensing data is possible [34]. However the survey area covered by COSMOS is too small to detect the most massive clusters (similar to Abell 2218 and 1689) and only future wide field space mission will be able to probe the full range of the cluster mass function directly with gravitational lensing.

33.3 Gravitational Telescope

In the most massive clusters the area covered by region with large magnification (say larger than $\sim$5 — or a gain of more than $\sim$1 mag) is very significant and allows a deeper look in the distant Universe also probing intrinsically fainter sources. The draw back of the magnification gain, is a reduction of the surveyed area in the source plane. Although one could think that because of its complexity the magnification gain is compensated by the reduction of surveyed area, there is more to gain from the lensing magnification: (i) the ability to probe much fainter the luminosity function of the targeted population, (ii) the gain in spatial resolution, (iii) the gain in time in follow-up observation of strongly magnified objects, (iv) the gain in the confusion limit for the long wavelength observations, allowing to explore uncharted regions.

One of the classical strongly magnified galaxies that has been studied in great detailed is the object called "CB58" at $z = 2.72$. First recognized as a booming primeval galaxy [35], it was thereafter identified as a typical Lyman-break galaxy (LBG) which was highly magnified by a cluster [36]. "CB58" has been studied in great detailed taking benefit of its very large magnification, using high resolution spectroscopy to study the inter-galactic medium in LBGs [37] and measuring the UV to FIR flux ratio of typical LBGs [38].

With the advent of SCUBA on the JCMT a major lens survey was undertaken on a sample of clusters [39] targeting the dust-obscured population of strongly star-

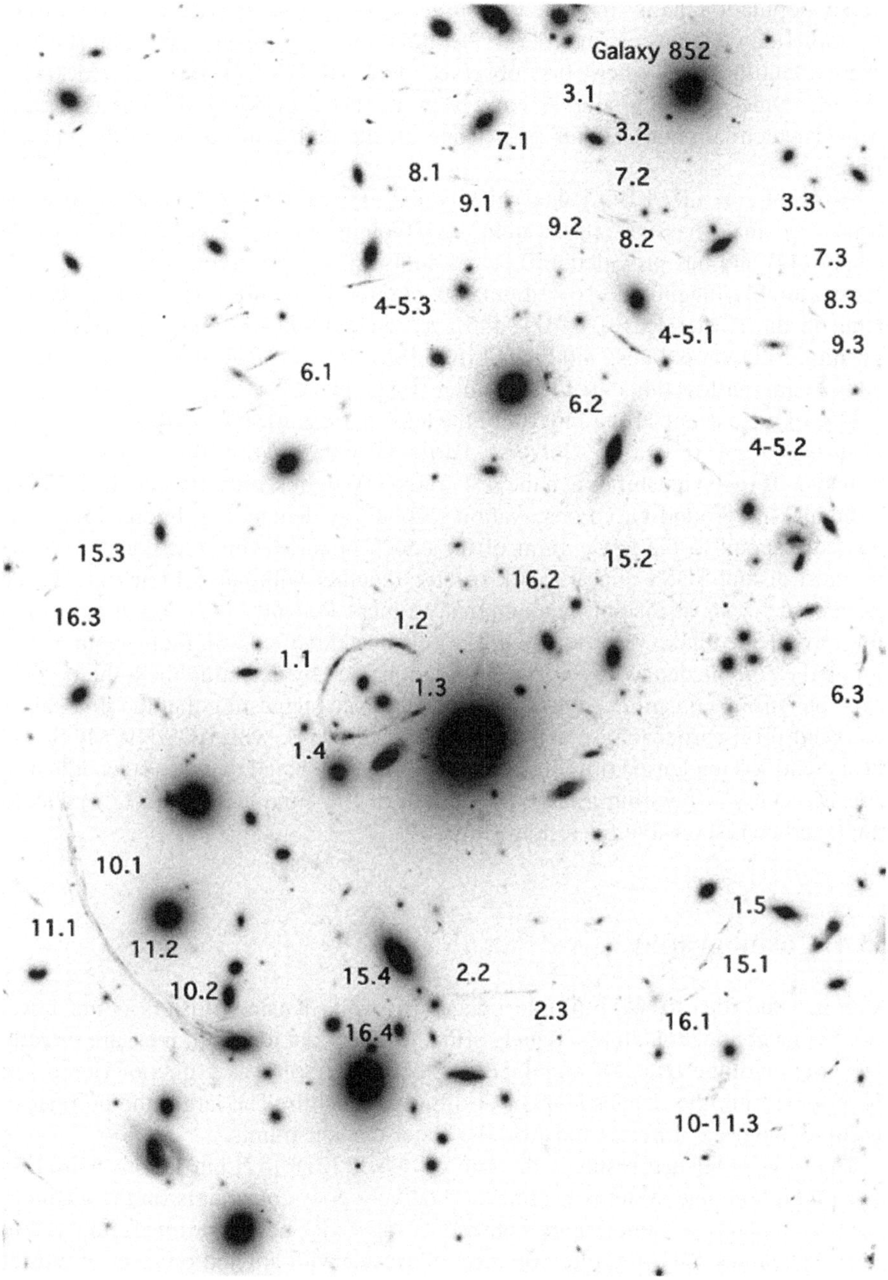

Fig. 33.1 (Color online) Deep ACS color image of the cluster Abell 1703 ($z = 0.28$) with the identification of the multiple-images [26]. The central ring like arc is at $z = 0.888$

forming galaxies in the distant Universe. Much have been learned on the (sub)mm-galaxy population thanks to the lensed surveys [40]. However, we had to wait to survey more than ten clusters in order to identified multiply-imaged (sub)mm sources. These identifications where first observed in Abell 2218 [41], and in MS0451-03 [42]. Other multiply sources have been identified in Abell 1689 as described in [43] which summarizes our knowledge on the (sub)mm counts to the faintest limits.

A recent $z = 3.07$ LBG galaxy nicknamed the "cosmic-eye" has been recently discovered in a massive cluster field — although mainly magnified by a single galaxy [44] and has provided 100 pc resolution image of a typical L* LBG. Similarly, a highly magnified ($16\times$) Lyman-α emitter at $z = 4.88$ has been studied in detail in the cluster RCS0224-002 [45]. A detailed spectroscopic analysis show: redshifted Lyman-α lines, and blue-shifted ISM lines, typical of a strong starburst galaxy with outflow that extend more than 10 kpc in size.

One of the current strong interest of the lensing magnification is the search of the most distant galaxies in the Universe. Particularly those at redshift $z > 6.5$ which corresponds to the redshift when the rest-frame UV light is redshifted to the NIR domain, making ground-based observation particularly demanding. Interestingly, this also correspond to the lower limit of the epoch of reionization as constrained by the most distant SDSS quasars [46]. *Hubble* together with natural telescope broke the record for most distant galaxies many times! Recently [47] found a $z \sim 6.8$ multiply-imaged galaxy in the Abell 2218 cluster identified as a i-z dropout. Combining the *Hubble* data with *Spitzer* observation, [48] by detecting the light of "old" stars, put strong constraints on the age of this object, suggesting that the first galaxies could form earlier than $z \sim 9$–10. More recent work with ACS/NICMOS [49] have identified ten lensed dropout candidates at $z \sim 7$–8 and two $z \sim 10$ candidates. This discovery is favoring a large population of low-luminosity (sub-L*) galaxies that could be responsible for re-ionization.

33.4 Cosmography

Although the idea to use multiple images in strong lensing clusters is not novel [50, 51] its application on real data is difficult. The main idea is to measure directly the variation of the D_{LS}/D_S angular diameter distance ratio as a function of redshift by probing multiple images sources at different redshift. The larger the number of multiple images at different redshift, the better the constraints.

The only published result is the one on Abell 2218 [52] but is only based on four multiple image systems leading to relatively poor constraints on the cosmography parameters. Cosmography constraints may only be interesting compared to other techniques only if applied on a set of cluster with applied on a set of cluster with more than about ten set of multiple images with known spectroscopic redshift. Nevertheless, aiming to reach cosmography constraints is probably the best way in ensuring that the understanding of the mass model is as best as it can be. Such motivation is motivating new technique to reconstruct cluster mass models [15], as well to identify more multiple images (which will ultimately need dedicated computer-assisted technique) and determine their redshift spectroscopically.

33.5 Conclusion and the Future

Cluster Lensing was NOT a science driver to build and launch *Hubble*. Indeed giant arcs were basically found at the time of the *Challenger* disaster that delayed by four years *Hubble*'s launch. Nevertheless, *Hubble* has revolutionized the study of cluster lenses by providing high resolution observation leading to: (i) unique identification of MANY multiple images (a major contribution to strong lensing), (ii) accurate shape (identification of the morphology of the strongly magnified sources) and shear measurements (with unique weak lensing measurements, including the one of COSMOS), (iii) accurate photometric redshift of the faintest galaxies (which benefit both strong and weak cluster lensing measurements).

These *Hubble* cluster lenses observations have allowed to address important questions in clusters physics (distribution and nature of the dark matter) and cosmology (sources of reionization, and constraints on cosmography). With the forthcoming servicing mission (particularly with the WFC3 camera and the ACS repair), *Hubble* will certainly continue exciting discoveries using cluster lenses.

Acknowledgements *Hubble* is one of the great observatory that has transformed our views on the Universe. I would like to thank here Duccio Macchetto for his excellent work at STSci that has contributed in improving our knowledge of this fascinating Universe.

References

1. Soucail, G., et al., 1987. Astron. Astrophys. 172, L14.
2. Kneib, J.-P., et al., 1996. Astrophys. J. 471, 643.
3. Ebbels, T., et al., 1996. Mon. Not. R. Astron. Soc. 295, 75.
4. Seitz, C., et al., 1996. Astron. Astrophys. 314, 707.
5. Smith, G., et al., 2005. Mon. Not. R. Astron. Soc. 359, 417.
6. Smith, G., et al., 2002. Mon. Not. R. Astron. Soc. 333, L16.
7. Natarajan, P., Kneib, J.P., 1997. Mon. Not. R. Astron. Soc. 287, 833.
8. Natarajan, P., et al., 1998. Astrophys. J. 499, 600.
9. Treu, T., et al., 2003. Astrophys. J. 591, 53.
10. Kneib, J.-P., et al., 2003. Astrophys. J. 598, 804.
11. Czoske, O., et al., 2002. Astron. Astrophys. 386, 31.
12. Broadhurst, T., et al., 2005. Astrophys. J. 621, 53.
13. Diego, J.M., 2005. Mon. Not. R. Astron. Soc. 362, 1247.
14. Jullo, E., et al., 2007. New J. Phys. 9, 447.
15. Coe, D., et al., 2008. Astrophys. J. 681, 814.
16. Zekser, K.C., et al., 2006. Astrophys. J. 640, 639.
17. Limousin, M., et al., 2007. Astrophys. J. 668, 643.
18. Halkola, A., et al., 2007. Mon. Not. R. Astron. Soc. 372, 1425.
19. Halkola, A., et al., 2007. Astrophys. J. 656, 739.
20. Leonard, A., et al., 2007. Astrophys. J. 666, 51.

21. Broadhurst, T., et al., 2005. Astrophys. J. Lett. 619, 143.
22. Bradac, M., et al., 2006. Astrophys. J. 652, 937.
23. Jee, M.J., et al., 2007. Astrophys. J. 661, 728.
24. Eliasdottir, A., et al.., 2009. Astrophys. J. 697, 1725.
25. Sharon, K., 2005. Astrophys. J. Lett. 629, 73.
26. Limousin, M., et al., 2008. Astron. Astrophys. 489, L23.
27. Ebeling, H., Edge, A., Henry, P., 2001. Astrophys. J. 533, 668.
28. Scoville, N., et al., 2007. Astrophys. J. Suppl. Ser. 172, 38.
29. Koekemoer, A., et al., 2007. Astrophys. J. Suppl. Ser. 172, 196.
30. Scoville, N., et al., 2007. Astrophys. J. Suppl. Ser. 172, 150.
31. Leauthaud, A., et al., 2007. Astrophys. J. Suppl. Ser. 172, 219.
32. Rhodes, J., et al., 2007. Astrophys. J. Suppl. Ser. 172, 203.
33. Massey, R.J., et al., 2007. Nature 445, 286.
34. Leauthaud, A., 2007. PhD thesis.
35. Yee, H., et al., 1996. Astrophys. J. 111, 1783.
36. Seitz, S., et al., 1998. Mon. Not. R. Astron. Soc. 298, 945.
37. Pettini, M., et al., 2000. Astrophys. J. 528, 96.
38. Sawicki, M., 2001. Astron. J. 121, 2405.
39. Smail, I., et al., 1998. Astrophys. J. Lett. 507, 21.
40. Blain, A.W., et al., 2002. Phys. Rep. 369, 111.
41. Kneib, J.-P., et al., 2004. Mon. Not. R. Astron. Soc. 349, 1211.
42. Borys, C., et al., 2004. Mon. Not. R. Astron. Soc. 352, 759.
43. Knudsen, K., et al., 2008. Mon. Not. R. Astron. Soc. 384, 1611.
44. Smail, I., et al., 2007. Astrophys. J. Lett. 654, 33.
45. Swinbank, M., et al., 2007. Mon. Not. R. Astron. Soc. 376, 479.
46. Fan, X., et al., 2003. Astron. J. 125, 1649.
47. Kneib, J.-P., et al., 2004. Astrophys. J. 607, 697.
48. Egami, E., et al., 2005. Astrophys. J. Lett. 618, 5.
49. Richard, J., et al., 2008. Astrophys. J. 685, 705.
50. Link, R., Pierce, M.J., 1998. Astrophys. J. 502, 63.
51. Golse, G., et al., 2002. Astron. Astrophys. 387, 788.
52. Soucail, G., et al., 2004. Astron. Astrophys. 417, L33.

Chapter 34
HST Observations of Gravitationally Lensed QSOs

Jean-François Claeskens, Dominique Sluse, and Jean Surdej

Abstract Thanks to its sharp view, HST has significantly improved our knowledge of tens of gravitationally lensed quasars in four different respects: (1) confirming their lensed nature; (2) detecting the lensing galaxy responsible for the image splitting; (3) improving the astrometric accuracy on the positions of the unresolved QSO images and of the lens; (4) resolving *extended* lensed structures from the QSO hosts into faint NIR or optical rings or arcs. These observations have helped to break some degeneracies on the lens potential, to probe the galaxy evolution and to reconstruct the true shape of the QSO host with an increased angular resolution.

34.1 Introduction

Strong lensing, i.e. the splitting of the image of a background source into several, magnified but distorted images, occurs each time an intervening massive object lies nearly on the same line-of-sight, provided its surface mass density is large enough (typically larger than 0.5 $\mathrm{gr/cm^2}$ at cosmological distances). The angular resolution of the telescope must be sufficient to resolve the images ($\Delta\theta \simeq 1'' \sqrt{M_E/10^{11}\ \mathrm{M_\odot}}$). This is a consequence of the curvature of space–time around massive objects, as predicted by General Relativity.

However, a massive object such as an isolated galaxy can be considered as an imperfect optical lens (like the foot of a glass of wine). Instead of a focus, a generic diamond-shaped closed curve is produced — the caustic. All observed configurations of multiply imaged QSOs as well as giant arcs can be explained by the exact relative position of the (possibly extended) source with respect to that caustic

J.-F. Claeskens (✉) · J. Surdej
Institut d'Astrophysique et de Géophysique, Université de Liège, Allée du 6 Août, 17, B5C, 4000 Liège, Belgium
e-mail: claesken@astro.ulg.ac.be

J. Surdej
e-mail: surdej@astro.ulg.ac.be

D. Sluse
Laboratoire d'Astrophysique, Ecole Polytechnique Fédérale de Lausanne (EPFL) Observatoire, 1290 Sauverny, Switzerland
e-mail: dominique.sluse@epfl.ch

F.D. Macchetto (ed.), *The Impact of HST on European Astronomy*, Astrophysics and Space Science Proceedings, DOI 10.1007/978-90-481-3400-7_34, © Springer Science+Business Media B.V. 2010

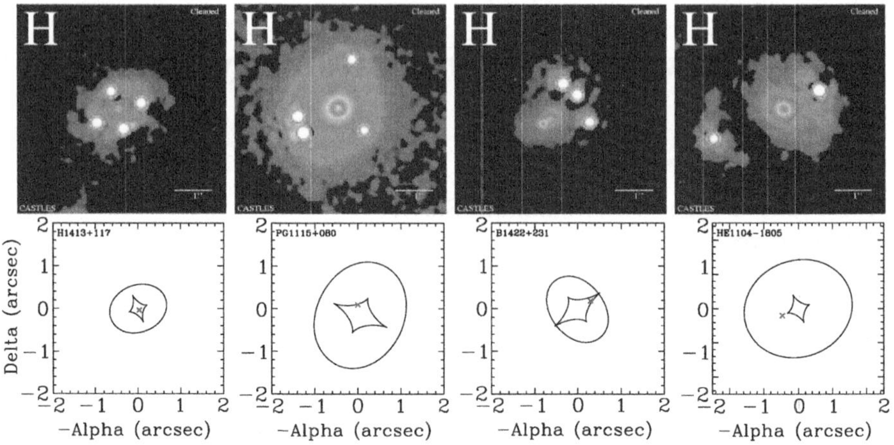

Fig. 34.1 (Color online) *Top*: Deconvolved HST NICMOS observations of four gravitationally lensed QSOs (CASTLES project, http://cfa-www.harvard.edu/glensdata); *bottom*: fitted source position w.r.t. the caustics of the Singular Isothermal Ellipsoid model. From *left* to *right*: four image cross, fold and cusp configurations, two image configuration

(see Fig. 34.1). Of course, besides being an exotic curiosity, this phenomenon betrays the nature of the deflector and can be used as an astrophysical tool. Reviews on gravitational optics and its astrophysical applications may be found in [1] and references therein. Here, we concentrate on selected results derived from HST observations of lensed QSOs.

In 1979, the first gravitationally lensed source was discovered serendipitously. The source was the radio-loud QSO Q0957+561A&B. However, as demonstrated later [2] with HST, only relying on *chance* to discover lensed sources leads to preferentially select *faint* ones, which are then more difficult to follow up and to characterize and which are further embedded in complicated observational biases.

However, as first mentioned in the 1980s [3], *luminous* quasars are excellent candidates to look for multiple images produced by isolated galaxies: the optical depth for lensing is high towards those distant sources *and* their luminosity function is steep, introducing an amplification bias, which boosts the probability of lensing by a factor of 10 or more. Although about 100 such systems are presently known, HST has only contributed to one discovery through a QSO survey (i.e. Q1208+1011 [4])! Indeed, ground based optical and radio QSO surveys (and recently the SDSS survey) were much more efficient and faster. But starting from here, HST has played a crucial role.

34.2 Confirming the Lensing Nature

While a set of four QSO images is conspicuous of lensing, double image configurations are not a signature unique to lensing. In the case of small angular separations, spectroscopy with the Faint Object Spectrograph onboard HST was required to assess the true statistics of lensing in optical samples (i.e. Q1208+1011A&B [4] & J03.13A&B [5]).

34.3 Improving the Constraints on the Lens Model

The generic properties of the lensing potential can be described with seven parameters: its center position with respect to the lensed images (θ_x, θ_y), the total mass within the images M_E, the total radial mass profile (logarithmic slope β, core radius r_c) and the shape of the iso-density contours (ϵ, φ). Moreover, the influence of the direct environment of the lens (neighbours, cluster, ...) may be described by three extra parameters: an external shear γ with orientation φ_γ and a possibly associated matter density κ_c.

However, the insight we can really get into the lensing potential is related, firstly, to the number of observational constraints, secondly to the accuracy on each of them, and finally to the existence of internal degeneracies between different properties of the potential.

34.3.1 The Lens Position

Finding the faint distant galaxy in the vicinity of the bright QSO images is a challenge. HST high angular resolution and infra-red capabilities (NICMOS) have provided the direct detection, the precise astrometry and sometimes the morphology of many lenses (e.g. HE1104-1805 [6], Q1009+025 [7], PG1115+080 [8]), thus increasing the number of constraints in these systems.

34.3.2 The Image Astrometry

Reaching an astrometric accuracy of ~3 mas, HST observations of gravitationally lensed quasars have shown that simple lens models (such as the Singular Isothermal Ellipsoid) systematically fail in reproducing the exact image and lens positions. Adding an external shear drastically improves the fit [9], independently of the adopted exact radial profile (e.g. slope, core radius [10]). The origin of this external shear is not entirely clear. While in some cases, the external shear derived from the environment can explain the lensed image positions (e.g. in PG1118+008 [12]), it is usually not strong enough [11].

Fig. 34.2 (Color online) Reconstructed shape of the lensed Seyfert I galaxy RXS J1131-1231 [14]

34.3.3 Extended Structures

Because the source position is unknown, the number of astrometric constraints is $2(N-1)$ in a N image system. This is just sufficient to determine M_E, ϵ, φ, γ and φ_γ, if $N = 4$. However, a degeneracy is observed between ϵ, γ and $\varphi - \varphi_\gamma$, whose strength depends on the exact lens configuration [9]. There is also a degeneracy between κ_c and M_E or between κ_c and β [13].

The detection of *resolved* arcs and ring-like features (i.e. [part of] the lensed QSO host) may help in breaking some of these degeneracies if they provide extra azimuthal and/or radial constraints. Such detections are one of the best achievements of HST. The first ones were made in the NIR (e.g. PG1115+080 [8], ...) and have been recently supplemented by optical rings, such as RXS J1131-1231 [14].

34.4 Investigating the Nature of the Lens and the Source

Besides numerous studies of the individual lensing systems, the above mentioned HST observations have allowed *statistical* analysis of the lens population, which represents a *mass*-selected sample in low density environments. An interesting example is the evolution of their M/L as a function of redshift through the Fundamental Plane analysis: it is found to be comparable to that observed in galaxy clusters [15]. On the other hand, a moderate but measurable amount of patchy extinction in the lenses has also been reported [16].

Finally, the HST high angular resolution combined with the lens magnification is opening deeper insight into the QSO *host galaxy*. As an example, Fig. 34.2 shows the reconstructed image of a Seyfert I host galaxy, where spiral arms, star-forming regions and interaction with a companion are clearly seen at a redshift $z = 0.66$. On the other hand, a *statistical* study of the lensed host galaxies has shown the rise of $M_{\rm bh}/M_*$ with redshift as far as $z \sim 4$ [17].

Acknowledgements This work was supported by PRODEX PEA C90194 HST.

References

1. Kochanek, C.S., Schneider, P., Wambsganss, J., 2006. In: Meylan, G., Jetzer, P., North, P. (Eds.), Proc. of the 33rd Saas-Fee Advanced Course, Springer, Berlin.
2. Ratnatunga, K.U., Griffiths, R.E., Ostrander, E.J., 1999. Astron. J. 117, 2010.
3. Turner, E.L., Ostriker, J.P., Gott, J.R. III, 1984. Astrophys. J. 284, 1.
4. Bahcall, J.N., et al., 1992. Astrophys. J. Lett. 400, L51.
5. Surdej, J., et al., 1997. Astron. Astrophys. 327, L1.
6. Remy, M., et al., 1998. New Astron. 3, 379.
7. Claeskens, J.-F., et al., 2001. Astron. Astrophys. 367, 748.
8. Impey, C.D., et al., 1998. Astrophys. J. 509, 551.
9. Keeton, C.R., Kochanek, C.S., Seljak, U., 1997. Astrophys. J. 482, 604.
10. Witt, H.J., Mao, S., 1997. Mon. Not. R. Astron. Soc. 291, 211.
11. Faure, C., et al., 2004. Astron. Astrophys. 428, 741.
12. Schechter, P.L., et al., 1997. Astrophys. J. 475, L85.
13. Wucknitz, O., Refsdal, S., 2001. ASP Conf. Ser. 237, 157.
14. Claeskens, J.-F., Sluse, D., Riaud, P., Surdej, J., 2006. Astron. Astrophys. 451, 865.
15. van de Ven, G., van Dokkum, P.G., Franx, M., 2003. Mon. Not. R. Astron. Soc. 344, 924.
16. Falco, E.E., et al., 1999. Astrophys. J. 523, 617.
17. Peng, C.Y., et al., 2006. Astrophys. J. 649, 616.

Chapter 35
Study of Quasar Host Galaxies Combining HST/ACS Images and VLT Spectra

Yannick Letawe, Géraldine Letawe, and Pierre Magain

Abstract We present the results of a study of the host galaxies of a few bright low redshift quasars, based on a combination of high resolution HST/ACS images with deep VLT spectra. Using a version of the MCS deconvolution algorithm specially adapted to HST images, the light from the central point source can be efficiently removed, which allows to study the environment of the quasar down to a few hundreds of parsecs from the center. In a similar way, the spectra can be separated into a quasar contribution and a spatially resolved slit spectrum of the host galaxy. Combining the images and spectra allows to carry out a detailed analysis of the immediate environment of the quasar: excitation and ionization state of the gas, dynamics... Up to now, this method has been applied to six quasars, including the peculiar HE0450-2958 (Magain et al. in Nature 437:381, 2005), for which no host galaxy could be detected.

35.1 Introduction

The study of quasar host galaxies helps to understand the quasar phenomenon. By inspecting and comparing them to non active galaxies, we can identify which mechanisms favour the triggering and fueling of the central engine. The challenge in quasar host study is to remove the quasar light contribution from the observations, allowing to inspect the hosts to their near center. So far, these studies have been carried out using either spectroscopic data or high resolution images, but none of them used both to analyze the quasar host properties, what we present here for six quasars.

These six quasars form a subsample of the 20 bright low redshift quasars spectroscopically analysed in [2], for which we also have high resolution images obtained with HST. The parent sample was drawn out of the Hamburg-ESO Survey with well defined selection criteria ($z < 0.3$, $M_V < -23$). The present subsample is obviously too small to attempt any statistical analysis, and suffers from selection biases. Thus, the present analysis can be seen as (1) a demonstration of the interest of combining high resolution images and spatially resolved spectra to study the quasar hosts and (2) the focus on interesting cases, which can shed light on the relations between the quasars and their environment. The complete analysis is available in [3].

Y. Letawe (✉) · G. Letawe · P. Magain
Département AGO, Université de Liège, allée du 6 août 17, 4000 Liège, Belgium
e-mail: yletawe@ulg.ac.be

F.D. Macchetto (ed.), *The Impact of HST on European Astronomy*, Astrophysics and Space Science Proceedings,
DOI 10.1007/978-90-481-3400-7_35, © Springer Science+Business Media B.V. 2010

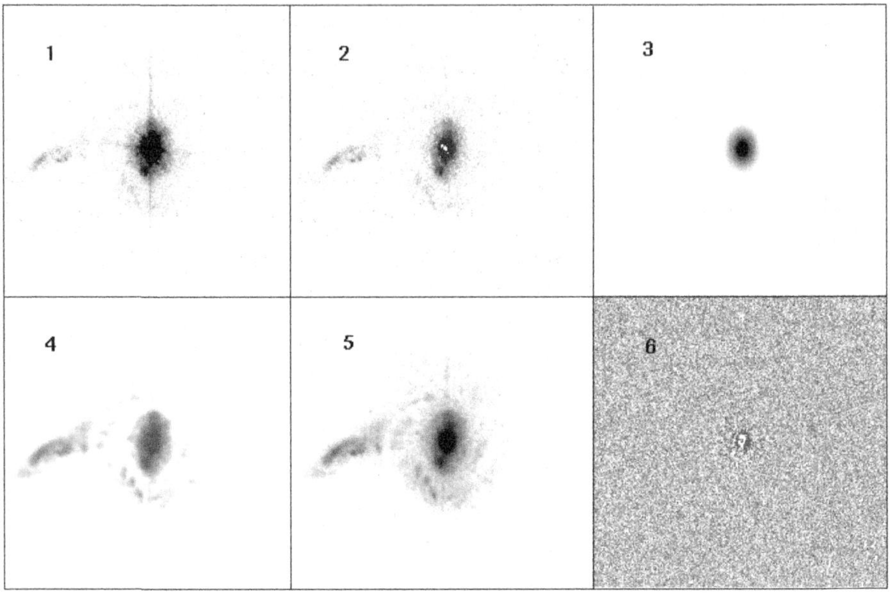

Fig. 35.1 Steps for quasar/host separation. *1*: Long exposure of HE0354-5500 ($z = 0.267$) after reduction. *2*: Removal of the point source. *3–5*: Fit of the galaxy: analytical model found for the bulge (*3*), numerical background (*4*) and total fit (*5*), the one used for the analysis. *6*: Residuals (raw data minus total fit) of the whole process

35.2 Observations and Data Processing

35.2.1 Images

The images have been obtained with the ACS/High Resolution Channel onboard HST with the broad V-band filter F606W in 2004 and 2005. For every quasar, long and short exposures have been taken during one HST orbit, short ones avoiding saturation at the center, and long ones providing a good S/N in the lower surface brightness regions. Moreover, a PSF star was observed in the same orbit with identical conditions. Reductions were performed with Pyraf tasks.

The HST/ACS images have been processed with the MCS deconvolution algorithm [4], whose principle is to obtain a result image with a resolution and sampling fixed by the user, provided that they conform to the sampling theorem. Since it allows to separate point sources from diffuse ones, it is particularly well suited for the study of quasar host galaxies. We processed in two steps, first removing the quasar light, then modeling the host galaxy by an analytical model (accounting for the bulge) plus a numerical background (for spiral arms and irregular features). An example is given in Fig. 35.1.

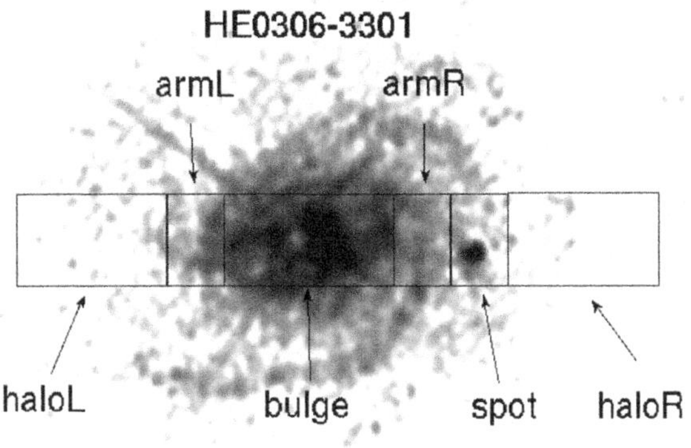

Fig. 35.2 Slit decomposition overlaid on deconvolved HST image of HE0306-3301 ($z = 0.247$), L and R standing for left and right

35.2.2 Spectra

The spectroscopic data, covering the optical range, were obtained with the ESO VLT/UT1 in April and November 2000 with FORS1 in Multi Object Spectroscopy (MOS) mode, with $19'' \times 1''$ slits. Reduction and deconvolution through the MCS method adapted to spectra [1] have been presented in detail in [2], resulting in spatially deconvolved spectra with the quasar spectrum separated from the spectrum of its host galaxy. We take here advantage of the spatial resolution of these spectra to separate the contributions from different sub-regions of the galaxies, guided by the higher resolution of the ACS images (see Fig. 35.2).

35.3 Results

In addition to the scrutiny of high resolution images of the hosts, we performed a detailed analysis of the gaseous content of the hosts, on 1D spectrum of each sub-regions of each host. We derived ionisation sources, radial velocity curves, and fitted blended lines. We present here only the main or most surprising results obtained for each galaxy.

HE0306-3301 Despite its apparently regular spiral structure (Fig. 35.2), the radial velocity curve (Fig. 35.3, left) reveals that this galaxy is strongly perturbed by the bright spot, being a probable tiny companion merging in the host and which could have ignited the quasar.

HE0354-5500 Spiral galaxy colliding with a smaller one (Fig. 35.1). The spectral fit of the Hα region far from the nucleus imposes the addition of a broad Hα line with similar shape as the quasar's (Fig. 35.3, right). We suggest this broad line is the quasar light scattered by dust or free electrons, as seen through polarisation in obscured AGN. This would be the first direct observation of such scattering in Type 1 quasar.

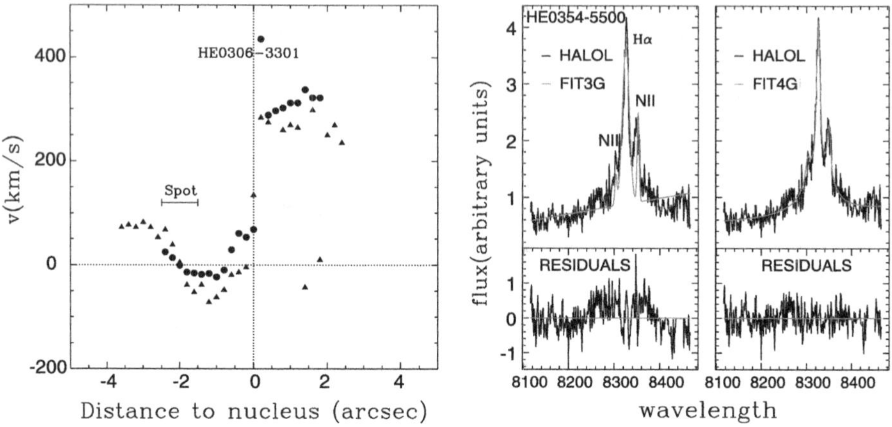

Fig. 35.3 (Color online) *Left*: radial velocity curve of HE0306-3301, strongly disturbed by the tiny companion called 'spot'. *Right*: fit of the Hα region far from HE0354-5500 nucleus. The addition of a broad component (on the *right*) is mandatory to obtain acceptable residuals

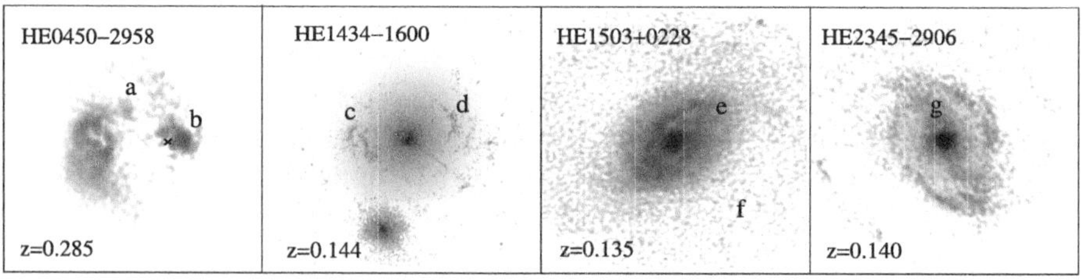

Fig. 35.4 HST/ACS deconvolved image of the four remaining hosts. The *black cross* indicates the position of the quasar in HE0450-2958 system. Intensity levels are logarithmic to improve contrasts. All images are $5.7 \times 5.7''$, except HE1434-1600 that is $11 \times 11''$ to include the companion galaxy. Regions labelled *a* to *g* are commented in the text

HE0450-2958 This object, with its undetected host galaxy, was analysed in detail in [5]. Several gaseous emissions, faint (labelled *a* in Fig. 35.4) or intense (*b*), ionized by the nucleus itself, are found on each side of the quasar (Fig. 35.4). Broad Hα is observed in the faint emission between the companion galaxy and the quasar. Additional data on this system, including NICMOS observations, are currently under analysis.

HE1434-1600 This elliptical galaxy (well reproduced by the analytical model) contains filaments (*c* and *d*) of gas in expansion, ionized by the quasar (Fig. 35.4), plus a companion galaxy, apparently weakly disturbed by interaction.

HE1503+0228 The inclined spiral structure of the central part of the host (*e* in Fig. 35.4), is surrounded by a fainter spherical component *f*, made of stars and gas. It could be a wider spiral arm, relaxed after some gravitational interaction. Broad nuclear reflected Hα is also observed in the spherical component.

HE2345-2906 The deconvolution of the ACS images reveals a prominent bar (*g* in Fig. 35.4), well known to facilitate matter inflows to the center of the galaxy and thus the fueling of the quasar.

35.4 Conclusions

1. A simultaneous analysis of VLT-spectra and HST-images, deconvolved with an MCS-based method, is an excellent way to extract precise and useful informations about quasar host galaxies to their near center and reveals interesting and sometimes unexpected properties;
2. The galactic Hα spectral region has, in some loci faraway from the quasar, a broad component which arises from a reflection of the nucleus light by dust or free electrons. Similar broad lines were only observed for obscured quasars, and thus our study supplies a generalisation of the phenomenon to Type 1 quasars;
3. This study re-emphazises the gravitational interaction as an efficient release mechanism of the quasar activity.

We are currently extending the imaging/spectral combination to integral field spectrosocpy, to achieve a better spatial resolution not only along the slit but over the whole fields of the hosts.

Acknowledgements This work has been supported by ESA and the Belgian Federal Science Policy Office under contract PRODEX 90195.

References

1. Courbin, F., Magain, P., Kirkove, M., Sohy, S., 2000. Astrophys. J. 539, 1136.
2. Letawe, G., et al., 2007. Mon. Not. R. Astron. Soc. 378, 83.
3. Letawe, Y., et al., 2008. Astron. Astrophys. 679, 967.
4. Magain, P., Courbin, F., Sohy, S., 1998. Astrophys. J. 494, 452.
5. Magain, P., et al., 2005. Nature 437, 381.

Chapter 36
Local Lyman α Emitters and Their Relevance to High Redshift Ones

Daniel Kunth, Hakim Atek, Göran Östlin, Matthew Hayes, Miguel Mas-Hesse, Claus Leitherer, Artashes Petrosian, and Daniel Schaerer

Abstract The Lyα line is an important diagnostic of star formation and activity in galaxies. The analysis of Lyα is complicated due to the resonant nature of the line and radiative transfer effects. High spectral resolution studies of local starburst galaxies with the unprecedented UV capabilities of the HST have shown that this line is either seen in absorption or in emission and in the latter case with a P Cygni profile indicative of a large scale outflow of neutral gas. Moreover, HST imaging obtained with the ACS has revealed that a substantial fraction of the Lyα photons are diffused far away from the emissive knots. Since the importance of Lyα for tracing large scale structure, correlation functions, and galaxy formation is recognized, Lyα will remain a very important probe of the distant universe for the foreseeable future,

D. Kunth (✉) · H. Atek
Institut d'Astrophysique de Paris, Paris, France
e-mail: kunth@iap.fr

G. Östlin · M. Hayes
AlbaNova U. Centre, Stockholm Observatory, Stockholm, Sweden

G. Östlin
e-mail: ostlin@astro.su.se

M. Mas-Hesse
Centro de Astrobiologia, LAEFF, Madrid, Spain
e-mail: mm@laeff.inta.es

C. Leitherer
STScI, Baltimore, USA
e-mail: leitherer@stsci.edu

A. Petrosian
Byurakan Astronomical Observatory, Erevan, Armenia
e-mail: artptrs@yahoo.com

D. Schaerer
Observatoire de Genève, Geneva, Switzerland
e-mail: daniel.schaerer@obs.unige.ch

F.D. Macchetto (ed.), *The Impact of HST on European Astronomy*,
Astrophysics and Space Science Proceedings,
DOI 10.1007/978-90-481-3400-7_36, © Springer Science+Business Media B.V. 2010

and it is therefore imperative to acquire a better understanding of what mechanisms regulate our ability to detect this line.

36.1 The Observational Puzzle

Lyα has been recognized in the late 1960s for its potential importance in searching for primeval galaxies. Reference [15] predicted its intensity to reach 10% of the L_{bol}. At very low metallicity the Lyα line is expected to be even stronger as emphasized by [17] rendering the line potentially detectable at very high redshift. The complexities of using Lyα as a cosmological tool result from the fact that it is a resonant line and have motivated a number of studies of star-forming galaxies at low-z and theoretical models of resonant line radiative transfer. UV spectra of nearby starbursts taken with *IUE* combined with optical spectra showed that the relative line intensity Lyα/Hβ was much below the expected case B value while similarly the $W_{\mathrm{Ly}\alpha}$ was smaller than expected. Among possible explanations was the combined fact that Lyα is a resonant line that can be easily destroyed by dust in an HI-rich medium [1], the inhomogeneity of the ISM [4], a possible underlying stellar Lyα absorption or the simple fact that a metallicity dependant extinction law should be applied ([23], see [18] for a review and references). However, the dust alone could obviously not be controlling by itself the visibility of the Lyα line that was in many cases seen in absorption in some of the most metal-poor local starburst known [8, 22]. In front of these unconclusive results deeper imaging and spectroscopic studies were undertaken of local starbursts using the UV capabilities of the HST in imaging and spectroscopic modes.

36.2 HST Spectroscopy and Imaging

When Lyα is seen in emission, it frequently exhibits a P Cygni profile and systematic velocity offset from low ionisation state metal absorption features in the neutral ISM traced by OI or SiII (see [9]), suggesting that an outflowing medium is an essential ingredient in the formation of the line profile and escape physics. Similar results have also been found at high-z [19, 20]. STIS 2-D spectroscopy brought more insight into the wide variety of possible Lyα profiles allowing a comparison with predictive models [13]. These results suggested that outflows and superwinds are important factors for Lyα escape. Furthermore ACS imaging at high spatial resolution was obtained and led to unprecedented studies of the coupling between the nebular emission, the dust and the stellar complex [10]. The first detailed imaging study of a local starburst, ESO 338-IG04 [5] found emission and absorption on very small scales in the central starburst regions, and little or no correlation with the FUV morphology. No clear correlations between stellar ages and the Lyα emission was found. One of the most noticeable feature revealed by these unprecedented HST images is that this starburst is surrounded with a large, diffuse, low surface brightness

Lyα halo. This component contributes ~70% to the global Lyα luminosity, resulting from the resonant decoupling and diffusion of Lyα. The total escape fraction was found to be just ~5%, implying that any global values (e.g.. SFR) that would be estimated from Lyα alone would be seriously at fault. In another case ESO 350-IG038 young star-forming only the dustiest knot had Lyα in emission while in the two others Lyα is in absorption. Being also a diffuse Chandra X-ray source it supports the view that the Lyα emission is associated to the presence of winds [6]. Both these objects are Irrs, metal deficient by a factor ~10 and ~$L*$.

The first unifying scenario to explain the observed diversity of Lyα profiles was given by [21] in terms of an evolutionary sequence of a starburst driven supershell/superwind and was confronted with local observations in depth by [13]. Further studies have introduced more realistical properties of the ISM in terms of its geometry. Full 3-D codes are now becoming available treating various gas distribution; ionisation temperatures, dust and kinematics ([24] and references therein) depending on the velocity of the HI moving shell driven by the starburst, the emission profile might appear double peaked, with P Cygni profile of even with broad absorption.

Lyman Break Galaxies (LBGs) which are distant spatially unresolved starburst galaxies show similar variety of Lyα [19, 20] indicating that superwinds of neutral gas are ubiquitous and testifying the presence of massive stars. Although early searches for distant Lyα emitters have been rather disappointing the situation has dramatically improved about ten years ago thanks to larger collecting area and fainter flux limits reached by modern detectors and using more appropriate strategies such as narrow band imaging techniques [7], long slit blind-searches or integral field units surveys. In this case most detected Lyman alpha emitters (LAEs) have P Cygni profiles indicative of powerful superwinds. Typical luminosities now can reach below that of LBGs [3, 12, 14]. Very distant LAEs have been found with the SUBARU telescope using narrow-band technique. However the number of high-z emitters is still an order of magnitude smaller than predicted by models [16]. LAEs can be used to infer the ionisation fraction of the IGM at different redshift [2, 11]. While Lyα surveys remain difficult to interpret, it is clear that this line should be a competitive tool for detecting distant primeval galaxies at large redshift from the ground. On the other hand the pronounced progress on very high redshift galaxies in the last few years is largely based on Lyα detection while redshift confirmations almost also exclusively depend on Lyα. It is therefore likely that the importance of Lyα will remain also when JWST is launched and when the next generation of 30 meter-class ground-based telescopes have their first light.

References

1. Charlot, S., Fall, S.M., 1993. Astrophys. J. 415, 580.
2. Dijkstra, M., Wyithe, S., Haiman, Z., 2007. Mon. Not. R. Astron. Soc. 379, 253.
3. Fynbo, J.U., Möller, P., Thomsen, B., 2001. Astron. Astrophys. 374, 443.
4. Giavalisco, M., Koratkar, A., Calzetti, D., 1996. Astrophys. J. 466, 831.

5. Hayes, M., Östlin, G., Mas-Hesse, J.M., Kunth, D., Leitherer, C., Petrosian, A., 2005. Astron. Astrophys. 438, 71.
6. Hayes, M., Östlin, G., Atek, H., Kunth, D., Mas-Hesse, M., Leitherer, C., Jimenez-Bailon, E., Adamo, A., 2007. Mon. Not. R. Astron. Soc. 382, 1465.
7. Hu, E.M., McMahon, R.G., 1996. Nature 382, 231.
8. Kunth, D., Lequeux, J., Sargent, W.L.W., Viallefond, F., 1994. Astron. Astrophys. 282, 709.
9. Kunth, D., Mas-Hesse, J.M., Terlevich, E., Terlevich, R., Lequeux, J., Fall, S.M., 1998. Astron. Astrophys. 334, 11.
10. Kunth, D., Leitherer, C., Mas-Hesse, J.M., Östlin, G., Petrosian, A., 2003. Astrophys. J. 597, 263.
11. Le Delliou, M., Lacey, C.G., Baugh, C.M., Morris, S.L., 2006. Mon. Not. R. Astron. Soc. 365, 712.
12. Malhotra, S., Rhoads, J.E., 2002. Astrophys. J. Lett. 565, L71.
13. Mas-Hesse, J.M., Kunth, D., Tenorio-Tagle, G., Leitherer, C., Terlevich, R.J., Terlevich, E., 2003. Astrophys. J. 598, 858.
14. Ouchi, M., et al., 2003. Astrophys. J. 582, 60.
15. Partridge, R.B., Peebles, P.J.E., 1967. Astrophys. J. 147, 868.
16. Pritchet, C.J., 1994. Publ. Astron. Soc. Pac. 106, 1052.
17. Schaerer, D., 2003. Astron. Astrophys. 397, 527.
18. Schaerer, D., 2007. In: Cepa, J. (Ed.), The Emission Line Universe, XVIII Canary Islands Winter School of Astrophysics. Cambridge Univ. Press, Cambridge, in press. arXiv:0706.0139.
19. Shapley, A.E., Steidel, C.C., Pettini, M., Adelberger, K.L., 2003. Astrophys. J. 588, 65.
20. Tapken, C., Appenzeller, I., Noll, S., Richling, S., Heidt, J., Meinkoehn, E., Mehlert, D., 2007. Astron. Astrophys. 467, 63.
21. Tenorio-Tagle, G., Silich, S.A., Kunth, D., Terlevich, E., Terlevich, R., 1999. Mon. Not. R. Astron. Soc. 309, 332.
22. Thuan, T.X., Izotov, Y.I., 1997. Astrophys. J. 489, 623.
23. Valls-Gabaud, D., 1993. Astrophys. J. 419, 7.
24. Verhamme, A., Schaerer, D., Maselli, A., 2006. Astron. Astrophys. 460, 397.

Chapter 37
The HST View of Low Luminosity AGN

Marco Chiaberge

Abstract We have studied the nuclear emission detected in HST data of carefully selected samples of low luminosity AGN (LLAGN) in the local universe. Although these objects lie on the so called "fundamental plane of black hole activity", we show that such a plane is not indicative of their physical nature. We find faint unresolved nuclei in a significant fraction of the objects. The nuclear emission is as low as 10^{-8} times the Eddington luminosity, indicating extremely low radiative efficiency for the accretion process and/or an extremely low accretion rate. When the Eddington ratio is plotted against the nuclear "radio-loudness" parameter, sources divide according to their physical properties. It is thus possible to disentangle between nuclear jets and accretion disks of different radiative efficiencies. This new diagnostic plane allows us to find objects that are the best candidates for hosting (and showing) radiatively inefficient accretion and determine in which ones we cannot see it. The (extremely limited) information available in the HST archive to derive the nuclear SEDs strongly supports our results.

37.1 Introduction

Low luminosity AGN (LLAGN) are the class of extragalactic objects in which RIAFs are most likely to be at work. RIAFs should lack both the "big blue-bump" and the IR (reprocessed) bump, which instead characterize optically thick, geometrically thin accretion disk emission and the surrounding heated dust (e.g. [6]). However, because of their intrinsic low luminosities, it is extremely difficult to study the emission from the accretion flow in LLAGN.

While it is now possible to routinely disentangle the nuclear emission from that of the host thanks to HST (e.g. [2–4, 9]), emission from the accretion flow cannot be seen when it is swamped by other nuclear radiation processes, or obscured by dust. Reference [10] have shown that among a sample of 17 LLAGN, 15 of them show variability over a timescale of a few months, which demonstrate their non-stellar origin. However, it was still unclear from their work whether the nuclear radiation is from a jet or from the accretion flow.

M. Chiaberge (✉)
Space Telescope Science Institute, 3700 San Martin Dr., Baltimore, MD 21210, USA
e-mail: marcoc@stsci.edu

M. Chiaberge
INAF, Istituto di Radioastronomia, Via P. Gobetti 101, 40129 Bologna, Italy

F.D. Macchetto (ed.), *The Impact of HST on European Astronomy*,
Astrophysics and Space Science Proceedings,
DOI 10.1007/978-90-481-3400-7_37, © Springer Science+Business Media B.V. 2010

In this paper, we review the current understanding of the nuclear emission in LLAGN, and we describe our method to discriminate between non-thermal jet emission and RIAF.

We consider the following samples of low luminosity AGN: (1) A complete sample of 33 FR I Radio Galaxies [3]; (2) 32 Seyferts [9]; (3) a complete sample of 21 LINERs from the Palomar Survey [8]; (4) 51 nearby "radio-bright" early-type galaxies. A detailed description of this sample is given in [2]. (5) The five broad-line radio galaxies with $z < 0.1$ included the 3CR catalog.

Note that being selected according to different criteria, these objects do not constitute a complete sample. However, they perfectly represent the overall properties of low power active nuclei in the local universe, where RIAFs are most likely to be found.

37.2 Is There a "Fundamental Plane of Black Hole Activity"?

LLAGN have been found to lie on the so-called "fundamental plane of black hole activity" [7, 11]. Here we want to test whether such a plane can be easily used for discriminating between different radiation processes associated to inflows and/or outflows around black holes.

We perform a simple test, to check whether objects with different emission processes and of different physical origin are located in distinct regions. The "fundamental plane" is based on the radio and X-ray luminosity, combined with the black hole mass, which is the ultimate "energy source". However, if we take the radio and X-ray fluxes of solar system objects, together with their masses, and we over plot them onto the "fundamental plane of black hole activity" we find something intriguing. Unexpectedly, the Sun, the Moon, Jupiter and Saturn do fall on the same plane, although they are tens of orders of magnitudes fainter and less massive than all "black hole" sources (Fig. 37.1). This implies that such a "fundamental plane" cannot be easily used as a diagnostic to discriminate between different radiation processes. On the other hand, the apparent correlation might only be a result of mere plotting luminosity vs. luminosity, and "artificially" rescaling the quantities with the mass. As we show in the following section, in order to discriminate different radiation processes for the nuclei (and eventually find RIAF candidates) we should use diagnostics that enhance the differences.

37.3 Low Luminosity Radio Galaxies

In the large majority of them ($\sim$85%), the radial brightness profiles of the nuclear regions is clearly indicative of the presence of a central unresolved nucleus (Fig. 37.2). The optical nuclei show a tight correlation with the radio core emission, both in flux and luminosity, which extends over more than 4 orders of magnitude [3]. The fact that the correlation is found both in flux and luminosity gives us confidence that

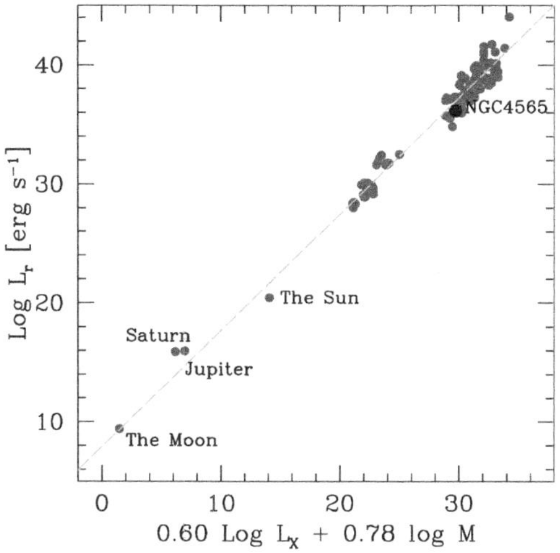

Fig. 37.1 (Color online) The so-called fundamental plane of black hole activity, to which we over-plotted solar system sources (adapted from [11]). The location of NGC 4565 (which we refer to in Sect. 37.4) along the correlation is also shown

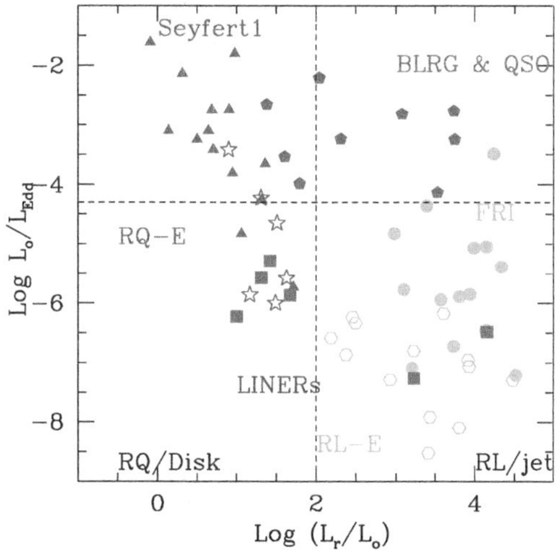

Fig. 37.2 (Color online) The diagnostic plane for the nuclear emission of LLAGN. Seyfert 1 are *triangles*, BLRG are *pentagons*, FR Is are *circles*, LINERs are *squares*, *hexagons* are early-type "core-galaxies" and *stars* are early-type "power-law" galaxies

it is not induced by a common redshift dependence. As the radio core emission is certainly non-thermal synchrotron radiation, the presence of a tight linear correlation between radio core and optical nuclear emission is a strong clue that the optical nuclei are also synchrotron. The detection of optical nuclei indicates that we have a direct view of the innermost nuclear regions of FR I. The high detection rate suggests that if obscuring material is present in these objects, it has to be distributed in a geometrically thin structure.

In order to rule out the presence of strong nuclear emission obscured at optical wavelengths (from e.g. the accretion disk, as in Type 2 AGN) we observed the 3C FR Is with HST/NICMOS at 1.6 μ. While most Type 2 AGN show a bright nuclear component in the near-IR [12], FRI I do not show any significant IR excess with respect to the optical emission. Furthermore, if VLBI radio data are used instead of VLA data, the correlation between IR nuclei and radio cores becomes slightly tighter and more linear.

37.4 How to Find RIAFs: Radio Loudness and Eddington Ratio

In [4] we showed that the nuclear properties of LLAGN are best understood when the ratio between optical luminosity and Eddington luminosity L_o/L_{Edd} is plotted against the "nuclear radio-loudness parameter" R, defined as the ratio between the *nuclear* radio (core) luminosity at 5 GHz ($F_{5\,GHz}$) as measured from high-resolution (VLA or higher resolution) data and the *nuclear* optical luminosity L_o as measured from HST images. This allows us to discriminate between jets and accretion disks of different radiative efficiency (and/or different accretion rates).

Sources separate into four quadrants: Seyfert 1s occupy the top-left quadrant, FR I radio galaxies are in the bottom-right quadrant. These two classes define the regions characterized by "radio-quiet" high radiative efficiency accretion and "radio-loud" jet emission, respectively. LINERs split into two sub-samples: both are in the region of low L_o/L_{Edd}, some of them being radio quiet and some radio loud. Nuclei of ellipticals separate according to the properties of the radial brightness profile of the host [1], and behave similarly to LINERs: core-galaxies are radio-loud, power-law galaxies are radio-quiet. Broad line radio galaxies, instead, are found at the top of the plot, at high values of L_o/L_{Edd}, while their location on the radio loudness axis most likely depends on the relative importance of the jet and disk emission in the optical band.

Let us now focus on the left side of the plot. In that region we find nuclei that show an optical excess with respect to the optical counterpart of the radio (synchrotron) radiation. For Seyfert 1s, such an optical excess is readily interpreted as emission from the accretion disk. On the other hand, objects with Eddington ratios as low as 10^{-6} cannot be anyhow reconciled with "standard" accretion disk models. Therefore, they are the best candidates to host radiatively inefficient accretion disks.

We thus predict that RIAFs can be detected and studied in the following objects: M 81, NGC 3245, NGC 3414, NGC 3718, NGC 3998, NGC 4143, NGC 4203, NGC 4565 and NGC 4736. In [5] we have derived the SED of NGC 4565 and we showed that it is unusual, lacking the typical signatures of radiatively efficient accretion disks. Unfortunately, the data currently available in the HST archive are not suitable to derive the nuclear SED of a substantial number of objects that are located in the bottom-left quadrant. However, for M 81, NGC 3998, and NGC 4203, we were able to derive the nuclear SED in the IR-to-UV spectral region. It is clear that also these objects do not show the SED typical of standard accretion disks.

References

1. Capetti, A., Balmaverde, B., 2006. Astron. Astrophys. 453, 27.

2. Capetti, A., Balmaverde, B., 2006. Astron. Astrophys. 440, 73.
3. Chiaberge, M., Capetti, A., Celotti, A., 1999. Astron. Astrophys. 34, 77.
4. Chiaberge, M., Capetti, A., Macchetto, F.D., 2005. Astrophys. J. 625, 716.
5. Chiaberge, M., Gilli, R., Macchetto, F.D., Sparks, W.B., 2006. Astrophys. J. 651, 728.
6. Elvis, M., et al., 1994. Astrophys. J. Suppl. Ser. 95, 1.
7. Falcke, H., Körding, E., Markoff, S., 2004. Astron. Astrophys. 414, 895.
8. Ho, L.C., Filippenko, A.V., Sargent, W.L.W., 1997. Astrophys. J. Suppl. Ser. 112, 315.
9. Ho, L.C., Peng, C.Y., 2001. Astrophys. J. 555, 650.
10. Maoz, D., Nagar, N.M., Falcke, H., Wilson, A.S., 2005. Astrophys. J. 625, 699.
11. Merloni, A., Heinz, S., di Matteo, T., 2003. Mon. Not. R. Astron. Soc. 345, 1057.
12. Quillen, A.C., et al., 2001. Astrophys. J. 547, 129.

Chapter 38
The Black Hole Masses in Galactic Nuclei

D.J. Axon

Abstract We review recent progress in dynamically measuring the black hole masses in nearby galactic nuclei. Some of the outstanding issues for both the gas disk and stellar dynamical methods are outlined and the impact of high S/N integral field spectrograph studies is emphasized. The importance of obtaining reverberation mapping measurements at higher redshift is discussed.

38.1 Introduction

Over the last ten years since the first HST long-slit spectroscopy [22] established the mass of the black hole ($M_{\rm BH}$) in the center of M87 a concerted effort has been made to directly measure the $M_{\rm BH}$ in many nearby galaxies [11]. The majority of the secure $M_{\rm BH}$ measurements in the local universe have been obtained in elliptical galaxies, through a combination of stellar and gas dynamical studies [11]. Taken together with a few pivotal measurements in spiral galaxies (e.g. the Milky Way [15, 16] and NGC 4258 [32]), these studies led to two critical insights; namely that almost all local galaxies harbor supermassive black holes (SMBHs) in their centers (e.g. [21]) and that an intimate relationship exists, between $M_{\rm BH}$ and the host galaxy bulge (or maybe dark matter halo) properties in the form of the $M_{\rm BH}$–$\sigma_\star$ and $M_{\rm BH}$–$L_{\rm bul}$ relations (e.g. [12, 13, 23, 25]). The existence of these correlations has important implications for theories of galaxy formation and evolution implying that the growth of SMBHs and the formation of bulges are closely linked [17, 43]. They have also opened a widely exploited statistical pathway to investigate the cosmological processes at work without having to directly measure $M_{\rm BH}$ in every galaxy (e.g. [30, 45]). On the basis of this work it now seems that the growth SMBHs is anti-hierarchical [28] in the sense that smaller black holes ($M_{\rm BH} < 10^7$ M$_\odot$) grow at lower redshifts ($z < 1$) with respect to more massive ones ($z \sim 1$–3).

These correlations, however powerful, show considerable scatter and are based on very few accurate $M_{\rm BH}$ determinations [11] and until recently the coverage at the low-mass end has been poor. In this review I will focus on the recent progress

D.J. Axon (✉)
Department of Physics, Rochester Institute of Technology, Rochester, NY 14623, USA
e-mail: djasps@rit.edu

D.J. Axon
Centre for Astrophysics Research, Science & Technology Research Institute, University of Hertfordshire, Hatfield AL10 9AB, UK

F.D. Macchetto (ed.), *The Impact of HST on European Astronomy*, Astrophysics and Space Science Proceedings,
DOI 10.1007/978-90-481-3400-7_38, © Springer Science+Business Media B.V. 2010

in directly measuring M_{BH} using gas kinematics (GK) and stellar dynamics (SD), starting with an overview of the state-of-art of both methodologies, contrasting their strengths and weaknesses, before turning to the new results themselves. The importance of IR studies using Integral Field Spectrographs (IFU) on large ground based telescopes with Adaptive Optics (AO) will emerge as an important development. Significant progress has also been made in anchoring broad line region Reverberation Mapping (RM) M_{BH} estimates [39] onto the M_{BH}–$\sigma_\star$ and M_{BH}–L_{bul} relations [35, 38] thus tying Active Galactic Nuclei (AGN) M_{BH} in with those of quiescent galaxies. In the process a third indirect way of measuring M_{BH} based on the virial relationship for the Broad Line region ($R_{BLR} \approx L^{0.50}$) has been established [20, 30]. Since the RM results has recently been reviewed [38] we will confine ourselves to a few short comments on the future role of RM in directly measuring M_{BH} beyond the local universe.

38.2 Black Hole Mass Determinations from Gas Kinematics

The basis of modeling emission line velocity fields is to assume that the gas is rotating in circular orbits within a thin disk in the galaxy potential. The gravitational potential is assumed to comprise of two components: the stellar potential, characterized by its mass-to-light ratio Υ and a dark mass concentration (the black hole), spatially unresolved at HST resolutions and characterized by its total mass M_{BH}. Standard χ^2 minimization is then used to match synthetic rotation curves to the observations and derive M_{BH}. Over time the modeling has become increasingly sophisticated, allowing for the spatial resolution (point spread function, psf), finite slit and pixel sizes and incorporating weighting over the line surface brightness, including the introduction of weights derived from synthetic models to represent the unknown line surface brightness distribution with the psf. Further refinements in the physical assumptions underlying the models have also taken place. In some of the early models ([2], and references therein) it was assumed that the gas remains coplanar with the observed large scale gas distribution. In practice there is no real physical justification for assuming the gas settles in the same plane at all radii; in many galaxies e.g. NGC 4258, Centaurus A, we know it certainly does not. Allowance for warping of the gas disk and radial variations of Υ have become routine. As a consequence of these relaxations the error estimates on M_{BH} is widened. The state of the art as we apply it is encapsulated in [37] and adopts the following recipe. Firstly, the observed surface brightness distribution is inverted to determine the stellar luminosity density and estimate the contribution of stars to the gravitational potential in the nuclear region. Our approach assumes that the gravitational potential is an oblate spheroid, with the oblate spheroid density distribution parameterized as [49]. Unfortunately, this inversion procedure is not unique when the gravitational potential is not spherically symmetric and a range inclinations (i) must be considered. The nature of any compact nuclear source is also crucial for the estimate of the stellar mass distribution. If, as in NGC 4041 it is due to a nuclear star cluster [27] it must be retained in the mass budget, whereas if it is due to light produced by an AGN it does not correspond to a stellar mass contribution and needs to be removed [10, 22, 26, 37]. Both cases are easily handled by adding a nuclear point source to the surface brightness modeling.

Secondly, the observations themselves are parameterized by fitting the observed line profiles to produce maps of flux, velocity and width. In order to allow for non-gaussian wings we model the line profile with a Hermite polynomial expansion [48] via a penalized likelihood technique [6, 31].

The observed kinematical quantities are averages over apertures defined by the slit width and the detector pixel size along the slit weighted by the intrinsic line surface brightness distribution (ISBD). An assumed ISBD(s) is thus a further fundamental ingredient in the modeling. While large scale emission line and reddening maps are useful for matching the micro-structure in the large scale rotation curves, these have little effect on the derived M_{BH} which is dominated by the high leverage points within the unresolved central emission spike [29]. In practice many different assumed ISBDs are investigated. To build the synthetic kinematical models the ISBD(s) has to be obtained for each disk inclination under consideration.

As well as the key physical parameters i, Υ and M_{BH} our modeling includes free parameters which take into account a variety of data related factors, such as any slit configurations and placements. In general the solutions have $\chi^2_{\mathrm{reduced}} > 1$ due to a combination of "miss fit" and an additional term arising because individual kinematic data points are not statistically independent. To allow for this second term the error bars in the velocity measurements are rescaled [2] by adding a constant error in quadrature, such that the overall best fitting model provides $\chi^2_{\mathrm{reduced}} = 1$. This is a quite conservative approach as it has the effect of increasing the final uncertainty on M_{BH}. Finally a grid search of M_{BH} vs. Υ as a function of i (with the remaining optimization parameters free) is performed to derive confidence contours on M_{BH} and Υ. Usually acceptable fits can be found for a wide range of inclinations, but many of the low inclinations are eliminated as they correspond to values of Υ larger than predicted by stellar population synthesis. In this context using H or K images is best as the allowed range of $\Upsilon_{K,H}$ are more tightly constrained by the stellar populations models (e.g. [24]) than $\Upsilon_{V,B}$.

38.2.1 Who Will Rid Me of This Turbulent Gas?

Several groups (e.g. [18, 33, 50]) have reported galaxies which have "regular" rotation curves but have a central gas velocity dispersion which exceeds the stellar velocity dispersion and which is not matched by the models. In Cygnus A for example (Fig. 38.1) additional turbulence $\sim$220 km s^{-1} is needed to match the nuclear line profiles. The origin of this broadening is currently not understood. It could be due to either the gravitational potential of a black hole or turbulent shocks in the gas. Another possibility is that we are seeing the effects of gas falling into the center in combination with geometrical warps.

However, a mismatch between the observed and model velocity dispersion is not necessarily an indication of non-circular motions or kinematically hot gas.

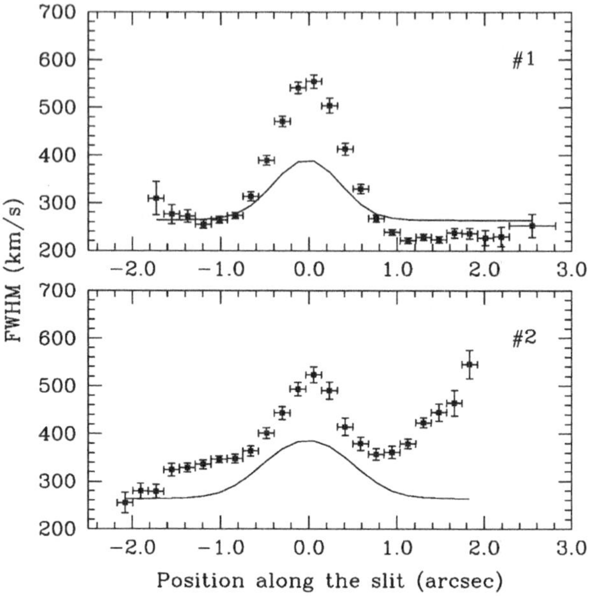

Fig. 38.1 The nuclear spectra of Cygnus A from Keck/NIRSPEC infrared data [44] showing the large FWHM in the emission lines compared to the model predictions (*solid lines*)

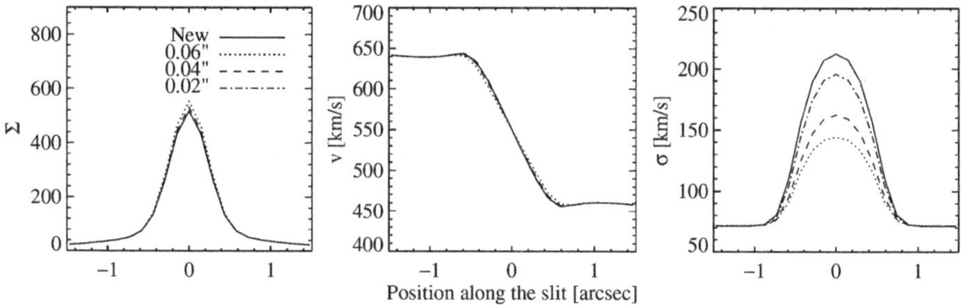

Fig. 38.2 Simulated instrument convolved surface brightness (*left panel*), radial velocity (*middle panel*) and velocity dispersion (*right panel*) computed with different ISBDs. The ISBDs have been chosen to provide very similar observed surface brightnesses and velocities along the slit. Without any significant variations in the real dynamical parameters large (factor of 2) changes in the observed velocity dispersion [29]

Detailed simulations [29] of the effect of the ISBD within the psf show that large variations in observed FWHM could be produced (Fig. 38.2) due to an inaccurate computation arising from too course a model grid, or adoption of an intrinsic brightness distribution which is too smooth. Thus while a FWHM rise is an essential signature of the presence of a SMBH the FWHM is not an accurate measure of $M_{\rm BH}$ in itself.

38.3 Stellar Dynamical Black Hole Measurements

The current state of the art of SD modeling recovers $M_{\rm BH}$ from the Line of Sight Stellar Velocity Dispersion (LOSVD) using three integral Schwarzschild dynamical (3I) realizations, drawn from a library of star orbits built under the assumption of an axisymmetric potential and some knowledge of phase space distribution function $f = f(E, L_z)$. In all the results published so far the nuclear Υ has been constrained to that measured on the large scale, and usually $i = 90°$ is assumed. Just as in GK modeling (Sect. 38.2) relaxing these two assumptions broadens the uncertainties on $M_{\rm BH}$. The low surface brightness of the stellar bulge makes application of the method difficult, as high S/N (>30) is needed to recover the higher order moments from the LOSVD and this simply cannot be obtained with HST. Obscuring material even in ellipticals can distort the LOSVD making the reconstruction and hence $M_{\rm BH}$ ambiguous. Of course the GK method is not immune from this problem either. Moreover, in broad line AGN the nuclear light is dominated by the AGN, swamping the stars and making it hard to measure the nuclear LOSVD and Υ. Traditionally the literature argues that application of this complex procedure, lead to conservative error bounds on $M_{\rm BH}$. However, a heated technical debate about the uniqueness and accuracy of the determined values of $M_{\rm BH}$ and Υ obtained from the 3I models at the achieved data S/N's has erupted in the literature [7, 14, 46]. Critical to the issue is whether the SBH sphere of influence is resolved or not. In the best case scenario the critics argue that the error bounds are unrealistically small towards the low mass end, and in the worst that $M_{\rm BH}$ is totally indeterminate. Certainly, in the original study of NGC 3379 a 'no-black hole solution' seems just as acceptable a fit to the data as the derived $M_{\rm BH}$ of $\sim 1.0 \times 10^8$ (see Fig. 11 of [13]). It is encouraging to see that obtaining high S/N 2D stellar dynamical maps from the ground with devices such as SAURON [41] may significantly improve the situation.

38.4 Recent $M_{\rm BH}$ Determinations

Table 38.1 gives a summary of the new $M_{\rm BH}$ measurements since the publication of [11]. Broadly speaking the new points are consistent with $M_{\rm BH}$–$\sigma_\star$ but with some significant outliers at the high mass end like the type 2 Seyfert NGC 5252. There are still only three galaxies outside the milky way in which the dynamics is unambiguously Keplarian, thus establishing beyond doubt the presence of a SBH; NGC 4258 [32], M87 [22] and Centaurus A [27, 29]. Until recently none of these had a $M_{\rm BH}$ determination by SD methods. The Silge et al. (2005) NIR CO band-head study [42] of Centaurus A made it the first external galaxy with reliable $M_{\rm BH}$ measurements from both techniques. Encouragingly the $M_{\rm BH}$ GK and SD measurements are in good agreement with each other (Table 38.1). The other crucial test NGC 4258 turns out to be as challenging as it gets for gas studies. Not only does the small scale gas disk show a ~90° warp compared to the large scale disk, but even worse the emission lines are highly centrally concentrated (dominated by the psf) and an extra broad-blueshifted kinematic component, presumably caused by

Table 38.1 Recent direct M_{BH} determinations. The additional abbreviations used are LG = Local Group, N = NAOS-CONICA, OA = OASIS, SA = SAURON

Galaxy	Type	D (Mpc)	Log[M_{BH}]	σ_c	Method	Telescope	Ref.
NGC 205	E5	LG	<4.34	23	SD (MgII)	STIS	[47]
NGC 224	SA(s)b	LG	8.04–8.36	170	SD (CaT)	STIS	[3]
NGC 1300	SB(rs)bcr	18.8	7.5–8.1	90	GK	STIS	[1]
NGC 1399	Cd,E1	18.3	8.6–9.6	317	SD IR(CO)	N	[19]
NGC 2748	SAbc	23.2	6.9–7.9	79	GK	STIS	[1]
NGC 3310	SAB(r)bc	17.4	<7.6	101	GK	STIS	[37]
NGC 3227	SAB(s)	20.4	6.85–7.3	155	SD IR(CO)	SINFONI	[8]
NGC 3379	E1	17.3	7.6–8.6	156	GK, SD(MgII)	STIS/SA/OA	[41]
NGC 3998:	SA:(s:)b	42	7.8–8.7	305	GK	STIS	[10]
NGC 4151	Sbc	17.1	7.66	156	RM		[4]
			<7.6		SD (CaT)	MMT	[36]
NGC 4303	SAB(rs)bc	16.1	6.4–6.8	108	GK	STIS	[37]
NGC 4258	SAB(s)bc	7.2	7.4–8.4	120	GK	STIS	[37]
NGC 4486a	E2	6.57	6.7–7.3	105	SIR(CO)	SINFONI	[34]
NGC 4593	RSB(rs)b	42	6.9–7.1	92	RM		[9]
			7.75–8.1	101	GK IR(H_2)	SINFONI	
NGC 5128	E	3.8	8.1–8.4	120	SD IR (CO)	GNIRS	[42]
			8.0–8.8		GK IR (Paβ +[SIII])	STIS/ISAAC	[29]
NGC 5252	SO S2	92.0	8.7–9.4	192	GK	STIS	[5]

the interaction with the jet, is clearly present in the line profiles. Despite these serious complications the GK estimate [37] is within 1σ of the maser mass estimate (4×10^7 $M_\odot$).

Even with the addition of several new GK M_{BH} measurements, in spiral galaxies RM M_{BH} estimates in Seyfert type 1 (36 AGNs in total) still predominate. There are only three AGN (NGC 3227, NGC 4151 and NGC 4593) for which the RM result can be compared with one of the direct dynamical methods. For NGC 3227 the IFU stellar dynamics measurement [8] (7–20) $\times 10^6$ $M_\odot$ is in acceptable agreement with the RM mass of $42 \pm 21 \times 10^6$ $M_\odot$ [40]. The situation is less satisfactory for NGC 4151 as here longslit stellar dynamics: only yields an upperbound on M_{BH} of $\leq 70 \times 10^6$ $M_\odot$ [36] (cf. the RM result $(46 \pm 5) \times 10^6$ $M_\odot$ [4]). In the case of NGC 4593 preliminary analysis AO assisted SINFONI IFU emission line cubes yields a M_{BH} 8 larger than the RM result [9]. On this basis I would argue that an agreement between the RM and dynamical mass estimates is plausible but not yet beyond question.

IFU data obtained on AO equipped 8m class telescopes (e.g. VLT, GEMINI, Keck) is now important and arguably the preferred choice for all methods (Table 38.1) since the extensive 2D coverage and high S/N GK and SK obtainable tightly constrain mass models. Furthermore, since non gravitational motions are usually spatial localized they can be recognized and excised from GK data. IR spectrographs enable measurements of the CO band heads at 2.3μ [8, 19, 34, 42] in the K band, or a suitable emission line, thereby reducing the effects of dust. Unfortunately at low redshift there are no emission lines with high equivalent width available. The best choice, $Pa\alpha\ \lambda 1.87\mu$, is in the atmospheric absorption band, but becomes accessible at moderate redshift [44], the reverse being true for the CO band heads. However H band measurements still produce reliable GK $M_{\rm BH}$ determinations [27, 29]. We are also now starting to see the first studies [41] which combine GK and SD observations. It is likely that AO assisted NIR spectroscopy will make an increasing impact on $M_{\rm BH}$ determinations as laser guide stars widen the choice of targets and attain resolutions comparable to that of the best STIS data from HST (0.1–0.2 arcseconds).

38.5 Determining $M_{\rm BH}$ Beyond the Local Universe

A detailed understanding of the co-evolution of galaxy bulges and nuclear black holes, requires the variations of $M_{\rm BH}$ and $\dot{M}$ in AGN to be mapped out as a function of luminosity and redshift and host galaxy properties. For example, the peak of quasar activity occurs at $z \sim 2.5$–3 and coincides with the time when the first deep potential wells assemble in plausible variants of hierarchical CDM models. Realistically RM is the **only** possible method of direct $M_{\rm BH}$ measurements beyond the local universe (at $z = 1\ 0.1'' \sim 0.8$ kpc). The advent of 8(30)m telescopes operating in queue-scheduled mode makes RM practical via long term monitoring campaigns out to high redshift. Extrapolating variability timescales from low-redshift AGN (taking into account time dilation) LAGs between 360–1200 day are predicted. Observations at $\sim$1 month intervals over 3–10 year base-lines should give adequate sampling of the light curves to determine BLR sizes and hence $M_{\rm BH}$. Observing in the NIR is best as this allows a direct comparison with current low-z studies (which are all based partly or entirely on the Balmer lines) and also avoids contamination by intervening absorption line systems (which seriously affects the UV continuum at high z). The JWST will also have a key role at high z.

Acknowledgements In addition to my collaborators I would like to thank Ross McLure, Brad Peterson, Monica Valluri and Alister Graham for stimulating discussions and for sharing their latest results and papers with me.

References

1. Atkinson, J., et al., 2005. Mon. Not. R. Astron. Soc. 359, 504.
2. Barth, A.J., et al., 2001. Astrophys. J. 555, 685.
3. Bender, R., et al., 2005. Astrophys. J. 631, 280.
4. Bentz, M., et al., 2006. Astrophys. J. 651, 775.

5. Capetti, A., et al., 2005. Astron. Astrophys. 431, 465.
6. Cappellari, M., Emsellem, E., 2004. Publ. Astron. Soc. Pac. 116, 138.
7. Cretton, N., Emsellem, E., 2004. Mon. Not. R. Astron. Soc. 347, 31.
8. Davies, R.I., et al., 2006. Astrophys. J. 646, 754.
9. Denney, K., et al., 2006. Astrophys. J. 653, 152.
10. de Francesco, G., Capetti, A., Marconi, A., 2006. Astron. Astrophys. 460, 439.
11. Ferrarese, L., Ford, H., 2005. Space Sci. Rev. 116, 523.
12. Ferrarese, L., Merritt, D., 2000. Astrophys. J. 539, L9.
13. Gebhardt, K., et al., 2000. Astrophys. J. 119, 1157.
14. Gebhardt, K., et al., 2003. Astrophys. J. 583, 92.
15. Genzel, R., et al., 2000. Mon. Not. R. Astron. Soc. 317, 348.
16. Ghez, A.M., et al., 2000. Nature 407, 349.
17. Haehnelt, M.G., Kauffmann, G., 2000. Mon. Not. R. Astron. Soc. 318, L35.
18. Ho, L., 2004. astro-ph/0401527.
19. Houghton, R., et al., 2006. Mon. Not. R. Astron. Soc. 367, 2.
20. Kaspi, S., et al., 2005. Astrophys. J. 629, 61.
21. Kormendy, J., Richstone, D., 1995. Annu. Rev. Astron. Astrophys. 33, 581.
22. Macchetto, F.D., et al., 1997. Astrophys. J. 489, 579.
23. Maggorian, J., et al., 1998. Astrophys. J. 489, 579.
24. Maraston, C., 1998. Mon. Not. R. Astron. Soc. 300, 872.
25. Marconi, A., Hunt, L.K., 2003. Astrophys. J. 589, L21.
26. Marconi, A., et al., 2001. Astrophys. J. 549, 915.
27. Marconi, A., et al., 2003. Astrophys. J. 586, 868.
28. Marconi, A., et al., 2004. Mon. Not. R. Astron. Soc. 351, 169.
29. Marconi, A., et al., 2006. Astron. Astrophys. 448, 921.
30. McLure, R.J., Dunlop, J.S., 2002. Mon. Not. R. Astron. Soc. 331, 795.
31. Merritt, D., 1997. Astron. J. 114, 228.
32. Miyoshi, M., et al., 1995. Nature 373, 127.
33. Noel-Storr, J., et al., 2003. Astrophys. J. Suppl. Ser. 148, 419.
34. Nowak, N., et al., 2007. Mon. Not. R. Astron. Soc. 379, 909.
35. Onken, C.A., et al., 2004. Astrophys. J. 615, 645.
36. Onken, C.A., et al., 2007. Astrophys. J. 670, 105.
37. Pastorini, G., et al., 2007. Astron. Astrophys. 469, 405.
38. Peterson, B., 2007. Astron. Soc. Pac. Conf. Ser. 373, 3.
39. Peterson, B., Bentz, M., 2006. New Astron. Rev. 50, 796.
40. Peterson, B., et al., 2004. Astrophys. J. 613, 682.
41. Shapiro, K., et al., 2006. Mon. Not. R. Astron. Soc. 370, 559.
42. Silge, J., et al., 2005. Astron. J. 130, 406.
43. Silk, J., Rees, M.J., 1998. Astron. Astrophys. 331, L1.
44. Tadhunter, C., et al., 2003. Mon. Not. R. Astron. Soc. 342, 861.
45. Treu, T., et al., 2007. Astrophys. J. 667, 117.
46. Valluri, M., Merritt, D., Emsellem, E., 2004. Astrophys. J. 602, 66.
47. Valluri M., et al., 2005. Astrophys. J. 628, 137.
48. van der Marel, R.P., Franx, M., 1993. Astrophys. J. 407, 525.
49. van der Marel, R.P., van den Bosch, F.C., 1998. Astron. J. 116, 2220.
50. Verdoes Kleijn, G., et al., 2002. Astron. J. 124, 2524.

Chapter 39
The HST/ACS Coma Cluster Treasury Survey

David Carter and the ACS Coma Cluster Treasury Survey Team

Abstract We describe here the HST/ACS Coma Cluster treasury survey, which aims to provide an unparalleled database of high spatial resolution images of a sample of cluster galaxies. The survey was designed to cover an area of 740 arcmin^2 in regions of different density of both galaxies and intergalactic medium within the cluster, to a depth for point sources of $B = 27.3$ mag in the F475W filter, and $I = 25.1$ mag in the F814 filter (in each case for 10σ detections).

The ACS failure of January 27th 2007 leaves the survey 28% complete, but with data reduction procedures, specifications of data products and distribution mechanism in place.

39.1 Introduction

The Coma cluster is the nearest rich, relaxed and dense cluster of galaxies, and provides a unique environment in which to study the effect of interactions with galaxies and the intracluster medium on the evolution of galaxies. Coma is also near enough that it is possible to study the properties of galaxies well into the dwarf regime with HST. The study of dwarf galaxies is important, because the popular hierarchical clustering scenario predicts that dwarfs in clusters are the first galaxies to form [9]; regions in which galaxy formation is initiated first are ones whose small-scale density peaks are superimposed on underlying large-scale density peaks, which evolve into rich clusters. Dwarf galaxies also have a shallower potential well than giants and, therefore, efficiently probe a wide range of environmental processes.

The ACS Coma Cluster Treasury Survey was proposed by an international consortium of 32 scientists from nine countries, and awarded 164 orbits of HST time in Cycle 15. The main scientific motivations of the survey are:

(1) Measurement of the luminosity function (LF) down to $M_V = -9$ for comparison with other more nearby clusters. This is critical for detecting dwarf galaxies at luminosities that overlap with the Local-group dwarf spheroidals, and testing whether the faint-end LF slope in clusters is universal.

D. Carter (✉)
Astrophysics Research Institute, Liverpool John Moores University, Twelve Quays House, Birkenhead, CH41 1LD, UK
e-mail: dxc@astro.livjm.ac.uk

F.D. Macchetto (ed.), *The Impact of HST on European Astronomy*, Astrophysics and Space Science Proceedings,
DOI 10.1007/978-90-481-3400-7_39, © Springer Science+Business Media B.V. 2010

(2) Determination of detailed morphologies using bulge-disk decomposition techniques for a wide range of galaxy luminosities, including those at the faint end of the cluster LF, thus revealing compact nuclei, nuclear bars, disks, and dust rings.
(3) Derivation of global colors and colour gradients at a range of luminosity and environment to investigate the origin of cluster galaxies, and to understand the environmental process acting upon them.
(4) Investigation of the bright and faint ends of the global scaling laws among structural parameters (e.g., luminosity–radius, luminosity–velocity dispersion, fundamental plane) and assess any environmental dependence.
(5) Study of the Globular Cluster (GC) systems of galaxies at a range of luminosity and morphological type, to see whether phenomena such as colour bimodality and the colour–luminosity relation for blue GCs are universal.
(6) To search for indicators of morphological transformation and star formation in the core and outskirts of the cluster, and how indicators of current merger activity (shells, ripples, tails) depend upon environmental density.
(7) Construction of samples of galaxies selected for the probability of cluster membership for more detailed followup with the largest ground-based telescopes.
(8) To provide a zero redshift benchmark for studies of high redshift clusters, and cluster and galaxy evolution.

39.2 Survey Design

For our passbands we choose F814W, as the passband which will give the deepest data for structural and luminosity function studies, and F475W, which is a compromise between colour baseline and speed. The aim of the survey is to provide a large sample of galaxies for study in a high density environment, and at the same time a comparison sample in a lower density region of the cluster for the specific science goals of examining the effect of environment upon morphology, structural parameters and stellar populations. Because of the density of confirmed cluster galaxies in the core [3] we have adopted a tiling strategy in the core region, tiling a region of approximately 18×21 arcmin, using 42 ACS pointings in a 7×6 pattern, with some overlap between adjacent tiles. An HST orientation of 282 degrees was chosen for these tiles in order to maximise the time for which the observations could be scheduled in two gyro mode [8]. A tile of the central mosaic was moved to the southern edge of the mosaic area, away from the star HD112877 ($V = 7.17$) which would have a negative impact on ACS observations near its position.

In the outer part of the cluster the density of known members is less than one per ACS tile, and we decided to target known members which can be used to address some of our primary science goals. The best studied region outside the core is the infall region around NGC4839 [6] where there have been a number of photometric [4] and spectroscopic [1, 5, 7] surveys. Forty further ACS pointings were defined, each containing one, or in most cases more, cluster members from these spectroscopic surveys. At the time of the ACS failure 21 of our 82 fields have been fully

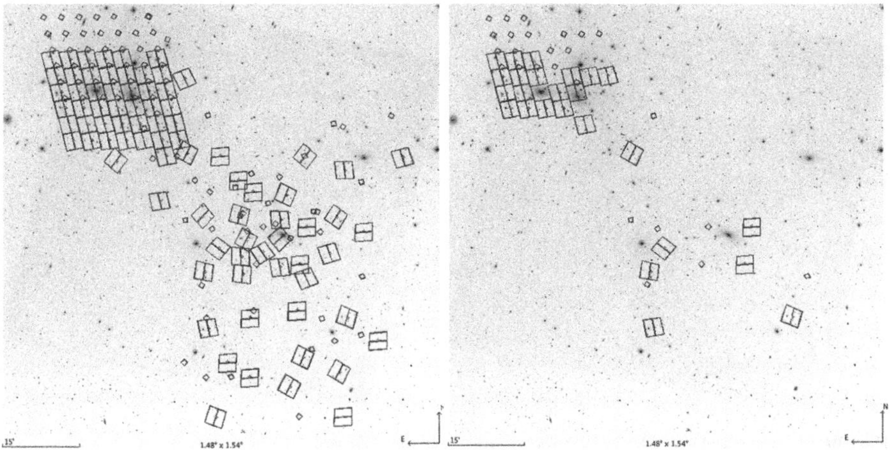

Fig. 39.1 (Color online) *Left*: Positions of ACS tiles for the survey proposed, overlayed on a DSS image of the core and infall region of the Coma cluster. The *small squares* represent the fields of NICMOS parallel exposures. *Right*: The survey as at the ACS failure of January 2007. *Tiles* shown have some or all of the proposed observations complete by this date

observed, and further four have either two or three dither positions observed. The survey is thus 28% complete, the locations of the 25 wholly or partially completed tiles are shown in the right panel of Fig. 39.1.

39.3 Data Products

As a Treasury survey, public data products as described in Table 39.1 will be made available on timescales of between 3 and 12 months via MAST in the USA, and Astrowise in Europe, and via a dedicated project web site. Description of the procedures for pipeline processing and initial catalogue generation will be described in the first two team papers [2]. In Fig. 39.2 we show representative pipeline processed images of a region from an outer pointing and the cluster centre.

Table 39.1 HST data products and data release schedule

Data product	Description	Time from observations
Raw data	HST pipeline calibrated images	Immediate
Processed image data	Co-added, CR-cleaned, drizzled	Three months
Version 2	Reprocessed with best reference files	Six months
Source lists	Positions and magnitudes	Six months
Full catalogs	Catalogs including morphological analysis	One year
External data	Redshifts and ground-based colors	One year

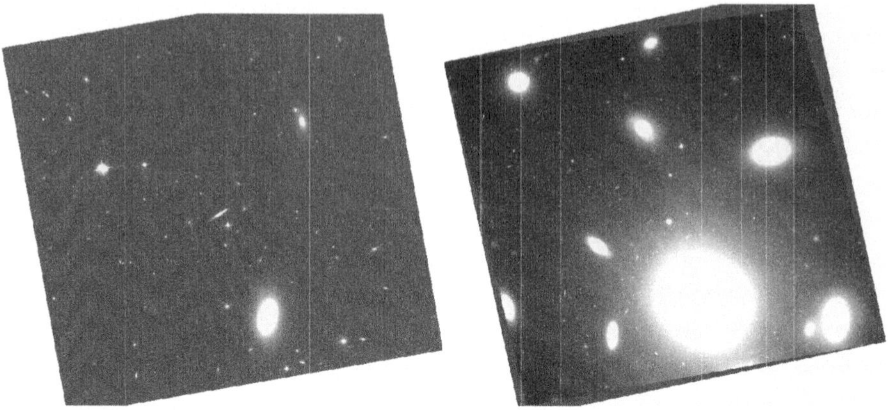

Fig. 39.2 Processed ACS F814W band images for an outer field (*left*) and the central field around NGC4874 (*right*)

39.4 Future Plans

The future of this project depends critically on the success of SM4. The survey can clearly be completed in 120 further orbits with ACS if it is repaired, otherwise it will be possible to complete it in around 160 orbits with WFC3 with slightly reduced areal coverage around the outer field centres. Extension of the wavelength baseline by a survey with WFC/IR of a subset of the central ACS area will greatly enhance the project, particularly the aspects involving the LF and stellar populations.

References

1. Caldwell, N., Rose, J.A., Sharples, R.M., Ellis, R.S., Bower, R.G., 1993. Astron. J. 106, 473.
2. Carter, D., et al.., 2008. Astrophys. J. Suppl. Ser. 176, 424.
3. Colless, M., Dunn, A.M., 1996. Astrophys. J. 458, 435.
4. Komiyama, Y., et al., 2002. Astrophys. J. Suppl. Ser. 138, 265.
5. Matković, A., Guzmán, R., 2005. Mon. Not. R. Astron. Soc. 362, 289.
6. Neumann, D.M., et al., 2001. Astron. Astrophys. 365, L74.
7. Poggianti, B.M., et al., 2001. Astrophys. J. 562, 689.
8. Sembach, K.R., et al., 2006. HST Two-Gyro Handbook, Version 3.0. STScI, Baltimore.
9. White, S.D.M., Frenk, C.S., 1991. Astrophys. J. 379, 52.

Chapter 40
Discovery of a Population of Evolved and Massive Galaxies at High Redshift

Bahram Mobasher and Tommy Wiklind

Abstract We present a new technique for identifying massive and evolved galaxies at $z \sim 5$. The technique is based on broad-band deep photometry, using the strength of Balmer Break features, which is shifted between the K-band and $m_{3.6}$ wavelengths at this redshift. The main criterion we use to identify these galaxies is therefore the K–$m_{3.6}$ colors. We found a total of 11 galaxies in the GOODS-S field at $z \sim 5$, with stellar masses between 10^{10} and 5×10^{11} $M_{\odot}$ and ages >100 Myr. The existence of such a population of galaxies strongly constrains current galaxy formation scenarios. The implications of this population for formation of galaxies is discussed.

40.1 Introduction

An important goal of observational cosmology is to understand how stars are assembled into galaxies and how this is related to the evolution of dark matter halos. In prevailing hierarchical models, star formation starts out in low mass systems, which build more massive galaxies through sequential merging (e.g. [4, 6]). In this picture, the most massive galaxies are found at relatively low redshifts through mergers of smaller sub-units, while the high redshift universe is populated by galaxies of modest stellar mass. Therefore, the very existence of massive systems at high redshifts will put strong constraints on the current galaxy formation scenarios.

Recent observations, done at wavelengths stretching from the UV to mid-infrared, have revealed a population of old galaxies at $z > 5$, with stellar mass $\sim 10^{11}$ $M_{\odot}$ [7]. Broad band photometric data combined with stellar population synthesis models show that many of these galaxies contain stars formed within ~1 Gyr after the Big Bang. However, it is not known how these stars were assembled into their present host galaxies, whether this was done during multiple merger events, as proposed in hierarchical models, or if the stars and their host galaxy are co-eval.

B. Mobasher (✉)
Physics and Astronomy Department, University of California, 900 University Ave, Riverside, CA 92508, USA
e-mail: mobasher@ucr.edu

T. Wiklind
ESA/STScI, 3700 San Martin Drive, Baltimore, MD 21218, USA
e-mail: wiklind@stsci.edu

F.D. Macchetto (ed.), *The Impact of HST on European Astronomy*, Astrophysics and Space Science Proceedings,
DOI 10.1007/978-90-481-3400-7_40, © Springer Science+Business Media B.V. 2010

The existence of massive and evolved galaxies at redshifts $z \geq 5$, when the universe was ≤ 1.2 Gyr old, is surprising and poses a fundamental problem for hierarchical models if their number density exceeds that of correspondingly massive dark matter halos (e.g. [4]). The number density of old and massive galaxies at $z \sim 5$ is therefore among the most fundamental parameters constraining galaxy formation models. In this communication, we first review the technique for selecting massive and old galaxies at $5 < z < 7$ and then apply the technique on real data to identify galaxies with such characteristics. Using number density of these sources, we then study the implication of these galaxies for galaxy formation scenarios.

40.2 Selection of the Population of Massive Galaxies at High-z

To identify the age of stellar population in galaxies at any redshift, we use the Balmer Break feature at 3,648 Å. This is an age-dependent diagnostic of the stellar population, most prominent in A-stars. The Balmer Break feature occurs for stellar populations in both passively evolving and continuously star forming systems, but on different time scales (cf. [1]) and for ages between 0.1 and 1.0 Gyr. Using the Balmer Breaks, we could therefore isolate the evolved and massive galaxies at high redshift by taking the following steps:

- For galaxies at $5 < z < 7$, the K- and 3.6 μm bands straddle the Balmer Break. Therefore, a red K–$m_{3.6}$ color is sensitive to the old stellar population in galaxies at this redshift range.
- At $z > 5$, the observed K–$m_{3.6}$ colors correspond to rest-frame J-band (1.2 μm), which is directly proportional to the stellar mass of galaxies. Therefore, candidates which are at $z > 5$ and are relatively bright at $m_{3.6}$ band, are also expected to be massive.
- A serious problem in identifying such galaxies is the contamination by nearby ($z \sim 2$) dusty and star-forming galaxies. To avoid these interlopers, we also use the J–H and H–$m_{3.6}$ colors to identify high redshift galaxies, with the additional requirement that they should not be detected in the B- and V-bands.

Using the optical, near- and mid-infrared data in the Great Observatories Origins Deep Survey (GOODS) southern field, we identify the candidates which satisfy the above criteria. Figure 40.1 presents the color–color diagrams which are used to isolate these candidates. In order to select the final candidates, we fit spectral energy distributions (SEDs), derived from population synthesis models, to the broad-band photometric data for the color selected galaxies. For each candidates we estimate its redshift, stellar mass, extinction and age. Because of the technique with which these objects are selected, we call them Balmer Break Galaxies (BBGs). Figure 40.2 shows the image displaying an example for a BBG. Note the non-detection of the source at shorter wavelengths and its brightness at mid-IR wavelengths.

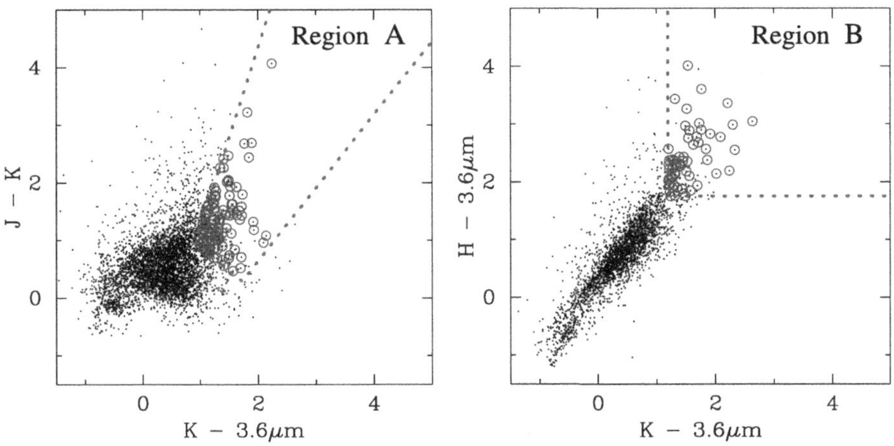

Fig. 40.1 (Color online) Color–color plots for selection of the BBGs. The *blue dots* show galaxy candidate at $z \sim 5$, with stellar populations more than 100 Myr old and stellar masses $10^{10} - 5 \times 10^{11}$ $M_{\odot}$. The primary selection criterion is the K–$m_{3.6}$ color, which straddles the redshifted Balmer Break. Two different secondary colors are chosen to exclude the outliers

40.3 Sources of Uncertainty

There are a number of uncertainties in selecting the BBGs. These include:

- Photometric errors — this will result in wrong classification of interlopers into the BBG class or, exclusion of true BBGs from the sample. The effect of photometric errors on sample selection is examined by performing detailed Monte Carlo simulations, simultaneously changing the magnitudes of the BBGs at different passbands, assuming Gaussian error distributions, and comparing the observed and template SEDs [2, 7]. After 1,000 trials, the distribution of redshift, stellar mass and extinction provides a measure of the sensitivity of the estimated parameters on the photometric errors (Fig. 40.2).
- Degeneracy — there is known degeneracy between redshift and extinction, resulting contamination of the high-z BBGs by dusty, star-forming galaxies at $z \sim 2$. This is clear from Fig. 40.2, which shows the dependence of redshift on reddening (E_{B-V}) as a result of photometric errors. To constrain the observed degeneracy, we need to acquire spectroscopic redshifts.
- Stellar Population Models — we have tested different population synthesis models and all produce consistent results. The absence of Asymptotic Giant Branch (AGB) stars from the population synthesis models could result in an overestimate of the stellar masses. However, including the contribution from the AGB population only marginally changes our estimated values for stellar mass.

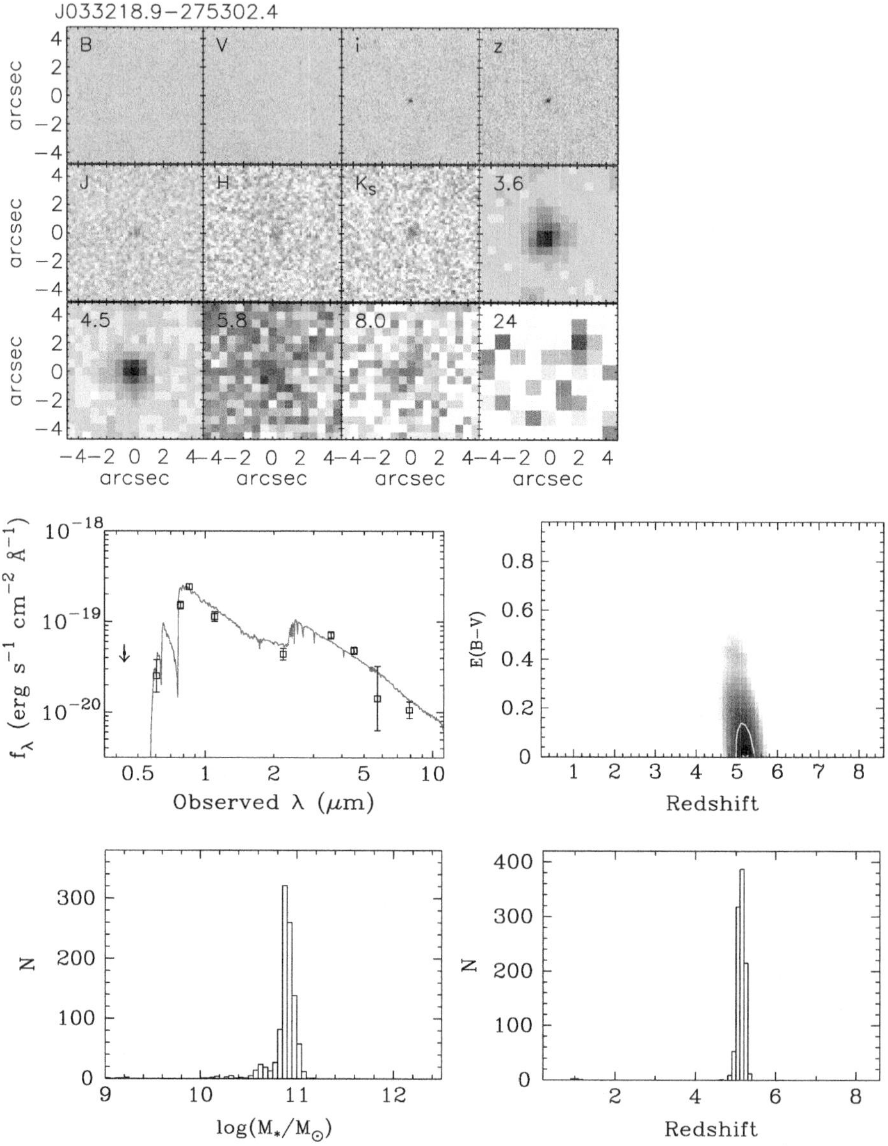

Fig. 40.2 (Color online) *Top*: images of one of the BBG candidates. Starting from *top-left*, the panels show: ACS (BViz); ISAAC (JHK); IRAC (3.6, 4.5, 5.8 and 8.00 μm); MIPS (24 μm). The *middle left panel* shows the observed data with the best-fit model SED corrected for dust extinction. The *middle right panel* shows the χ^2_ν contours for the best fits as a function of redshift and extinction E_{B-V}. The *bottom panel* shows results from 1,000 Monte Carlo realizations for redshift and stellar mass, indicating the probability distributions for the two parameters

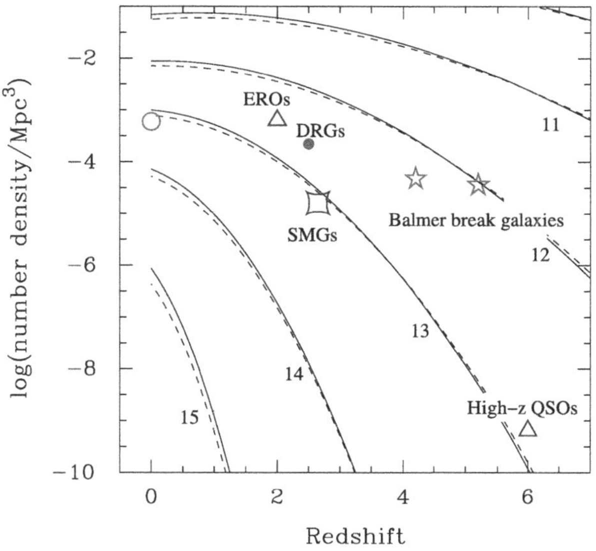

Fig. 40.3 The number density of cold Dark Matter halos as a function of redshift for several different halo masses $\log M/M_{\odot} =$ 15–11 [5]. The number densities for various other galaxy populations are shown, including Balmer Break Galaxies at $z \sim 5.2$ and $z \sim 4.4$ (*red asterix*); Sub-mm Galaxies (*blue box*); EROs at $z \sim 2$ and high-z QSOs at $z \sim 6$ (*blue triangle*); Dead and Red Galaxies (DRGs) (*red filled circle*)

40.4 Results

We find a total of 11 galaxies in the GOODS-S field with stellar masses $10^{10} < M/M_{\odot} < 5 \times 10^{11}$, ages in the range 0.6–1 Gyr and with redshifts $5 < z < 7$. An example of the BBGs is shown in Fig. 40.2. Since our target galaxies are dominated by the old stellar population, they lack emission lines and hence, it is difficult to confirm their redshifts via spectroscopy. Nevertheless, one of the BBGs in our sample has available spectroscopic redshift in excellent agreement with that estimated from its SED.

The existence of massive and evolved galaxies at redshifts $z \geq 5$, when the universe was $\leq$1–2 Gyr old, is surprising and poses a fundamental problem for hierarchical models if their number density exceeds that of correspondingly massive dark matter halos (e.g. [4]). In Fig. 40.3, we compare the number density of old and massive galaxies at $z \sim 5$, with those of dark matter halos. The corresponding halo mass is $\sim 10^{12}\ M_{\odot}$, which is too small, given the inferred stellar masses.

The main question here is how such massive and evolved galaxies at these high redshifts could be formed by mergers of smaller subunits? This requires intensive star formation activity over a short period of time. If this is the case, early evolution of these galaxies could produce significant UV radiation, making them serious candidates for re-ionization of the Universe. This has been investigated in [3] who found for some values of the compactness and filling factors, the BBGs could indeed contribute up to 20% of the re-ionization of the Universe. A confirmation of reality of the BBGs requires measurement of the spectroscopic redshifts for the candidates found in this study. We have undertaken a spectroscopic program on the Keck telescopes to obtain redshifts for these systems.

References

1. Leitherer, C., et al., 1999. Astrophys. J. Suppl. Ser. 123, 3.
2. Mobasher, B., et al., 2005. Astrophys. J. 635, 835.
3. Panagia, N., et al., 2005. Astrophys. J. 633, L1.
4. Somerville, R., Primack, J.R., Faber, S.M., 2001. Mon. Not. R. Astron. Soc. 320, 504.
5. Sheth, R.K., Torman, G., 1999. Mon. Not. R. Astron. Soc. 308, 119.
6. White, S.D.M., Rees, M.J., 1978. Mon. Not. R. Astron. Soc. 183, 341.
7. Wiklind, T., et al., 2008. Astrophys. J. 676, 781.

Chapter 41
Searching for High Redshift Galaxies Using Population Synthesis Models

Tommy Wiklind and Bahram Mobasher

Abstract We discuss the use of population synthesis models as a mean of selecting galaxies based on age/stellar mass. In particular, we apply SED fitting on a sample from the GOODS-S field.

41.1 Introduction

One of the fundamental goal of observational cosmology is to understand the formation of galaxies and its relation to the formation and evolution of dark matter halos, and the subsequent evolution leading to the universe we observe at $z \sim 0$. However, finding distant galaxies and deriving their characteristics is a challenging task, requiring the best telescopes and detectors as well as specialized techniques for identifying the target galaxies. A prerequisite for these kind of studies is deep large-scale, multi-wavelength surveys of 'empty fields'. The Hubble Space Telescope has been instrumental in opening up this new area, providing multiwavelength data sets of high quality. In addition, the HST and the associated institutions have ensured a high impact of the new data by pioneering the immediate release of the data and results [9]. The legacy of these early efforts has been further enhanced by a large observational effort using other telescopes, ranging from radio to X-ray wavelengths, and with spectroscopic follow-up observations.

The Hubble Deep Fields have been followed by similarly concerted efforts, going deeper, wider and with even more wavelength coverage. One such effort is the Great Observatories Origins Deep Survey (GOODS) [6]. There are two GOODS field, each covering $\sim$160 arcmin^2, with four optical bands. The same regions have been observed in the near-infrared using the VLT/ISAAC instrument and in the mid-infrared with the Spitzer Space Telescope. The GOODS fields are also covered by very deep X-ray [4] and radio observations [1]. Hence, they represent a very valuable data set for studying the distant galaxies and to follow their evolution.

T. Wiklind (✉)
ESA/STScI, 3700 San Martin Drive, Baltimore, MD 21218, USA
e-mail: wiklind@stsci.edu

B. Mobasher
Physics and Astronomy Department, University of California, 900 University Ave, Riverside, CA 92508, USA

F.D. Macchetto (ed.), *The Impact of HST on European Astronomy*,
Astrophysics and Space Science Proceedings,
DOI 10.1007/978-90-481-3400-7_41, © Springer Science+Business Media B.V. 2010

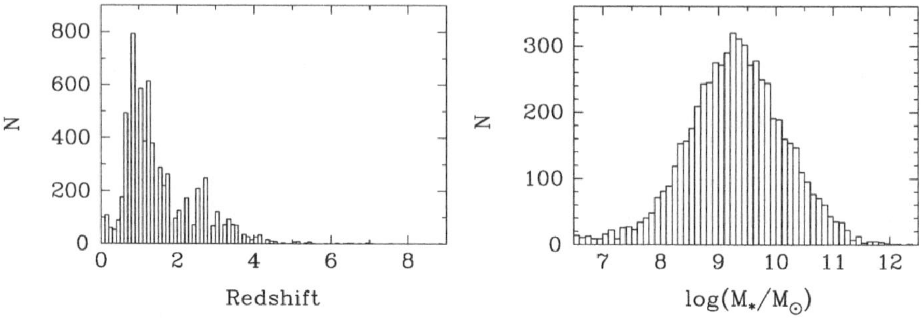

Fig. 41.1 Redshift (photometric) and stellar mass distribution for ~7,000 galaxies from the K-selected catalog in the GOODS-South field

41.2 Finding Distant Galaxies

Identification of high-redshift galaxies using the Lyman-break technique has been a well-established procedure for more than a decade [5]. The amplitude in flux density across the Lyman break, caused mainly by stellar atmospheric opacity, allows a clear distinction between foreground and high redshift galaxies. This method (the dropout technique) can be 'tuned' to select galaxies at different redshifts by choosing appropriate filter combinations. Several thousands of Lyman-break galaxies (LBGs) have been identified at redshifts from $z \approx 3$–7. While the dropout technique works well for actively starforming galaxies, it is less efficient for selecting passively evolving galaxies, where the UV part of the spectral energy distribution (SED) is depressed because of a lack of young and massive stars.

We recently applied a technique to select post-starburst galaxies at redshifts $z \geq 5$ using a combination of color selection and fitting population synthesis models to multi-wavelength broad photometric data from the GOODS-South field [8]. The main color selection is done using the redshifted Balmer break, which for $5 \leq z \leq 9$ falls between the VLT/ISAAC K-band and the Spitzer/IRAC 3.6 μm band. The color selection was applied to a K-selected sample consisting of ~7,000 galaxies from the GOODS-South field. In order to do the final selection, we fitted population synthesis models (BC03) [2], exploring a large part of the relevant parameter space. Fitting population synthesis models to broad band photometric data suffers from degeneracies among some of the parameters, most notably between age, metallicity and extinction. In order to assess the confidence of the SED fits and to explore the impact of the degeneracies, we submitted each galaxy to a Monte Carlo simulation, were each photometric data point was allowed to vary stochastically according to its photometric uncertainty. The best-fit parameters were derived for each new photometric data set. The distribution of parameters (photometric redshift, age, extinction, metallicity and star formation history) is thus a probability distribution. The results from this study is presented in [8] (see also Chap. 40 by Mobasher and Wiklind in this book).

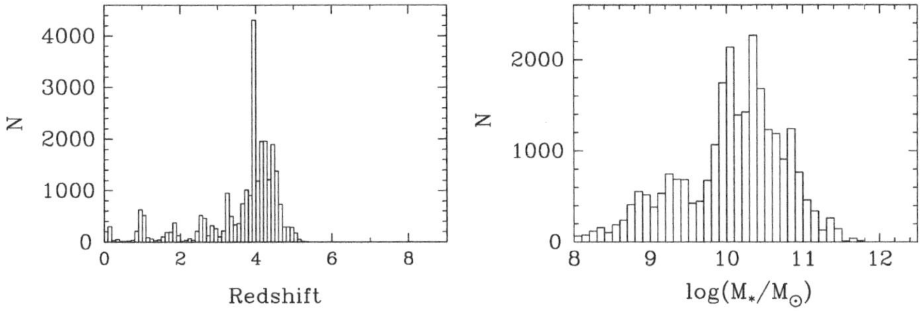

Fig. 41.2 Results from Monte Carlo simulations of 26 Balmer break galaxy candidates in the redshift range $4 \leq z \leq 5$. Shown are the photometric redshift and stellar mass distributions

41.2.1 Age Selected Balmer Break Galaxies

Finding the best-fit SED using population synthesis models is a powerful method to characterize large samples of galaxies, allowing the definition of mass- or age-selected samples of galaxies rather than luminosity selected ones. In Fig. 41.1 we show the distribution of photometric redshift and stellar mass for all ~7,000 galaxies in our K-selected sample from the GOODS-South field. The median redshift is $z = 1.2$ and the median stellar mass is $\log(M_*/\mathrm{M}_\odot) = 9.36$. The redshift distribution has a high-z tail, extending to $z > 6$ while the stellar mass distribution is fairly well represented by a normal distribution, extending over $\log(M_*/\mathrm{M}_\odot) = 7$–12. From this sample we selected all galaxies in the redshift range $4 \leq z \leq 5$, with stellar ages in excess of 200 Myr. This resulted in 26 galaxies which can be characterized as post-starburst galaxies, similar to the Balmer break galaxies discussed earlier [8] (see also Chap. 40 by Mobasher and Wiklind in this book). We performed Monte Carlo simulations on these 26 galaxies and part of the result is shown in Fig. 41.2. The median photometric redshift is $z = 4.2$, with a tail of $z < 4$ solutions (note, no $z > 5$ tail). The fraction of Monte Carlo realizations resulting in a redshift lower than the one used for initially selecting the galaxies is ~25%. Hence, it is likely that only 19–20 of our 26 galaxies are truly in the target redshift range. The probability distribution for the stellar masses is also shown in Fig. 41.2, showing a relatively large dispersion. About 70% of the realizations give a stellar mass in excess of 10^{10} $\mathrm{M}_\odot$. Hence, we find ~19–20 galaxies in the redshift interval $z = 4$–5, with a stellar population older than 200 Myr and a stellar mass $>10^{10}$ $\mathrm{M}_\odot$. The completeness of this selection is presently undetermined, and the values are therefore lower limits.

41.2.2 Other Systematic Effects

The influence of stellar evolution is another source of systematic effects for population synthesis models. In particular, the late AGB phase is a major contributor of luminosity at red optical and near-infrared wavelengths, but theoretical modelling of

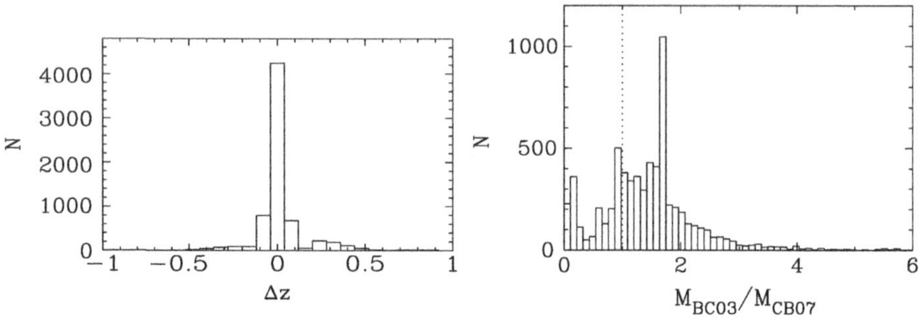

Fig. 41.3 Difference in results for the BC03 [2] models and the new CB07 [3] models when applied to the ~7,000 galaxies in our K-selected GOODS-S sample

this stellar evolutionary phase is difficult (cf. [7]). For stellar populations characterized by a single age parameter, the effect of the late AGB phase is most prominent in the age range 0.2–1.0 Gyr, which is the range for our Balmer break galaxies at $z > 4$ [8].

Recently, [7] introduced a model including the effects of thermally-pulsating AGB stars. The effect on the SED used to fit photometric data is significant and, for a given set of observed data points, will result in a lower stellar mass needed to account for the luminosity. In addition, more subtle effects on the fitting results, such as the redshift and age may come into play as well. A second set of population synthesis models, with an updated and improved treatment of AGB stars, is being released by Charlot and Bruzual (CB07) [3]. We used these models to look for systematic effects on our results for the K-selected sample of galaxies from the GOODS-South field. We did two set of fitted models, one using the Bruzual and Charlot [2] models, with a Salpeter initial mass function, and the new Charlot and Bruzual [3] models, using a Chabrier initial mass function. Both models were applied to all ~7,000 galaxies in our sample. We then compared the parameters on a galaxy-to-galaxy basis. The results for photometric redshift and stellar mass is shown in Fig. 41.3. While the difference is negligible for redshift, there is indeed a systematic difference for the estimated stellar mass. On average, the stellar mass obtained with the new CB07 models are 1.6 times smaller than the one obtained with the BC03 models. Part of the difference is due to the different initial mass functions, and part on the additional luminosity contributed by the late-phase AGB stars.

References

1. Afonso, J., et al., 2006. Astron. J. 131, 1216.
2. Bruzual, G., Charlot, S., 2003. Mon. Not. R. Astron. Soc. 344, 1000.
3. Charlot, S., Bruzual, G., 2007. Private communication.
4. Giacconi, R., et al., 2002. Astrophys. J. Suppl. Ser. 139, 369.
5. Giavalisco, M., 2002. Annu. Rev. Astron. Astrophys. 40, 579.
6. Giavalisco, M., et al., 2004. Astrophys. J. Lett. 600, 93.

7. Maraston, C., 2005. Mon. Not. R. Astron. Soc. 362, 799.
8. Wiklind, T., et al., 2008. Astrophys. J. 676, 781.
9. Williams, R.E., et al., 1996. Astron. J. 112, 1335.

Chapter 42
The ACS Grism Mode and ACS Grism Observations of Deep Fields and High z Ly-α Galaxies

Nor Pirzkal, Sangeeta Malhotra, James Rhoads, Chun Xu, and the GRAPES Team

Abstract The ACS grism mode offers an unprecedented opportunity to obtain slitless, highly multiplexed, spectroscopic observations down to very faint magnitudes. In this paper, we show that by selecting objects in a deep field based in the spectra of these objects rather than their broad band colors, a more accurate object selection can be performed. The ACS grism mode enabled us to identify faint, young, and low mass ($\approx 10^6$–10^8 $M_\odot$) galaxies at $4.0 < z < 5.7$, that are likely to be the building blocks leading to the assembly of the more massive galaxies that we observed in the present universe.

42.1 The ACS Grism Mode

The ACS grism mode enables slitless spectroscopy over the entire ACS field of view of 11 arcmin^2. This mode is sensitive to light from 5,500 Å $< \lambda <$ 9,500 Å. While the spectral resolution achieved by the ACS grism is low at $R = 100$ (40 Å/pixel), the advantage of this mode is that every object in the field produced to a spectrum in the corresponding grism image. Objects therefore do not need to be pre-selected, as it is the case with normal slit spectroscopy where objects are pre-selected based on their colors or morphology. An example of a field observed with the ACS grism is shown in Fig. 42.1. Details on the ACS grism and related data reduction are given in details in [3].

As shown in Fig. 42.2, selecting an object based on its apparent broad band colors or morphology can be misleading. Selecting a very red object, as measured by its

N. Pirzkal (✉)
Space Science Department of the European Space Agency, Space Telescope Science Institute, Baltimore, MD 21218, USA
e-mail: npirzkal@stsci.edu

S. Malhotra · J. Rhoads
School of Earth and Space Exploration, Arizona State University, Tempe, USA

C. Xu
Shanghai Institute of Technical Physics, Shanghai, China

F.D. Macchetto (ed.), *The Impact of HST on European Astronomy*, Astrophysics and Space Science Proceedings,
DOI 10.1007/978-90-481-3400-7_42, © Springer Science+Business Media B.V. 2010

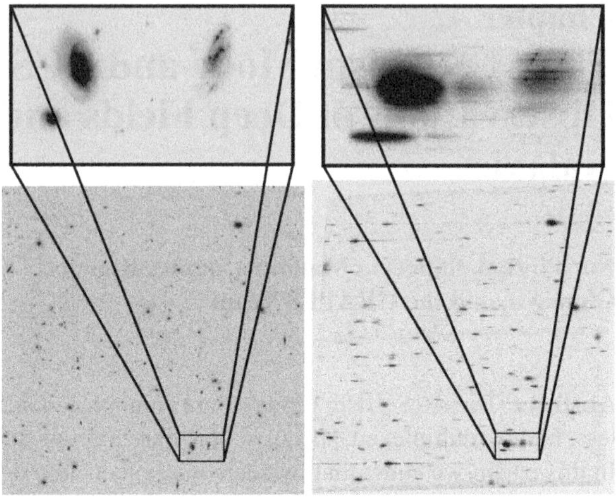

Fig. 42.1 A example of an ACS field (*left panel*) observed using the ACS grism (*right panel*). Clearly, every object, including more complex ones, produce a spectrum when imaged using the ACS grism

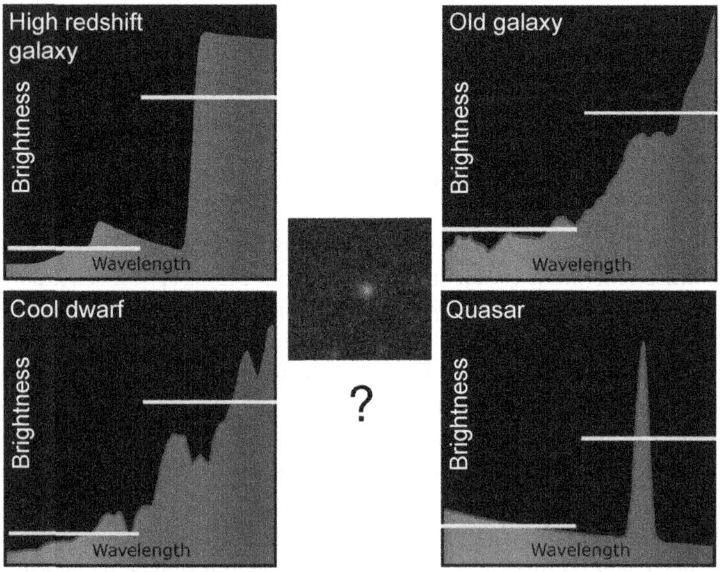

Fig. 42.2 (Color online) The color of an object is not necessarily a good fail proof indicator of the nature of that object

i and *z* band broad band fluxes, does no guaranty that this object is a high redshift object. It might as well be a red M dwarf star in our own galaxy, or a much larger and nearer galaxy whose stellar content is old and dusty, and therefore very red, or an object with one of the broad band flux is contaminated by a broad emission line. Hence, even low resolution spectra can substantially help in identifying the nature of an object.

42.2 The GRAPES Survey

The GRAPES (GRism ACS Program for Extragalactic Science, PI: S. Malhotra; see description in [3]), obtained grism observations of the Hubble Ultra Deep Field (HUDF, see [1]), using a total of 40 Hubble Space Telescope (HST) orbits, and using four different position angles to medicate the effect of source contamination. This field was also observed using HST/NICMOS, VLT/ISAAC and Spitzer/IRAC and ancillary imaging data ranging from the rest-frame UV to optical are therefore available for nearly all the objects in the HUDF. Out of the more than 10,000 objects in the field, GRAPES successfully obtained spectra for 1,500 of the brightest objects. Sources with continuum level down to $I_{AB} \approx 27.2$, or about two magnitude fainter than the deepest ground based surveys, produced spectra with good continuum signal-to-noise. The GRAPES data allowed us to identify emission line objects in the un-interrupted redshift ranges of $0.3 < z < 1.5$ using [OII]λ3727, [OIII]$\lambda\lambda$4959,5007, and Hα, and $3.5 < z < 6.0$ using Ly-α (see [4]), nine which were Ly-α galaxies at $4.0 < z < 5.7$.

42.3 Ly-α Sources in the HUDF

42.3.1 Properties

Table 42.1 summarize the morphological properties of these objects. The half-light radius (R_{50}), Concentration (C), Asymmetry (A) and absolute magnitude ($M_{1,500\,\text{Å}}$) were estimated at a rest-frame wavelength of 1,500 Å. The redshifts and the line fluxes are from [4]; the line fluxes from [4] shown in parenthesis were corrected using an aperture correction that we derived taking into account the size of each

Table 42.1 GRAPES LAEs and their morphological measurements

UID	R.A. J2000	Dec. J2000	Redshift z	Flux 10^{-18} ergs s^{-1} cm^{-2}	$M_{1,500\,\text{Å}}$	R_{50} kpc	C	A
631	3:32:40.09	−27:49:01.19	4.00	58. (20)	−20.21	0.67	2.74	0.097
712	3:32:42.81	−27:48:58.51	5.20	17. (6)	−21.31	0.52	2.53	0.11
4442	3:32:39.38	−27:47:39.17	5.76	24. (7)	−19.54	0.61	2.89	0.22
5183	3:32:34.51	−27:47:27.11	4.78	32. (11)	−19.82	0.54	2.84	<0.1
5225	3:32:33.26	−27:47:24.76	5.42	22	−22.77	1.86	2.18	0.31
6139	3:32:37.95	−27:47:10.99	4.88	60. (22)	−22.14	1.35	3.19	0.41
9040	3:32:41.08	−27:46:42.45	4.90	36. (13)	−22.41	1.47	2.65	0.12
9340	3:32:40.67	−27:45:56.11	4.71	49. (15)	−20.52	0.66	2.69	0.89
9487	3:32:40.17	−27:46:00.55	4.10	55. (17)	−21.98	0.62	2.36	0.11

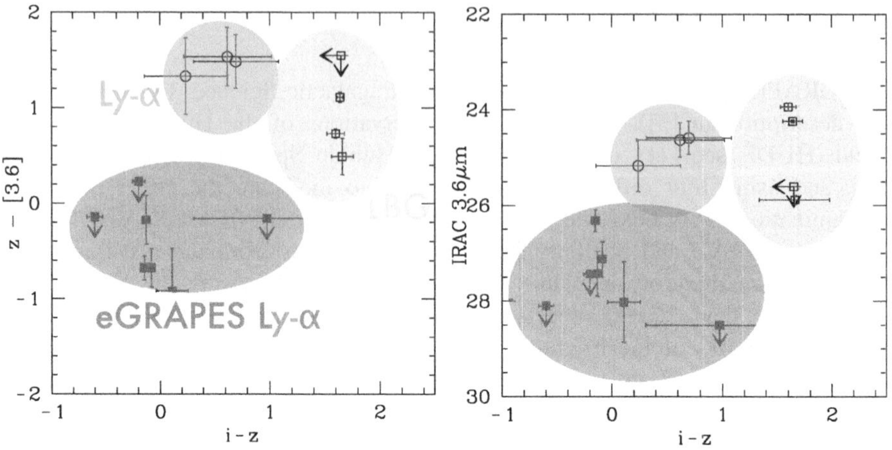

Fig. 42.3 (Color online) Colors of the GRAPES LAEs (*filled squares*) in z–3.6 μm vs. i–z space. These objects appear to be extremely blue, potentially being very young, low-extinction objects. Similarly, assuming low values of extinction (*right*), these sources appear to be fainter in the rest-frame optical (3.6 μm), and to have potentially low masses

individual source. When plotting rest-frame UV colors and rest-frame optical luminosities, as shown in Fig. 42.3, these sources are shown to be significantly bluer and fainter intrinsically than previously observed high redshift Ly-α sources or Lyman break galaxies.

42.3.2 SED Fitting and Results

We fitted the available broad band photometry available for these objects to an extensive series of Bruzual and Charlot (BC03, [2]) models. The redshifts of the sources were fixed to the spectroscopically determined redshift but extinction, metallicity, age, and mass were allowed to vary. We investigated several star formation histories and relied on the Spitzer observations to constraint the amount of older stars in these objects, and hence the mass contribution of these stars. The overwhelming conclusion is that all of these sources all have masses consistent with young (few 10^6 yr), low mass objects, with masses on the order of a few times 10^8 $M_\odot$.

42.4 Conclusion

As part of the GRAPES slitless spectroscopic survey of the HUDF, we have identified a sample of faint LAE galaxies at $4.0 < z < 5.7$. These objects were selected solely based on the detection of Ly-α emission in their low-resolution ACS slitless spectra. These spectra allowed us to identify LAEs down to extremely faint Ly-α

luminosities ($> 2.3 \times 10^{42}$ ergs s^{-1}). The GRAPES LAEs, with derived star formation rates of $\approx$8 $M_\odot$ yr^{-1} and stellar mass estimates that are likely to be no larger than a few times 10^8 $M_\odot$, must have formed quickly, unless their SFR was significantly lower earlier on. The LAE sources described in this paper are potentially the youngest and least massive galaxies observed to date, and at a time when our universe was just about $\approx$1 Gyr old.

References

1. Beckwith, S.V.W., et al., 2006. Astron. J. 132, 1729.
2. Bruzual, G., Charlot, S., 2003. Mon. Not. R. Astron. Soc. 344, 1000.
3. Pirzkal, N., et al., 2004. Astrophys. J. Suppl. Ser. 154, 501.
4. Xu, C., et al., 2007. Astron. J. 127, 213.

Chapter 43
The Role of *HST* in the Study of Near- and Mid-infrared-selected Galaxies

Karina I. Caputi

Abstract Because of their unique quality, *Hubble Space Telescope* (*HST*) data have played an important complementary role in studies of infrared (IR) galaxies conducted with major facilities, as *VLT* or *Spitzer*, and will be as well very valuable for future telescopes as *Herschel* and *ALMA*. I review here some of the most recent works led by European astronomers on IR galaxies, and discuss the role that *HST* has had in the study of different IR galaxy populations. I particularly focus the analysis on the GOODS fields, where the multiwavelength data and unique *HST* coverage have enabled to jointly put constraints on the evolution of star formation activity and stellar-mass growth with cosmic time.

43.1 Introduction

Studies of galaxy populations much benefit from the coordinated efforts of multiwavelength observations, which allow for a deeper insight into different galaxy properties and evolution. One of the best examples of these coordinated efforts is the *Great Observatories Origins Deep Survey* (*GOODS*) (P.I.: M. Giavalisco), with 320 arcmin^2 of the sky fully covered with deep observations from X-rays through radio wavelengths. Both North and South GOODS fields have been designed to be a unique case of *HST* coverage: they have deep and homogeneous maps in four broad-band *Advanced Camera for Surveys* (*ACS*) filters B, V, I_{775} and z_{850}. The large amount of science produced over the last years using GOODS data demonstrates the importance of *HST* Treasury and other major-telescope Legacy programs.

Near and mid-IR selected galaxies constitute quite unbiased tracers of two different astrophysical properties: stellar mass and star-formation/AGN activity, respectively. Although, in general, these galaxies populations are studied separately, a joint analysis of results allows to better understand the interplay between star formation, AGN activity and stellar-mass growth at different cosmic times. In the study of near- and mid-IR-selected galaxies, deep *HST* data have had a major role in the following aspects: (1) the determination of optical spectral energy distributions (SEDs), (2) the computation of precise ($\sigma \approx 3$–5%) photometric redshifts, as a complement

K.I. Caputi (✉)
Institute for Astronomy, Swiss Federal Institute of Technology (ETH Hönggerberg), Wolfgang-Pauli-Strasse 16, 8093, Zurich, Switzerland
e-mail: caputi@phys.ethz.ch

F.D. Macchetto (ed.), *The Impact of HST on European Astronomy*, Astrophysics and Space Science Proceedings,
DOI 10.1007/978-90-481-3400-7_43, © Springer Science+Business Media B.V. 2010

to existing spectroscopic redshifts and (3) galaxy morphology. The two former aspects have been of key importance to conduct complete studies of galaxy luminosity evolution and stellar mass assembly.

43.2 Near-IR-selected Galaxies

An important landmark in the study of near-IR-selected galaxies has been the K20 survey [8], which included part of the GOODS-South among their fields. The first science results obtained by the K20 survey have soon been extended after the progressive public releases of the deeper *ISAAC-VLT* J_s, H and K_s-band data in the GOODS-South (P.Is.: C. Cesarsky, E. Giallongo). Finally, after the launch of *Spitzer* in 2003 (P.I.: M. Werner), both GOODS fields have been observed with *IRAC* at 3.6, 4.5, 5.8 and 8.0 μm (P.I.: M. Dickinson).

At the typical depths of the GOODS data (21-22 Vega mag), K_s-selected galaxies mostly span the redshift range $z = 0$–1.5, but ~20% of these galaxies are found at higher $z = 1.5$–5.0 redshifts [4, 6, 17]. The latter include Extremely Red Galaxies (ERGs), which have been the subject of many studies looking for the progenitors of present-day massive ellipticals [3, 4, 23, 24, 27].

The evolution of the rest-frame K_s-band luminosity function (LF) has been studied up to $z \approx 1.5$ with the K20 survey [20], and then extended up to redshift $z \approx 2.5$ using GOODS and other datasets [1, 4, 6, 26]. All these works agree in a mild but still significant increase of the bright end of this LF from $z \approx 0$ to $z \approx 2.0$–2.5. At the same time, the overall number density of galaxies decreases, making the near-IR luminosity density to be nearly constant with redshift (cf. Fig. 43.1).

Multiwavelength SED modelling has also allowed to obtain stellar-mass estimates for near-IR-selected galaxies. The redshift evolution of galaxy number densi-

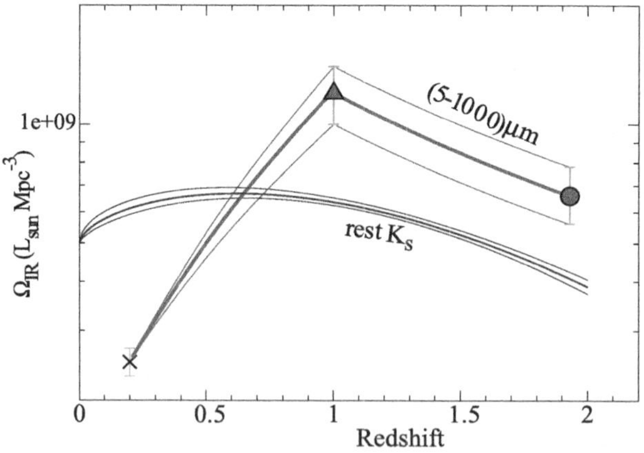

Fig. 43.1 The evolution of the IR luminosity densities associated with the stars already locked in galaxies (rest-K_s) and with on-going star formation (5–1,000 μm). The cosmology adopted has $H_0 = 70$ km s^{-1} Mpc^{-1}, $\Omega_M = 0.3$ and $\Omega_\Lambda = 0.7$

ties is differential with stellar mass: while the density of moderately massive galaxies ($M \sim 1 \times 10^{11}$ $M_{\odot}$) continuously increases from redshift $z \sim 4$ down to $z \sim 0.5$, virtually all of the most massive ($M > 2$–3×10^{11} $M_{\odot}$) galaxies appear to be in place by $z \sim 1.5$–2.0 [6, 14, 21, 22, 25, 28]. The formation epoch of the most massive galaxies seems to be constrained to the redshift range $z \sim 2$–4, as no evidence of such galaxies has been found at $z > 4$ in the GOODS-South field, indicating that these objects must be very rare (if any exists) at very high redshifts [12].

Moreover, during the last years, a general consensus has been achieved on the redshift evolution of the total stellar mass density: half of the stellar mass has been assembled in galaxies before the Universe was 30–40% of its present age [4, 6, 15].

43.3 Mid-IR-selected Galaxies and Their Link to Near-IR Galaxies

The study of mid-IR selected galaxies has been a subject of major interest in European astronomy since the launch of the *Infrared Space Observatory* (*ISO*), e.g. [2, 13, 16]. The amount of energy contained in the extragalactic IR background ($\lambda \approx 5$–$1{,}000$ μm) is comparable to that in the optical background [11], which emphasizes the importance of understanding the role of mid- and far-IR galaxy populations in the general context of galaxy evolution.

After *ISO* set the first constraints on the mid-IR Universe up to redshift $z \sim 1$, *Spitzer* has allowed us to prove that mid-IR galaxies have in fact a significant contribution up to higher $z \sim 2$–3 redshifts, e.g. [5]. A sort of 'downsizing' effect is observed in the typical stellar masses of the galaxies hosting the bulk of IR activity: at $z \sim 2$, most of the IR activity is found in $\gtrsim 10^{11}$ $M_{\odot}$ galaxies, while at $z = 0.5$–1.0, an important fraction of the IR activity is concentrated in intermediate-mass galaxies, with $10^{10} < M < 10^{11}$ $M_{\odot}$ [5, 18].

A result that might have come as a surprise is the large fraction of massive galaxies that are experiencing an ultra-luminous IR phase at high redshifts [6, 9]. Very recent results suggest that the AGN contribution to this IR phase at $z \sim 2$ might be much more important than previously thought [10].

The mid-IR LF strongly evolves from $z = 0$ to 1 [7, 19]. But, after removing to our best current knowledge the AGN contribution at high redshifts, the mid-IR LF associated with star formation shows only a modest evolution from $z = 1$ to $z = 2$ [7]. This is reflected in the observed evolution of the bolometric IR (5–1,000 μm) luminosity density, as shown in Fig. 43.1. From this figure, it is clear that the amount of energy associated with on-going star formation is significantly larger than the light produced by the stars already locked in galaxies at $z \gtrsim 0.6$–0.7, while the balance is reversed at lower redshifts. The (5–1,000) μm luminosity related to star formation can be interpreted as the gradient of the evolution of the assembled stellar mass. By $z \sim 0.7$, 80% of the present-day stellar mass is already in place [6], so there is only a minor amount of star formation still needed afterwards to finish assembling our local Universe.

References

1. Arnouts, S., Walcher, C.J., Le Fèvre, O., et al., 2007. Astron. Astrophys. 476, 137.
2. Aussel, H., Cesarsky, C.J., Elbaz, D., et al., 1999. Astron. Astrophys. 342, 313.
3. Caputi, K.I., Dunlop, J.S., McLure, R.J., Roche, N.D., 2004. Mon. Not. R. Astron. Soc. 353, 30.
4. Caputi, K.I., Dunlop, J.S., McLure, R.J., Roche, N.D., 2005. Mon. Not. R. Astron. Soc. 361, 607.
5. Caputi, K.I., Dole, H., Lagache, G., et al., 2006. Astrophys. J. 637, 727.
6. Caputi, K.I., Dole, H., Lagache, G., et al., 2006. Astron. Astrophys. 454, 143.
7. Caputi, K.I., Lagache, G., Yan, L., et al., 2007. Astrophys. J. 660, 97.
8. Cimatti, A., Mignoli, M., Daddi, E., et al., 2002. Astron. Astrophys. 392, 395.
9. Daddi, E., Dickinson, M., Chary, R., et al., 2005. Astrophys. J. 631, L13.
10. Daddi, E., Alexander, D.M., Dickinson, M., et al., 2007. Astrophys. J. 670, 173.
11. Dole, H., Lagache, G., Puget, J.-L., et al., 2006. Astron. Astrophys. 451, 417.
12. Dunlop, J.S., McLure, R.J., Cirasuolo, M., 2007. Mon. Not. R. Astron. Soc. 376, 1054.
13. Franceschini, A., Aussel, H., Cesarsky, C.J., et al., 2001. Astron. Astrophys. 378, 1.
14. Franceschini, A., Rodighiero, G., Cassata, P., et al., 2006. Astron. Astrophys. 453, 397.
15. Fontana, A., Pozzetti, L., Donnarumma, I., et al., 2004. Astron. Astrophys. 424, 23.
16. Genzel, R., Cesarsky, C.J., 2000. Annu. Rev. Astron. Astrophys. 38, 761.
17. Grazian, A., Fontana, A., de Santis, C., et al., 2006. Astron. Astrophys. 449, 951.
18. Hammer, F., Flores, H., Elbaz, D., et al., 2005. Astron. Astrophys. 430, 115.
19. Le Floc'h, E., Papovich, C., Dole, H., et al., 2005. Astrophys. J. 632, 169.
20. Pozzetti, L., Cimatti, A., Zamorani, G., et al., 2003. Astron. Astrophys. 402, 837.
21. Pozzetti, L., Bolzonella, M., Lamareille, F., et al., 2007. Astron. Astrophys. 474, 443.
22. Renzini, A., 2006. Annu. Rev. Astron. Astrophys. 44, 141.
23. Roche, N.D., Dunlop, J.S., Almaini, O., 2003. Mon. Not. R. Astron. Soc. 346, 803.
24. Roche, N.D., Dunlop, J.S., Caputi, K.I., et al., 2006. Mon. Not. R. Astron. Soc. 370, 74.
25. Saracco, P., Longhetti, M., Severgnini, P., et al., 2005. Mon. Not. R. Astron. Soc. 357, L40.
26. Saracco, P., Fiano, A., Chincarini, G., et al., 2006. Mon. Not. R. Astron. Soc. 367, 349.
27. Smail, I., Owen, F.N., Morrison, G.E., et al., 2002. Astrophys. J. 581, 844.
28. Thomas, D., Maraston, C., Bender, R., et al., 2005. Astrophys. J. 621, 673.

Chapter 44
Large Scale Structure and Galaxy Evolution in COSMOS

Nick Scoville and COSMOS Team

Abstract Galaxy evolution and AGN growth in the early universe are believed to be strongly driven by merging (hierarchical growth) and galaxy dynamical interactions. Thus, a full exploration of the environmental influences is absolutely essential to understanding this early evolution. The Cosmic Evolution Survey (COSMOS; Scoville et al. in Astrophys. J. Suppl. Ser. 172:1, 2007) is specifically designed to probe the correlated coevolution of galaxies, star formation, active galactic nuclei (AGN) and dark matter (DM) large-scale structures (LSS) over the redshift range $z > 0.5$–3. The survey includes multi-wavelength imaging and spectroscopy from X-ray to radio wavelengths covering a 2 deg^2 equatorial field. Given the very high sensitivity and resolution of these datasets, COSMOS also provide unprecedented samples of objects at $z > 3$ and will be a fundamental resource for future cosmology studies. I review the characteristics of the COSMOS survey and show exciting initial results mapping large scale structures in galaxies and dark matter.

44.1 Introduction

Our understanding of the formation and evolution of galaxies and their large-scale structures (LSS) has advanced enormously over the last decade — a result of a phenomenal synergy between theoretical and observational efforts. Deep observational studies using the Hubble Space Telescope (HST) and the largest ground based telescopes have probed galaxy and AGN populations back to redshift $z = 6$ when the universe had aged less than 1 billion of its current 13 billion years. Just as remarkable is the enormous success of numerical simulations for ΛCDM models in reproducing many of the current LSS characteristics, all starting from an initial, nearly uniform, hot universe!

44.1.1 The Cosmic Evolution Survey — COSMOS

The COSMOS survey is the first survey encompassing a sufficiently large area that it can address the coupled evolution of LSS and galaxies, star formation and AGN.

N. Scoville (✉)
Astronomy Department, Caltech 105-24, Pasadena, CA 91125, USA
e-mail: nzs@astro.caltech.edu

F.D. Macchetto (ed.), *The Impact of HST on European Astronomy*,
Astrophysics and Space Science Proceedings,
DOI 10.1007/978-90-481-3400-7_44, © Springer Science+Business Media B.V. 2010

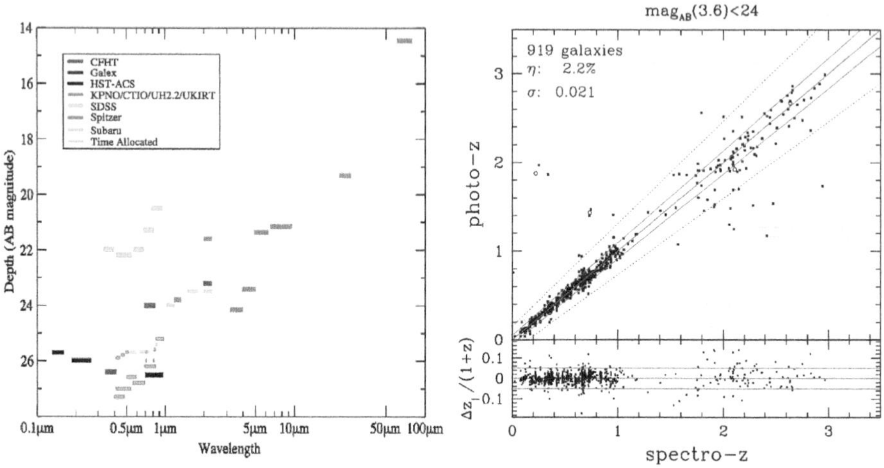

Fig. 44.1 (Color online) *Left*: The 5σ sensitivities are shown for the UV-optical and IR bands in COSMOS (3 arcsec apertures except ACS 0.15 arcsec — [2]). *Right*: Comparison of the newly derived COSMOS photo-z's using 20 bands [7] with the spectroscopic redshifts [9] for 2,100 galaxies with 3.6 μm < 24 mag_{AB} (some with I_{AB} down to 26 mag) indicates $\sigma_z/(1+z) < 0.02$

COSMOS is the largest HST survey ever undertaken — imaging an equatorial, 2 deg^2 field with single-orbit I-band exposures to a depth of $I_{AB} = 28$ mag (5σ) (see Fig. 44.1). Extensive multi-λ ground and space-based observations of this field have been gathered spanning the entire spectrum from X-ray, UV, optical/IR, mid-infrared, mm/submm and to radio with extremely high sensitivity imaging and spectroscopy. This full spectrum approach is required to probe the coupled evolution of young and old stellar populations, starbursts, the ISM (molecular and ionized components), AGN and dark matter. Each of these cosmic components may be best probed quite differently. The multi-λ approach is also required due to the differential redshifting of cosmic history and the presence of dust obscuration in many of the most rapidly evolving galactic regions. The large area coverage of COSMOS is motivated to sample the largest structures existing in the local universe — smaller area coverage can lead to severe cosmic variance problems.

COSMOS detects $\simeq 2 \times 10^6$ galaxies and AGN sampling a volume in the high redshift universe ($z > 0.5$–4) approaching that sampled locally by the Sloan Digital Sky Survey (SDSS). Subaru optical imaging and photometric redshifts have been determined for approximately 800,000 galaxies. The COSMOS spectroscopic surveys (VLT and Magellan) will yield 35,000 galaxies with accurate redshifts at $z = 0.5$–2.5, all having 0.05 arcsec HST imaging.

44.2 COSMOS Science Goals

The COSMOS survey addresses nearly every aspect of observational cosmology over the majority of the Hubble time at $z \geq 0.5$–6:

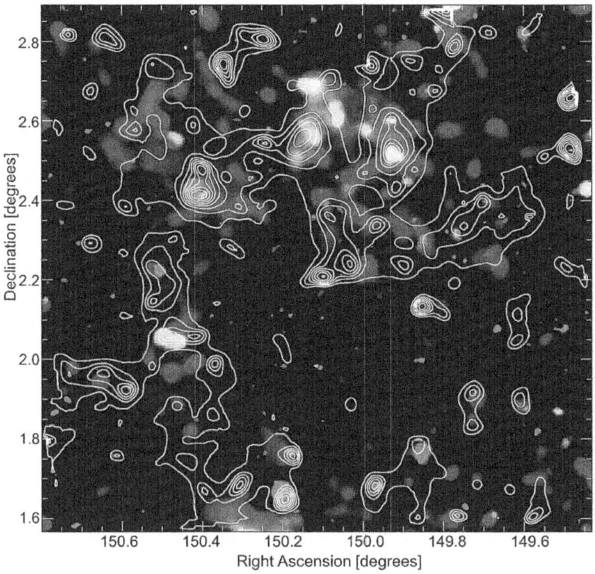

Fig. 44.2 (Color online) Comparison of the weak lensing dark matter mass map (*contours*) with galaxy LSS (*blue/grayscale*) and diffuse X-ray emission (*red/grayscale*) [10, 16]. This is the projected density distribution **obtained from the full 3-d distribution** derived using photometric redshifts for line of sight discrimination. Major structures shown here occur at $z = 0.25$, 0.5, 0.73 and 1.0 with sizes up to ~40 Mpc and total masses up to 10^{15} $M_\odot$, similar to that of the COMA cluster

- the assembly of galaxies, clusters and dark matter up to 2×10^{14} $M_\odot$;
- reconstruction of the dark matter distributions and characteristics out to $z \sim 1$ using weak gravitational lensing at $z < 1.5$;
- the evolution of galaxy morphology, galactic merger rates and star formation as a function of LSS environment and redshift;
- evolution of AGN and the dependence of black hole growth on galaxy morphology and environment; and
- the mass and luminosity distribution of the earliest galaxies, AGN and intergalactic gas at $z = 3$–6 and their clustering.

The growth of galaxies, AGN and dark matter structure is traced in COSMOS over a period corresponding to ~75% of the age of the universe. Galaxies in the early universe are built up by two major processes: dissipational collapse and merging of lower mass protogalactic and galactic components. Their intrinsic evolution is then driven by the conversion of primordial and interstellar gas into stars, with galactic merging and interactions triggering star formation and starbursts.

The need to sample very large scales arises from the fact that structure occurs on mass scales up to $\geq 10^{14}$ $M_\odot$ and existing smaller surveys are likely to be unrepresentative at $z \sim 1$. Earlier projects, such as GOODS and GEMS, adequately sample masses up to 3×10^{13} $M_\odot$, whereas COSMOS samples the largest known structures at $\sim 2 \times 10^{14}$ $M_\odot$ (dark and luminous matter). COSMOS measures all galaxy populations in their large-scale context as a function of redshift, providing essential guidelines for the next generation of theoretical models (see Fig. 44.2).

44.3 COSMOS Observational Components

44.3.1 Galaxy Evolution: HST Imaging and SEDs

The evolutionary status of galaxies can be analyzed from either their morphologies or their spectral energy distributions (SED, characterizing the stellar population). Morphological parameters for the galaxies are obtained from the HST imaging (e.g. bulge/disk ratios, concentration, asymmetry, size, multiplicity, clumpiness). The COSMOS I-band HST images have sufficient depth and resolution to allow classical bulge-disk decomposition for L^* galaxies at $z \leq 2$, while less detailed structural parameters such as compactness, asymmetry, clumpiness and size can be measured for all galaxies down to the spectroscopic limit ($I_{AB} \sim 25$), out to $z \sim 5$. In COSMOS, deep imaging (from Subaru, GALEX, UKIRT and NOAO, and SPITZER-IRAC) provides SEDs to characterize the integrated stellar populations of the 2 million galaxies detected with HST. The SEDs are derived self-consistently with the photometric redshift determinations.

44.3.2 Galaxy Redshifts: Photometric and Spectroscopic

Determining the redshifts or lookback time of individual galaxies is clearly one of the most difficult and time consuming aspects of any cosmological evolution survey. In COSMOS this is even more difficult since the redshifts are needed with sufficient precision not just to determine the cosmic epoch, but also to place the galaxies within or outside of LSS appearing along the line of sight. Without high precision, structures become 'blurred' due to scattering of galaxies to different distances in the line of sight and for specific galaxies, their environment cannot be determined.

In COSMOS, photometric redshifts are obtained from deep (mostly ground-based) imaging — from Subaru, CFHT, UKIRT, and NOAO [1, 11, 17]. At present the photometric-redshift accuracy is $\sigma_z/(1+z) \sim 0.02$ for approximately 4×10^5 galaxies at $z < 2.5$ [7, 11]. These photometric redshifts are based on 20 bands of optical/IR data and the accuracy is unprecedented for photo-z's. It is sufficient to enable initial definition of the LSS, especially for the denser environments [16].

Very large spectroscopic surveys are now ongoing as part of COSMOS at the VLT and Magellan telescopes [8, 9]. The spectroscopic sample will eventually include approximately 35,000 galaxies and 1,000 AGN down to limits of $I_{AB} = 24.5$ and $I_B = 25.5$ mag. These spectroscopic samples will provide very precise definition of the environment, albeit for smaller subsets of the overall COSMOS galaxy population.

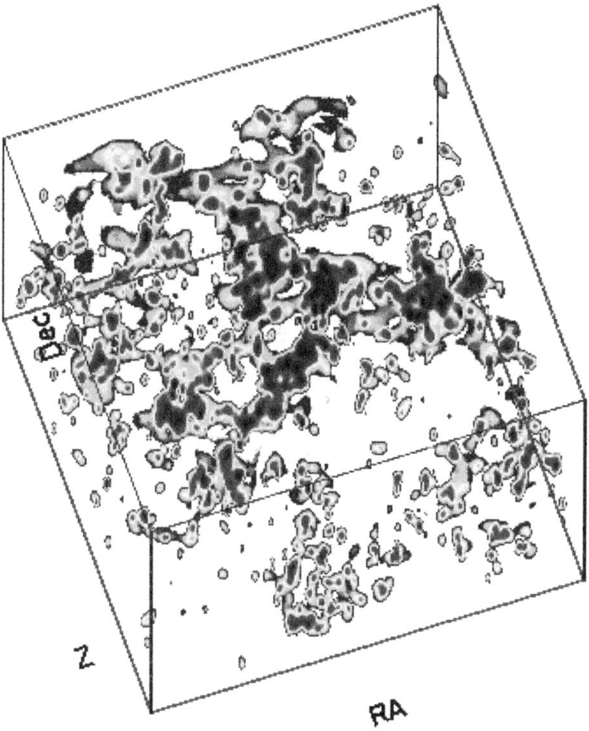

Fig. 44.3 (Color online) Three dimensional view (RA, DEC and redshift) of the large scale structures [16] seen in the galaxy overdensities obtained from 200,000 galaxies with COSMOS photometric redshifts at $z = 0.1$–1.5

44.3.3 Galaxy Overdensities and Weak Lensing

The environment or LSS in which a given galaxy resides is defined from the local number density of galaxies or from the DM density as determined from weak lensing or the local galaxy velocity dispersion. The COSMOS HST imaging provides measures of the close-in environment (from galaxy multiplicity and merger indicators such as tidal distortions) and larger-scale DM environment. As noted in Sect. 44.3.2, definition of the environment is critically dependent on moderately high accuracy redshifts (<2%).

The wide-area, uniform ACS and Subaru COSMOS surveys allow determination of spatial correlation functions as a function of type (morphological and SED) and luminosity and their evolution with redshift and environment. The enormous sizes of the samples which become available in COSMOS enable precision approaching that of SDSS but at much higher redshift.

A major goal of COSMOS has recently been realized — the first imaging of large scale structures (38 LSS on scales of 3–30 Mpc at $z = 0.2$–1.1) in both the dark matter from weak lensing [10] and in the baryons from galaxy overdensities [5, 16] and diffuse X-ray emission [4, 6]. In Fig. 44.3, a 3-d view of the large scale structures seen in the galaxy overdensities is shown as derived from the high accuracy photometric redshifts [16]. Large structures have also been seen in COSMOS at $z = 5.7$ in Lyα emitting sources [3, 13].

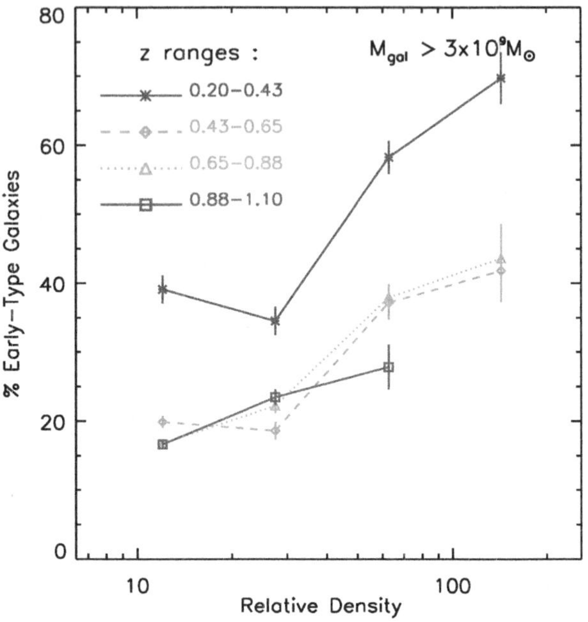

Fig. 44.4 (Color online) The percentage of early type galaxies is shown as a function of environmental density for four redshift bins at $z = 0.2$–1.1 [16]

The dependence of morphology and galactic spectral energy distribution (SED) on environment has been followed in the $z < 1.1$ structures and into the field population [2, 14, 16]. For example, Fig. 44.4 shows the percentage of galaxies with early type SEDs as a function of redshift and local environmental density (normalized relative to the mean at each redshift) [16]. With very small dispersions as a result of the large sample sizes and highly accurate SEDs, one clearly see the increase frequency of early types at low redshift and in denser environments. Similar correlations are seen in the median SFRs of galaxies (derived from the UV continuum) with respect to both redshift and epoch [12, 16].

44.4 COSMOS Multi-wavelength Surveys

The COSMOS field is located near the celestial equator ($\delta = 2°$) to ensure visibility by all astronomical facilities, especially unique instruments such as the next generation 20–30 m optical/IR telescope(s). The time requirements for deep imaging and spectroscopy over a total area of 2 deg^2, containing over a million galaxies makes it strategically imperative that the field be readily observable by all large optical/IR telescopes. For radio studies, high-declination fields such as Lockman Hole, HDF-North, Groth strip and CDF-South are ruled out — they can not be easily observed by *both* (E)VLA in the north and ALMA in the south.

The COSMOS field is accessible to essentially all astronomical facilities, enabling complete multi-λ datasets (X-ray, UV, optical/IR, FIR/submm to radio) as

Table 44.1 Multi-λ COSMOS data

Data	Bands, λ, Res.	AB mag 5σ pt. src	Investigators/Time
HST-ACS	814I	28.8	C12-13 581 o
HST-ACS	475g	28.15	C12 9 o
HST-NIC3	160W	25.6 (6% area)	C12-13 590 o
HST-WFPC2	300W	25.4	C12-13 590 o
Subaru-SCam	B, V, r′, i, z′, g′	27–26	Taniguchi et al. 10 n
Subaru-SCam	10 IB filters	26	Taniguchi et al. 16 n, Scoville
Subaru-SCam	NB816	25	Taniguchi et al. 2.5 n
CFHT-Megacam	u*	27	Sanders et al. 33 hr
CFHT-Megacam	u, i*	26	LeFevre et al. 12 hr
CFHT-LS	u–z		Deep LS Survey
LBT	Ugr	27	Giallongo et al. 48 hr
NOAO/CTIO	K_s	21	Mobasher et al. 18 n
CFHT/UKIRT	J, H, K	24.5–23.5	Sanders et al. 25 n
UH-88	J	21	Sanders et al. 22 n
GALEX	FUV, NUV	26.1, 25.8	Schminovich et al. 200 ks
XMM-EPIC	0.5–10 keV	10^{-15} cgs	Hasinger et al. 1.4 Ms
CXO	0.5–7 keV	0.9°	Elvis et al. 1.8 Ms
VLT-VIMOS sp.	$(R = 200)$	$\# = 3{,}000\ (I_{AB} < 23)$	Kneib et al. 20 hr
VLT-VIMOS sp.	$(R = 200, 600)$	$\# = 45{,}000\ (I_{AB} < 25, z \geq 0.8)$	Lilly et al. 540 hr
Mag.-IMAX sp.	$(R = 3{,}000)$	$\# = 2{,}000$	Impey, McCarthy, Elvis 30 n
Keck/GEMINI sp.	$(R = 5{,}000)$	$\# = 4{,}000\ (I < 24)$	10n Team Members
Spitzer-MIPS	160, 70, 24 μm	17, 1, 0.15 mJy (5σ)	Sanders et al. 450 hr
Spitzer-IRAC	8, 6, 4.5, 3 μm	11, 9, 3, 2 μJy (5σ)	Sanders et al. 170 hr
IRAM-MAMBO	1.2 mm	1 mJy (20 × 20 arcmin)	Bertoldi et al. 90 hr
CSO-Bolocam	1.1 mm	3 mJy	Aquirre et al. 40 n
JCMT-Aztec	1.1 mm	0.9 mJy (1σ)	Sanders et al. 5 n
VLA-A	20 cm	25 μJy (1σ)	Schinnerer et al. 60 hr
VLA-A/C	20 cm	10 μJy (1σ)	Schinnerer et al. 275 hr
VLA-A-1°	20 cm	7 μJy (1σ)	Schinnerer et al. 75 hr
SZA (full field)	9 mm	S-Z to $2 \times 10^{14}\ M_{\odot}$	Carlstrom et al. 2 mth

summarized in Table 44.1. The status of these observational programs is continuously updated on the COSMOS web-site:

http://www.astro.caltech.edu/~cosmos/.

The major COSMOS datasets become publicly available in staged releases (following calibration and validation) through the web site for IPAC/IRSA:

http://irsa.ipac.caltech.edu/data/COSMOS/

Acknowledgements I gratefully acknowledge the contributions of the entire COSMOS collaboration consisting of more than 100 scientists.The HST COSMOS Treasury program was supported through NASA grant HST-GO-09822.

References

1. Capak, P., et al., 2007. Astrophys. J. Suppl. Ser. 172, 99.
2. Capak, P., et al., 2007. Astrophys. J. Suppl. Ser. 172, 284.
3. Capak, P., et al., 2008. Astrophys. J. 681, L53.
4. Finoguenov, A., et al., 2007. Astrophys. J. Suppl. Ser. 172, 182.
5. Guzzo, L., et al., 2007. Astrophys. J. Suppl. Ser. 172, 254.
6. Hasinger, G., et al., 2007. Astrophys. J. Suppl. Ser. 172, 29.
7. Ilbert, O., et al., 2008. In: Kodama, T., Yamada, T., Aoki, K. (Eds.), Panoramic Views of Galaxy Formation and Evolution, ASP Conference Series, vol. 399, p. 169. ASP, San Francisco.
8. Impey, C.D., et al., 2007. In: Karas, V., Matt, G. (Eds.), Black Holes from Stars to Galaxies – Across the Range of Masses, IAU Symp., vol. 238, p. 287. Cambridge University Press, Cambridge.
9. Lilly, S., 2007. Astrophys. J. Suppl. Ser. 172, 70.
10. Massey, R., et al., 2007. Nature 445, 286.
11. Mobasher, B., et al., 2007. Astrophys. J. Suppl. Ser. 172, 117.
12. Mobasher, B., et al.., 2009. Astrophys. J. 690, 1074.
13. Murayama, T., et al., 2007. Astrophys. J. Suppl. Ser. 172, 523.
14. Scarlata, C., et al., 2007. Astrophys. J. Suppl. Ser. 172, 494.
15. Scoville, N., et al., 2007. Astrophys. J. Suppl. Ser. 172, 1.
16. Scoville, N., et al., 2007. Astrophys. J. Suppl. Ser. 172, 150.
17. Taniguchi, Y., et al., 2007. Astrophys. J. Suppl. Ser. 172, 9.

Chapter 45
Mass Estimations of Supermassive Black Holes in Brightest Cluster Galaxies

Elena Dalla Bontà, Laura Ferrarese, Enrico Maria Corsini, Jordi Miralda-Escudé, Lodovico Coccato, and Alessandro Pizzella

Abstract We observed a sample of three Brightest Cluster Galaxies (BCGs), Abell 1836-BCG, Abell 2052-BCG, and Abell 3565-BCG, with the Advanced Camera for Surveys (ACS) and the Imaging Spectrograph (STIS) on board the Space Telescope. For each target galaxy we obtained high-resolution spectroscopy of the Hα and [N II] λ6583 emission lines at three slit positions, to measure the central ionized-gas kinematics. ACS images in three different filters (F435W, F625W, and FR656N) have been used to determine the optical depth of the dust, stellar mass distribution near the nucleus, and intensity map. We present supermassive black hole (SBH) mass estimates for two galaxies which show regular rotation curves and strong central velocity gradients, and an upper limit on the SBH mass of the third one. For the SBHs of Abell 1836-BCG and Abell 3565-BCG, we derived $M_{\bullet} = 4.8^{+0.8}_{-0.7} \times 10^9$ M$_{\odot}$ and $M_{\bullet} = 1.3^{+0.3}_{-0.4} \times 10^9$ M$_{\odot}$ at 1σ confidence level, respectively. For the SBH of Abell 2052-BCG, we found $M_{\bullet} \leq 7.3 \times 10^9$ M$_{\odot}$.

The link between the evolution of SBHs and the hierarchical build-up of galaxies is imprinted in a number of scaling relations connecting SBH masses to the global properties of the host galaxies. The high SBH mass end of these relations has yet to be fully explored: the massive galaxies expected to host the most massive SBHs are generally at large distances, making a dynamical detection of SBHs observationally challenging. This is unfortunate, since black holes with mass in excess of 10^9 M$_{\odot}$ occupy an integral part in our understanding of the co-evolution of SBHs

E. Dalla Bontà (✉) · E.M. Corsini · A. Pizzella
Università degli Studi di Padova, Dipartimento di Astronomia, Vicolo dell'Osservatorio, 3, 35122 Padova, Italy
e-mail: elena.dallabonta@unipd.it

L. Ferrarese
Herzberg Institute of Astrophysics, Victoria, Canada

J. Miralda-Escudé
Institut de Ciencies de l'Espai (CSIC-IEEC)/ICREA, Bellaterra, Spain

L. Coccato
Max-Planck-Institut fuer extraterrestrische Physik, Garching bei Muenchen, Germany

F.D. Macchetto (ed.), *The Impact of HST on European Astronomy*, Astrophysics and Space Science Proceedings,
DOI 10.1007/978-90-481-3400-7_45, © Springer Science+Business Media B.V. 2010

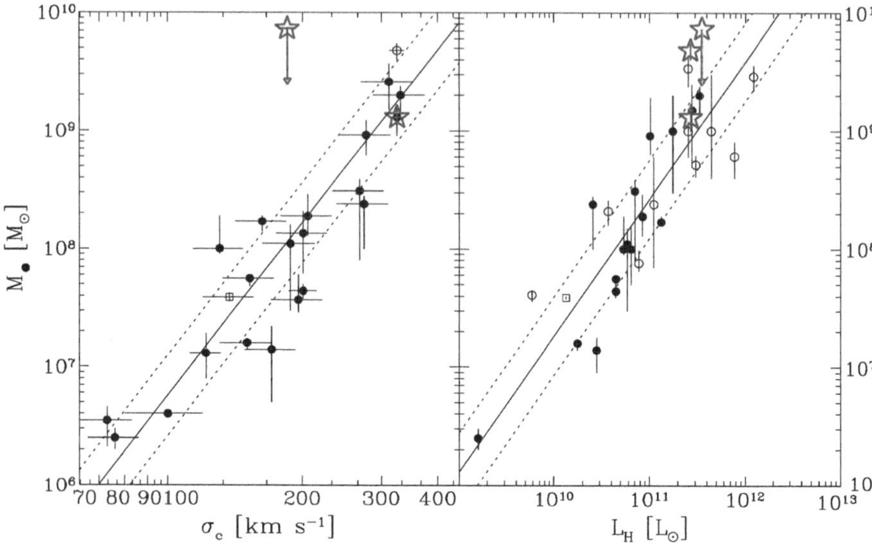

Fig. 45.1 (Color online) Location of the SBHs masses of our BCG sample galaxies (*star symbols*) with respect to the $M_\bullet$–σ_c relation of [2] (*left panel*), and near-infrared $M_\bullet$–$L_{\rm bulge}$ relation of [4] (*right panel*). No σ_c is yet available for Abell 1836-BCG

and galaxies: these are the systems that have undergone the most extensive and protracted history of merging; moreover, they represent the local relicts of the high redshift quasars detected in optical surveys. To investigate the high-mass end of the SBH mass function, we selected three BCGs from the sample of [3]. Their large masses, luminosities and stellar velocity dispersions, as well as their having a merging history which is unmatched by galaxies in less crowded environments, make these galaxies the most promising hosts of the most massive SBHs in the local Universe.

Following [1], a model of the gas velocity field for each galaxy is generated by assuming that the ionized-gas component is moving onto circular orbits in an infinitesimally thin disc centered at the galactic nucleus. The model is projected onto the plane of the sky for a grid of assumed inclination angles of the gaseous disk. Finally, the model is brought to the observational plane by accounting for the width and location (namely position angle and offset with respect to the disk center) of each slit, for the point spread function of the STIS instrument, and for the effects of charge bleeding between adjacent CCD pixels. The mass of the SBH is determined by finding the model parameters (SBH mass, inclination of the gas disk, and mass-to-light ratio of the stellar component) that produce the best match to the observed velocity curves. For Abell 1836-BCG, we derive $M_\bullet = 4.8^{+0.8}_{-0.7} \times 10^9$ M$_\odot$, the largest SBH mass to have been dynamically measured to-date. The best fit inclination angle is $i = 76^\circ \pm 1^\circ$, while only an upper limit to the stellar mass-to-light ratio is found ($M/L_I \leq 4.0$ M/L$_{I_\odot}$ at 1σ confidence level). For Abell 3565-BCG we determine $M_\bullet = 1.3^{+0.3}_{-0.4} \times 10^9$ M$_\odot$, with $i = 50^\circ \pm 1^\circ$ and $M/L_I = 9.0 \pm 0.8$ M/L$_{I_\odot}$

(1σ confidence level). In the case of Abell 2052-BCG we find $M_{\bullet} \leq 7.3 \times 10^9$ $M_{\odot}$, following the method by [5]. In Fig. 45.1 we show the location of our SBH masses determinations in the near-infrared $M_{\bullet}$–$L_{\rm bulge}$ relation of [4] and in the $M_{\bullet}$–σ_c relation, as given in [2].

References

1. Coccato, L., et al., 2006. Mon. Not. R. Astron. Soc. 366, 1050.
2. Ferrarese, L., Ford, H., 2005. Space Sci. Rev. 116, 523.
3. Laine, S., et al., 2003. Astrophys. J. 125, 478.
4. Marconi, A., Hunt, L.K., 2003. Astrophys. J. 589, L21.
5. Sarzi, M., et al., 2002. Astrophys. J. 567, 237; Mon. Not. R. Astron. Soc. 313, 469.

Chapter 46
Near-UV Study of Active Galactic Nuclei with Advanced Camera for Surveys

V.M. Muñoz Marín, R.M. González Delgado,
H.R. Schmitt, R. Cid Fernandes, and E. Pérez

Abstract During the last decade spectroscopic and high resolution HST observations have established a clear link between the starburst phenomenon and the activity of the central AGN. We report here on our investigation of the nature of the near-UV light in these objects, its relation to the circumnuclear starburst phenomenon, and the connection of this to the evolution and growth of the galaxy bulge and central black hole.

During the last decade, spectroscopic and high resolution HST observations have established a clear link between the starburst phenomenon and the activity of the central AGN (see [1] and references therein). Whether there is a causal link between them, or they just happen to coexist, as they are fed by the same mechanisms, is something to be determined. In addition, the physical processes operating in the central regions of Low Luminosity AGN (LLAGN) are still nowadays a matter of debate. Its ionizing source could either be a low efficiently accreting super-massive black hole, as well as a central ionizing massive star cluster. In a number of objects, both processes may coexist and play a similarly important role in the galactic nuclei evolution. Our goal is to investigate the nature of the near-UV light in these objects, its relation to the circumnuclear starburst phenomenon, and the connection of this to the evolution and growth of the galaxy bulge and central black hole.

We have constructed an useful image catalogue of the inner region of a large sample of 75 nearby Sy galaxies (14 Sy1; 14 Sy1.8-1.9; 47 Sy2) at a mean distance of 57 Mpc. These galaxies are imaged in F330W with ACS at High Resolution Configuration (HRC) (Fig. 46.1 shows some examples). At this wavelength we are able to best separate the contribution of the young and intermediate stellar population, from the redder host galactic bulge emission. In addition, the HRC configuration

V.M. Muñoz Marín · R.M. González Delgado · E. Pérez
Instituto de Astrofísica de Andalucía (CSIC), Granada, Spain

H.R. Schmitt
Remote Sensing Division, Naval Research Laboratory, Washington, DC, USA

R. Cid Fernandes
Depto. de Física-CFM, Universidade Federal de Santa Catarina, Florianópolis, SC, Brazil

F.D. Macchetto (ed.), *The Impact of HST on European Astronomy*,
Astrophysics and Space Science Proceedings,
DOI 10.1007/978-90-481-3400-7_46, © Springer Science+Business Media B.V. 2010

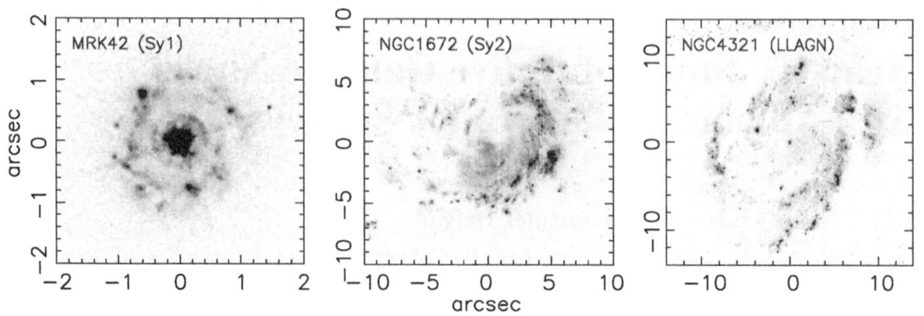

Fig. 46.1 HST ACS (F330W) images of Mrk42, a Sy1 galaxy, NGC 1672, a Sy2, and NGC 4321, a LLAGN

of ACS allows us to identify individual star clusters and separate their contribution from the diffuse background emission.

We estimate the size and the luminosity of the emitting regions and extract the luminosity profile, with magnitudes and surface brightness being tabulated at several radii. We also determine the presence of unresolved compact nuclei through PSF comparison. We calculate the asymmetry (A) and compactness (C) parameters. In addition, the circumnuclear stellar cluster population is identified, and the contribution of the stellar clusters to the total light, at this wavelength, is estimated.

We find that Seyfert 1 galaxies are completely dominated by its bright and compact nucleus, that remains point-like at this resolution, while we find almost no unresolved nucleus in Seyfert 2. The Seyfert types 1 and 2 are quite segregated in an A versus C plot, being Sy1 very compact and having low asymmetry values, while Sy2 show a very wide range of C and A. Stellar clusters are found somewhat more frequently in Sy2 (in ~70% of the galaxies) than in Sy1 (~57%), and contribute more to the total light in Sy2, but this seems to be mostly due to the large contribution of the compact nucleus in Sy1. When subtracting the contribution of the compact nucleus no clear difference is seen between different Sy type hosts. A more extended discussion of these results can be found in [1].

We have also constructed a LLAGN sample consisting in 26 objects (18 classified as LINER and eight as Transition Objects, or TO) at a mean distance of 19 Mpc. This sample has also been imaged with ACS-HRC and the filter F330W (see one example in Fig. 46.1). We are carrying out an aperture photometry analysis similar to the one described for Seyferts. In order to model better the extended emission we are also conducting a surface brightness analysis with elliptical isophotal fitting. In our preliminary results, LLAGN show similar values of C and absolute magnitudes in the inner 300 pc, to those of Sy2. No difference is seen, when considering TOs and LINER separately. We find a conspicuous unresolved nucleus in about 25% of the objects.

As the future goals of the project, we aim to determine which is the dependency of the frequency of compact unresolved nuclei, and the fraction of light coming from star clusters, on the nuclear activity classification. Also, we plan to characterize the properties of the circumnuclear cluster population (size, luminosity, mass, age, etc.).

This will allow us to study the possible existence of an evolutionary sequence among different activity types.

References

1. Muñoz Marín, V.M., González Delgado, R.M., Schmitt, H.R., Cid Fernandes, R., Pérez, E., Storchi-Bergmann, T., Heckman, T., Leitherer, C., 2007. Astron. J. 134, 648.

Chapter 47
A Quantitative Analysis of the Morphology of Star Formation in a Sample of GOODS-HST/ACS Galaxies

Ruymán Azzollini, J.E. Beckman, and Leonel Gutiérrez Albores

Abstract Here we present the methodology we have devised to study the spatial distribution of Massive Star Formation in galaxies. We show results of our analysis performed on a sample of 4,503 galaxies in a range of redshifts extending up to $z \sim 1.2$, in the Chandra Deep Field South. We use data from GOODS-HST and COMBO17. The aim is to produce techniques which can be used to help decide between different basic models of galaxy formation using the spatial distribution of star formation within galaxies as the test parameter. In situ knowledge of the star formation distribution as a function of redshift will be a valuable counterpart to population synthesis studies which at best integrate this kind of information over long periods.

We have used five parameters which describe the morphology of a distribution of light, and which have been applied before to the study of galaxy morphology: Concentration ("C", [3]), Asymmetry ("A", [1]), Clumpiness ("S", [5]), the Gini parameter ("G", [2]), and M20 [7]. We performed photometry on the GOODS-HST/ACS [6] images, in all four bands: B435w, V606w, i775w and z850LP, and matched the resulting catalog to the photometric redshifts derived from COMBO17 data, published in [9]. This is our parent sample, composed of 4,572 objects with measured redshift, of which, 4,323 (95%) have $z < 1.1$. We discard 69 objects (1.5%), as being suspect of hosting an AGN, or being QSOs of some kind, according to [8]. The final sample, with 4,503 objects, has been analysed morphologically, using our own software. We have also used data from COMBO17 and the KCORRECT software [4], to obtain rest-frame absolute magnitudes in several bands from 1,500 to 6,000 A and ultimately to estimate stellar mass (M*), Mean Surface Mass density (μ^*), Star Formation Rate (SFR) and specific SFR (sSFR) for all these objects.

In Fig. 47.1 we show the parameters for objects at different redshifts, but shifting the observation band, in such a way, that we keep track of an invariant region of the

R. Azzollini · J.E. Beckman · L. Gutiérrez Albores
Instituto de Astrofísica de Canarias, 38205 La Laguna, S/C de Tenerife, Spain

J.E. Beckman
Centro Superior de Investigaciones Científicas, Serrano, 117, 28006 Madrid, Spain

F.D. Macchetto (ed.), *The Impact of HST on European Astronomy*,
Astrophysics and Space Science Proceedings,
DOI 10.1007/978-90-481-3400-7_47, © Springer Science+Business Media B.V. 2010

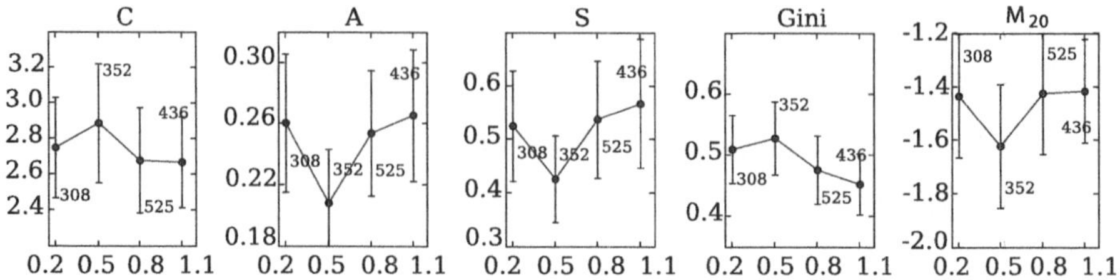

Fig. 47.1 (Color online) Morphological parameters as a function of redshift. *Points* are averages of values at redshift bins of $\Delta z = 0.2$ (*error bars* show σ, and *numbers* population of sample). In every bin, the band of observation has been shifted in order to track the same region of the rest-frame spectrum, at around 3,600 A

rest-frame spectrum, at around 3,600 A. In this way we avoid the need for morphological K-corrections in order to interpret this graph in terms of spatial distribution of SF. We see that the changes are not very significant, and in every case, compatible with no evolution, inside the error bars. We also present an analysis of the change of morphological parameters with Signal to Noise Ratio (SNR), and SNR with redshift, which, though not conclusively, indicates that a significative amount of the observed variations could be due to a mere effect of the variation of the SNR of objects with observing band and redshift. Considerable subsequent effort is needed to measure and decontaminate the results from effects of observational biases. In any case, it seems reasonable to state that if there is any change in the spatial distribution of the SF in the range of redshifts $0.2 \leq z \leq 2$, it is a small one, in comparison to the usual range within which these parameters vary for galaxies at longer wavelengths.

We also studied the five morphological parameters as a function of M^*, μ^*, SFR and sSFR, for objects in the range of redshifts $0.7 < z < 1.2$. No dependencies of the morphological parameters with the physical parameters are observed in B435w, which at this redshift is dominated by newborn stars.

References

1. Abraham, R.G., Tanvir, N.R., Santiago, B.X., Ellis, R.S., Glazebrook, K., van den Bergh, S., 1996. Mon. Not. R. Astron. Soc. 279, L47.
2. Abraham, R., van den Bergh, S., Nair, P., 2003. Astrophys. J. 588, 218.
3. Bershady, M., Jangren, A.A., Conselice, C.J., 2000. Astron. J. 119, 2645.
4. Blanton, M.R., et al., 2003. Astron. J. 125, 2348.
5. Conselice, C.J., 2003. Astrophys. J. Suppl. Ser. 147, 1.
6. Giavalisco, M., et al., 2004. Astrophys. J. 600, 93.
7. Lotz, J.M., Primack, J., Madau, P., 2004. Astron. J. 128, 163.
8. Szokoly, G.P., et al., 2004. Astrophys. J. 155, 271.
9. Wolf, C., et al., 2004. Astron. Astrophys. 421, 913.

Part IV
HST, H_0 and Dark Energy

Chapter 48
Visiting Hubble in Orbit

Claude Nicollier

Abstract Since its installation on orbit in 1990, the Hubble Space Telescope (HST) has been visited four times by Space Shuttle crews for correcting faulty devices, replacing failed or degraded components, and installing higher performance subsystems and scientific instruments. This servicing concept has proven extremely successful and has not only allowed the survival of the orbiting observatory despite significant problems, but has also allowed keeping HST as an up to date facility with state of the art instrumentation. This author participated as a Mission Specialist in two of the four servicing missions so far, the first in 1993 and the third in 1999. The last servicing mission, SM4, is now planned for 2009 and should extend the useful life of HST well into next decade.

48.1 Servicing HST in Orbit

Following years of preparation, training, and final integration of the required equipment in the Shuttle's payload bay, an HST servicing mission starts at lift off at such a time that, following Main Engines Cutoff (MECO) about 8:30 minutes after leaving the launch pad, the initial Shuttle orbit is in the plane of HST's orbit. A series of maneuvers follow MECO so as to complete the rendezvous with HST about 48 hours after lift off. The first few of these maneuvers are computed on the ground, on basis of the knowledge of both the target's (HST) and the chaser's (Shuttle) state vectors. Whenever the first measurements of HST's position relative to the Orbiter can be performed on board using various sensors (star tracker, then rendezvous radar), the maneuvers are computed on board and later executed by the crew. The final phase of the rendezvous is flown entirely manually by the mission commander or pilot, at least as far as all Orbiter translations are concerned. The spaceship's attitude is normally controlled by the Digital Autopilot following keyboard inputs by the crew. Typically, the orbiter is inertially stabilized during the final approach, matching HST's similar attitude control mode.

The Telescope is grappled with the shuttles robotic arm at the end of the rendezvous, during a final phase of precise formation flying of the Shuttle and HST,

C. Nicollier
ESA/European Astronaut Centre, Cologne, Germany

C. Nicollier (✉)
Space Center EPFL, Ecole Polytechnique Fédérale de Lausanne, Lausanne, Switzerland
e-mail: claude.nicollier@epfl.ch

F.D. Macchetto (ed.), *The Impact of HST on European Astronomy*,
Astrophysics and Space Science Proceedings,
DOI 10.1007/978-90-481-3400-7_48, © Springer Science+Business Media B.V. 2010

and then secured on a rotating platform at the rear end of the payload bay. Up to five spacewalks or Extravehicular Activities (EVAs) are then performed during the following days, according to a precise and well thought out plan, to accomplish all mission objectives.

Any servicing EVA involves two spacewalking astronauts, one normally using a foot restraint at the end of the robotic arm as a work platform, the other one being the "free floater", using various handholds and other appendages for translation or restraint, and making occasional use of a removable portable foot restraint for appropriate positioning at the worksite.

Special tools have been developed for HST servicing, in particular the "Pistol Grip Tool" or PGT, a multi-use EVA friendly motorized and microprocessor-controlled wrench, with selectable torque, rotation rate and number of turns. Most of HST bolts and fasteners are hexagonal 7/16 inch size so that PGT socket exchange is minimized during typical servicing scenarios.

A typical HST servicing EVA lasts 6 to 7.5 hours, with a maximum of just over 8 hours. Considering the preparation time before each EVA (about 2 hours), and the duration of the required activities post-EVA (about 4 hours for spacesuit reconditioning and preparation work for the next day's EVA), each EVA day is very busy, with virtually no time off! The use of two teams of two spacewalkers allows a sharing of the workload on these typically rather short and very full missions, and also provides manpower redundancy, both teams being fully cross trained on all planned tasks.

So far four HST servicing missions have been performed, and a fifth one is in preparation. Below is a quick look at the tasks performed on each of the past servicing missions.

48.2 The First Four Servicing Missions

The first Servicing Mission (STS-61 or HST SM1) was accomplished in December 1993. The most important objective of the mission was to restore the optical quality of the Telescope, significantly degraded by spherical aberration caused by incorrectly shaped primary mirror. A new Wide Field and Planetary Camera (WFPC-2) was installed, with its own optical correction, and a further optical corrector, COSTAR (Corrective Optics Space Telescope Axial Replacement) was put in place to serve the axial scientific instruments. The solar arrays were also exchanged, and one of the two solar array wings had to be jettisoned on-orbit because it could not be fully retracted and consequently not accommodated in the Orbiter's payload bay. Various other pieces of equipment were also replaced, including three Rate Sensor Units or RSUs, in a total of five spacewalks. The Telescope was also reboosted to its original orbital altitude.

The second Servicing Mission (STS-82 or HST SM2) took place in February 1997. Two new axial scientific instruments (NICMOS or Near Infrared Camera and Multi-Object Spectrograph, and STIS or Space Telescope Infrared Spectrometer) were installed in replacement of two original instruments. Some other critical pieces

Fig. 48.1 (Color online) Replacement of three Rate Sensor Units during the first EVA of HST SM3A (STS-103) in December 1999

of equipment were also replaced including one FGS or Fine guidance Sensor, and one of the Reaction Wheel Assemblies (RWA). A reboost of the Telescope was also successfully performed.

The third Servicing Mission (STS-103 or HST SM3A) was originally planned for mid-2000. At the beginning of 1999, it was cut short from a very ambitious program, designated 3A instead of 3, and moved to an earlier date, following the failure of a third (out of six) rate-sensing gyroscope. With only three functioning rate sensing gyroscopes, the Telescope had become zero failure tolerant for three-axis rate sensing, an essential feature in attitude and slew control. A fourth rate gyroscope failed before the mission took place in December 1999, preventing the Telescope from gathering any scientific data for several weeks. Main mission accomplishments were the replacement of the six rate sensing gyroscopes, of the Telescope's main computer, and of one FGS (see Fig. 48.1).

Servicing mission 3B or STS-109 was successfully accomplished in March 2002. The Telescope's solar arrays were replaced by a new rigid and more efficient type, another RWA was exchanged, and a new scientific instrument installed, the Advanced Camera for Surveys or ACS. A cooling system was also added to NICMOS, and a degraded Power Control Unit or PCU was replaced.

All was well and the ACS rapidly became the mostly used focal instrument, with remarkable sensitivity and resolution!

48.3 One More

A fifth and final servicing mission was going to take place in 2004. After a hesitation period following the loss of Columbia and its crew in February 2003, then a decision to cancel the mission, NASA finally decided to put it back on the manifest, as STS-125 or HST SM4, with the intent to fly it the fall of 2008 for an extension of HST's productive life until at least 2013.

Obviously, with this significant delay, HST has had a few extra problems and the intent to fix these problems added to the scope and complexity of this mission. The extra problems affected the STIS instrument, the ACS, and more recently one Science Data Formatter channel.

The mission has now been pushed to 2009 on orbiter Discovery. Again five EVAs planned for the replacement of all batteries and all rate sensing gyroscopes in three RSUs, installation of Wide Field Camera 3 or WFC-3 to replace WFPC-2, of a brand new scientific instrument designated COS or Cosmic Origins Spectrograph, and repairs of STIS and ACS. The failed Science Data Formatter will also be repaired, and a Soft Capture Mechanism will be installed for docking of a not yet designed and built robot intended to take with it the Telescope into a controlled deorbit some time in the future, way beyond the end of HST's useful life...

48.4 The Challenge of On-orbit Servicing

At first sight, it seems that the task of replacing failed or degraded components on HST should not be too difficult, and pretty standard operational procedures used in laboratories or workshops on Earth could be applied. Typically, the subtasks consist in demating connectors, loosening bolts or fasteners, removing the failed component, installing the new one, tightening the attachment bolts or fasteners, remating connectors, and that's it.

However, all of this is done from bulky spacesuits, inflated with circulating pure oxygen at a pressure of about 0.3 bar, with thick gloves, and in the absence of gravity! The spacesuit is quite rigid, and leg, arm, and especially hand fatigue is quite common. Visibility is also impaired with the design of the helmet/visor. Working in the absence of gravity means that all forces and torques on the body, while working, have to be compensated for somehow. Body stabilization while working is a must and, depending on the location, not always easy. Everyone outside is of course equipped with a safety tether, and all tools and pieces of equipment handled during the EVA are also at least temporarily secured with a safety tether. The geometry of all these tethers can become quite complicated, and care has to be taken to avoid a mess and wasted time untangling them... Spacewalking for work is certainly fun, but also demanding and fatiguing!

48.5 Final Words

The servicing concept has been a huge success for HST. Without it the orbiting observatory would have since long become useless. The success of the Servicing Missions so far was the result of the engagement, hard work, talent and huge motivation of all personnel involved, primarily at the Goddard, Johnson, Kennedy, and ESA centres, and within the involved industrial teams. The strong leadership present at all levels in the program, the attention to details, and the thorough preparation and training for each of the servicing tasks has paid off. Strict operational discipline, as is the case for all Shuttle missions, has been rigorously followed since STS-61 and carried over all following missions.

Personally, I feel tremendously privileged to have been part of this effort. The spirit of the First Servicing mission was extraordinary, and I was overwhelmed by the excellence of the team that put it together, trained us, and supported us during the on-orbit phase. It was the first time that a complex scientific instrument was going to be repaired and serviced by a combination of spacewalking and robotic means. The challenge was enormous, and we had success. Similar level of success was then repeated three times, and I have no doubt that SM4 will again impress us, and give us several more years of useful service from this amazing discovery machine!

Chapter 49
HST and JWST: Present and Future

Michael Hauser

Abstract The *Hubble Space Telescope* (*HST*) has had a transformative impact on science, the science community, and the public around the world. As *HST* completes its 17th year of operations, NASA is preparing to send astronauts to service *Hubble* a fifth time in order to refurbish critical systems, install a new generation of powerful instruments, and possibly repair instruments that have currently failed. This mission is planned to extend *Hubble*'s operational life until 2014, potentially overlapping the initial operations of the *James Webb Space Telescope* (*JWST*). In this paper I review the impact and status of *HST* and the promise of *JWST* and beyond.

In anticipation of the planned 1984 launch of *HST*, John Bahcall and Lyman Spitzer wrote [1]:

> The *Space Telescope* will help solve many astronomical puzzles. The greatest excitement, however, will come when the pictures returned from the satellite reveal things no one in this generation of astronomers has dreamed of, phenomena that only the next generation will be privileged to understand.

Though the launch was delayed until 1990, the *Hubble Space Telescope* (*HST*) has indeed had a transformative impact on science, the science community, and the public around the world. Before the *HST* launch, we did not know how old the Universe was. Black holes seemed more science fiction than fact. The only planets we knew were the nine (now eight!) in our solar system. Since then we have marveled at images and discoveries flowing from *Hubble*: our first glimpse of galaxies whose light has traveled across nearly 13 billion years of cosmic history, of iconic celestial landscapes of breathtaking beauty, and at the startling discovery that the Universe is pervaded by a mysterious 'Dark Energy' that causes the expansion discovered by Edwin Hubble to accelerate. We of the next generation are indeed privileged to witness the scientific revelations and awe-inspiring images from *Hubble*, which continue to emerge as evidenced in this Symposium. Figure 49.1 shows a recently released spectacular image of the Carina Nebula.

In keeping with the overall theme of this Symposium, "The Impact of *HST* on European Astronomy," I will focus on some of the broad impacts of the *Hubble* mission rather than its science, then discuss future plans for *HST*, the exciting *JWST* mission that will follow, and ideas of what might come beyond 2020.

M. Hauser (✉)
Space Telescope Science Institute, 3700 San Martin Drive, Baltimore, MD 21218, USA
e-mail: hauser@stsci.edu

F.D. Macchetto (ed.), *The Impact of HST on European Astronomy*, Astrophysics and Space Science Proceedings,
DOI 10.1007/978-90-481-3400-7_49, © Springer Science+Business Media B.V. 2010

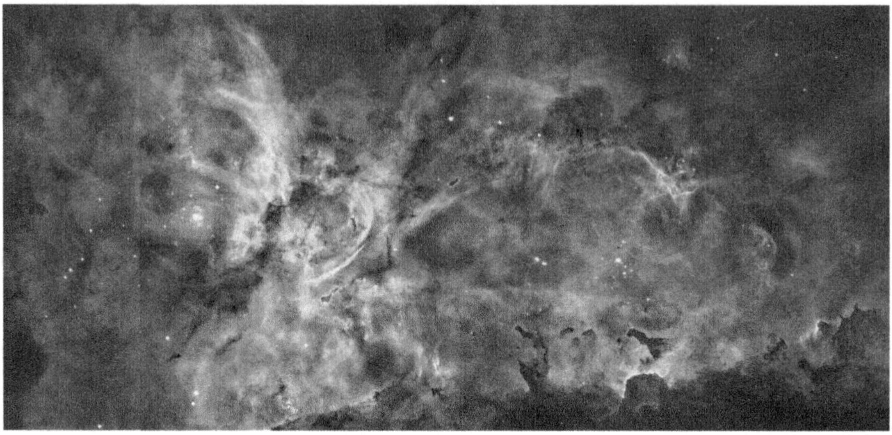

Fig. 49.1 (Color online) Carina Nebula, starburst in the extreme. This image was released April 24, 2007 to celebrate the 17th launch anniversary of the *HST*. This large mosaic was created from 48 ACS frames and additional data from CTIO

A major impact of *HST* has been a dramatic expansion of international collaboration on large, multi-wavelength research programs. This was spurred by the release of the first Hubble Deep Field image [5], which stimulated an unprecedented international collaborative effort employing leading-edge observatories on the ground and in space to understand the nature of the sources revealed. A more recent example is *Hubble*'s largest area survey, the Cosmic Evolution Survey, carried out by an international team of some 70 astronomers [3, 4], which yielded the remarkable first three-dimensional map of dark matter [2] described at this meeting by Scoville (see Chap. 44 in this book).

The *HST* has been a major research tool for European astronomers. Over the period 1997–2007, 19% of the 9,973 total *HST* observing proposals were submitted by astronomers from European institutions. Typically, some 20% of these proposals were selected for execution, utilizing about 15% of the observation time with *HST* (Fig. 49.2). The number of refereed publications based on *HST* observations has climbed steadily throughout the mission, reaching nearly 700 in 2006, with about 55% of the papers including European authors (Fig. 49.3). Twenty-two percent (1,368) of the 6,091 *HST* papers to date have been published in European journals, including A&A, MNRAS, and Nature. The unprecedented steady increase in the number of papers per year over a multi-decade time scale is the result of the continual evolution of the *HST* observatory, thanks to the human space flight program. The astronomical community owes great thanks to the NASA and ESA astronauts who have risked their lives to make this possible.

One of the strengths of the *HST* program has been the creation of research tools with wide applicability. The *HST* data archive, held jointly at STScI, the European Coordinating Facility (ECF) and the Canadian Astronomy Data Center has become a major research resource. Electronic access now satisfies the needs of researchers worldwide. There are now nearly 2,000 European archive users registered at STScI,

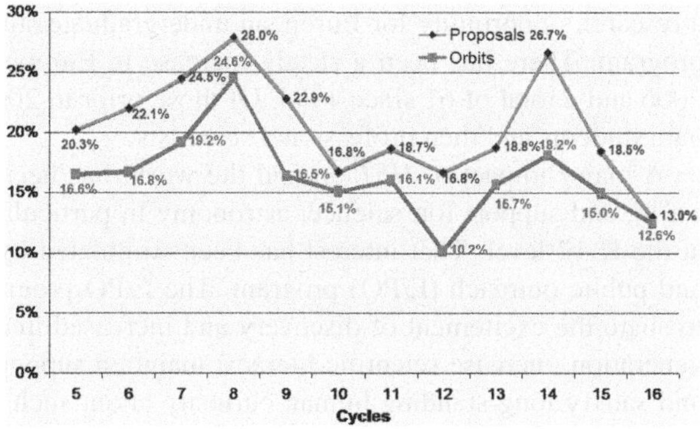

Fig. 49.2 (Color online) ESA proposal acceptance fraction for proposals (*upper curve*) and orbits (*lower curve*) from Cycle 5 to Cycle 16

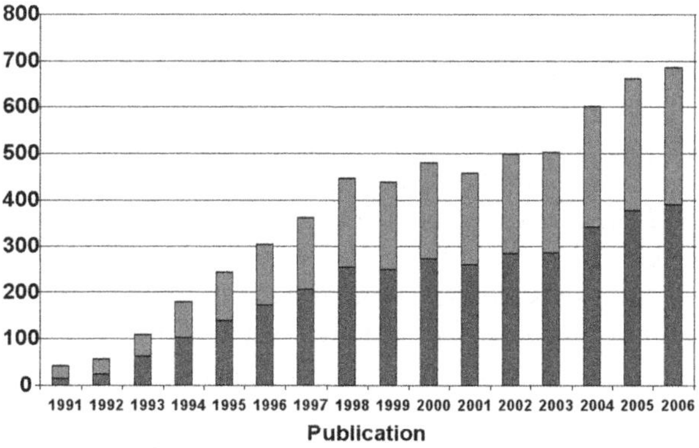

Fig. 49.3 (Color online) Number of refereed papers per year based on *HST* data. *Lower bars* indicate the number with European authors or co-authors

some 25% of all registered users. Currently, nearly half of all *HST* refereed publications use archival data. The Hubble Guide Star Catalog, created by scanning plates from ground-based observatories in both hemispheres, is now routinely used by observatories around the world: over the 6-month period ending 1 January 2007 there were 1.1 million requests from 143 countries. The *Hubble* observation scheduling software (SPIKE) and pipeline data processing system (OPUS) have been adopted by numerous US and European observatories on the ground and in space.

Another impact of *HST* has been to stimulate the development of human capital. An *HST* innovation was the introduction of the Hubble Fellow program, which provides a 3-year post-graduate fellowship to outstanding young scholars at an US institution of their choice. Of the 201 Fellows to date, 35 have been Europeans, many now prominent in the European and US communities. Per the NASA/ESA Memorandum of Understanding for the HST Program, 15 ESA staff (or ESA-supported staff) have been intimately involved in *Hubble* science operations at the STScI in Baltimore. Most have been astronomers, but several engineers have also served. Over the lifetime of *Hubble*, that has included 63 different individuals from 12 countries, including Austria, Belgium, Denmark, France, Germany, Ireland, Italy, Netherlands, Spain, Sweden, United Kingdom, and Switzerland. Of course, other Europeans participated in *HST* operations at the ECF. The STScI has also provided a research opportunity for European undergraduate students in its summer student program. There has been a steady increase in European participation, with 13 in 2006 and a total of 61 since 1994. Of those prior to 2004, most have become graduate students and then professional scientists.

A major impact of *HST* around the world has been an increasing public interest in, and support for, science, astronomy in particular, and on science education at the K-12 level. That interest has been stimulated by a vigorous *HST* education and public outreach (E/PO) program. The E/PO program is motivated by a desire to share the excitement of discovery and increased understanding, inspire the next generation, increase scientific literacy, maintain support for astronomical research, and satisfy long-standing human curiosity about such questions as where are we? Where did we come from? Where are we going? Are there others like us? At STScI we have a broad E/PO program, including news, informal education in venues such as science museums and planetariums, formal K-12 education programs providing curricular materials designed and evaluated by professional teachers, and a vigorous on-line outreach program. Combining all of these functions in a single, science-based organization has proven an efficient way to reach the public and the classroom with materials informed by the science of *HST*. In Europe there are also high-quality *HST* press releases and the innovative "Dr. J." podcasts to better reach the modern generation.

A very visible indication of the success of the *HST* education and outreach programs was the strong negative reaction from the public when NASA decided to cancel the planned fifth servicing mission (SM4) in the aftermath of the loss of the Shuttle *Columbia* in 2003. Partly as a result of that reaction, and after expert assessment of the benefits and risks of servicing *HST* again, NASA Administrator Michael Griffin announced on October 31, 2006, that a *Hubble Space Telescope* servicing mission would be added to the Shuttle's manifest. Mindful of the recommendations of the Columbia Accident Investigation Board to protect crew safety, he also directed that a second Shuttle must be ready at the time of launch in case a crew rescue mission would be needed.

Preparations for SM4 are now proceeding rapidly, aiming for launch in September 2008 on Shuttle *Atlantis*. The flight crew of seven astronauts, including three members with prior experience on *Hubble* servicing missions (Scott Altman, commander; John Grunsfeld and Mike Massimino, EVA team leads), is actively engaged in training. The mission objectives are to restore the health of observatory systems, including replacing all gyros and all batteries, replacing one fine guidance sensor (FGS2r, which is slowly degrading), and adding new outer blanket layers to protect degrading thermal blankets. In addition, two powerful new scientific instruments, Wide Field Camera 3 (WFC3) and the Cosmic Origins Spectrograph (COS), will be installed to replace, respectively, Wide Field and Planetary Camera 2 (WF/PC2) and the corrective optics device, COSTAR. Finally, the crew will attempt to repair the electronics in the Space Telescope Imaging Spectrograph (STIS), which failed in 2004, and the electronics for the Wide Field Camera and High Resolution Cameras (WFC, HRC) in the Advanced Camera for Surveys (ACS), which failed in 2006. These repairs are extremely challenging for the crew, since the instruments were not designed for servicing in flight. Successful accomplishment of the SM4 objectives will extend the life of *HST* to at least 2014, and put it at the historic peak of its observational capabilities. Until SM4 is accomplished, a full *HST* cycle 16 research program will continue using WF/PC2, the Near Infrared Camera and Multi-object Spectrometer (NICMOS), the Solar Blind Channel (SBC) of the ACS, and FGS1r for astrometry. A supplemental cycle 16 *Call for Proposals* will not be needed provided SM4 occurs before the end of January 2009.

Following SM4, the COS will provide high sensitivity, moderate- and low-resolution spectroscopy of astronomical objects in the 1,150–3,200 Å wavelength range using grisms for dispersion. The design is optimized for maximum FUV throughput, and will dramatically increase faint object discovery power and the spectroscopic efficiency of *HST*. Science themes to be pursued with the COS include the large-scale structure of matter in the Universe, formation of galaxies out of the intergalactic medium, galactic halos and out flowing winds from star-forming galaxies, formation of the chemical elements for life in massive stars and supernovae, formation of stars and planetary systems from dust grains in molecular clouds, and the composition of planetary atmospheres and comets in our solar system.

The WFC3 will provide wide-field imaging with continuous spectral coverage from the ultraviolet into the infrared, dramatically increasing both the survey power and the panchromatic science capabilities of *HST*. By combining a high-sensitivity UV/optical channel with a high-sensitivity near-infrared channel, WFC3 will be capable of direct, high-resolution imaging over the entire wavelength range from 200 to 1,700 nm. WFC3 is equipped with a comprehensive range of wide-, intermediate-, and narrow-band filters, providing broad applicability to a variety of new astrophysical investigations. Parallel observations with WFC3 and a repaired ACS will be a uniquely powerful imaging capability, with ACS most sensitive in the visible and far-red, and WFC3 in the UV and near-infrared. WFC3 will be used to study such topics as the nature of dark energy and dark matter, how and when galaxies first assembled, the universality of the processes of star formation in galaxies, the evolution of stars and the distribution of their masses, the dependence of star-formation and planetary disk formation on environmental conditions, and the composition of planets, comets, and minor planets in our solar system and beyond.

After the most recent *HST* servicing mission (SM3B), the *New York Times* observed in an editorial, "It (*HST*) has taught us to see the properties of the universe humans have been able, for most of their history, to probe only with their thoughts." There is no doubt that the new capabilities on the post-SM4 *HST* will allow our probing to reveal even more wonders.

I turn now to the next major observatory in space, the *James Webb Space Telescope* (*JWST*), currently under active development. *JWST* is also an international collaborative mission, involving NASA, ESA and the Canadian Space Agency (CSA). The *JWST* design has been driven by the capabilities needed to pursue four broad science themes:

1. The end of the dark ages: the first bright objects that formed and the ionization history of the Universe.

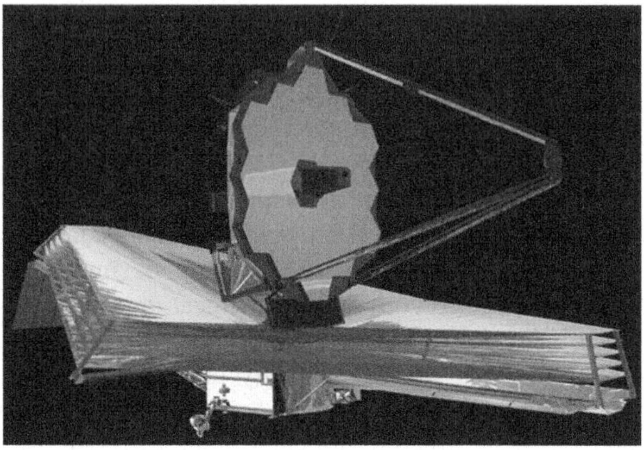

Fig. 49.4 (Color online) The *James Webb Space Telescope*. The large multi-layer sunshield keeps the telescope and instruments shielded from direct Sun and Earth radiation

2. Assembly of galaxies: evolution of galaxies and dark matter, including gas, stars, metals, physical structures (like spiral arms) and active nuclei.
3. The birth of stars and protoplanetary systems: birth and early development of stars and the formation of planets.
4. Planetary systems and the origins of life: physical and chemical properties of planetary systems (including our own) and the building blocks of life.

We have heard about many of these themes at this Symposium, but for all of these problems we eventually reach technology and photon-collection limitations. A capability to operate with high sensitivity and high angular resolution from near- to far-infrared wavelengths will take us beyond even what a 2.4-m, warm UV/opt/NIR observatory or the smaller ISO and *Spitzer* cryogenic infrared observatories in space can provide. The *JWST* (see Fig. 49.4) will be a cryogenic, infrared-optimized observatory with a segmented primary mirror 6.5-m across operating near the Earth–Sun Lagrange point L2. The infrared capability greatly advances our ability to deal with the cosmic redshift and cosmic dust which impede pursuit of all of the major themes at UV and optical wavelengths.

Management of *JWST* is led by the Goddard Space Flight Center (GSFC), with prime contractor Northrop Grumman Space Technology. The observatory will have four instruments: NIRCam, an imager with a large field of view and high angular resolution, covering a wavelength range of 0.6 to 5 μm, provided by NASA under the leadership of the University of Arizona; NIRSpec, a 1–5 μm, multi-object spectrometer that can obtain simultaneous spectra of more than 100 objects, provided by ESA with some components provided by NASA/GSFC; MIRI, a 5–27 μm imager and spectrograph, provided by a European consortium with ESA, and the NASA Jet Propulsion Laboratory; and FGS, a fine-guidance sensor that also incorporates a 1.6 to 4.9 μm Tunable Filter Imager, provided by the CSA. The science operations center for *JWST* will be at the STScI, with staff supported by ESA and the CSA, as well as by NASA.

The *JWST* mission incorporates many technology advances that required an extensive technology-development phase prior to commitment to a flight program. Ten critical technologies were identified, and earlier this year all ten were found by an independent review team to have advanced to the technology readiness level necessary to proceed with engineering design of *JWST*, a major mission milestone. The mission preliminary design review and programmatic non-advocate review of cost and schedule will occur in April 2008. Success at these reviews will lead to formal NASA commitment for flight development. In reality, a great deal of engineering design and development for the telescope and instruments has proceeded in parallel with the technology studies, and the mission is on track for launch by an ESA-supplied Ariane 5 rocket in June 2013. With a 5-year mission lifetime requirement and 10-year goal, *JWST* promises to serve the scientific community for a long time to come.

Though the *JWST* mission design is driven by four specific scientific themes, *JWST* will be, as is *HST*, a powerful general-purpose observatory enabling pursuit of known problems and of making new discoveries. It is not too early for the broad scientific community to begin thinking about the research they wish to do with *JWST*. To facilitate that, the *JWST* program (STScI and GSFC) is sponsoring a conference in Tucson, Arizona, in September 2007 "Astrophysics in the Next Decade: *JWST* and Concurrent Facilities." This meeting will focus on science enabled by *JWST* and concurrent orbital and ground-based facilities. A book containing a compilation of the talks and synopses of the discussion periods is planned, which will yield a reference text of lasting value for astronomers who will be developing investigations for the *JWST* and other concurrent facilities.

"Flagship" space observatories such as *HST*, *Chandra*, and *Spitzer* have achieved remarkable democratization of astronomical research, providing access to world-class facilities to the best investigators from any part of the world and any institution, large or small. A crucial element of this is time allocation by the community through peer review based solely upon scientific merit. The result has been engagement of a broad community in revolutionizing our view of the Universe. *JWST* will be run the same way, and serve a similar community of over 7,000 astronomers worldwide.

It is also not too early to begin thinking about what might lie beyond *JWST*. The US space policy announced by the President in 2004 (Vision for Space Exploration, VSE) put the NASA focus on human exploration beyond Earth orbit. NASA Administrator Griffin made it clear that scientists should look for opportunities compatible with this program. At STScI we took this challenge seriously, and in November 2006, STScI held a conference carefully titled "Astrophysics Enabled by the Return to the Moon." This scientific conference identified a few specific programs advantageously conducted from the surface of the Moon (if opportunities are available), but generated even more enthusiasm for science enabled by the infrastructure developed for the VSE program.

One of the most exciting elements of the new infrastructure is the Ares V launch vehicle. It is a cargo vehicle, with unprecedented lift and size. For example, it could lift 60,000 kg to L2 within a rocket shroud of perhaps 12-m diameter. With current space infrastructure, significant costs are incurred by needing to use lightweight materials and designs for large structures, such as *JWST*. Ares V enables a much more brute force approach—one could even imagine using ground-based mirror technology! One can also imagine 20–25 years of useful service via servicing just the instrument module with robots or humans, which would require much smaller launchers than the Ares V rocket. In order to pursue these ideas, STScI will host a workshop November 13–15, 2007 entitled "Astrophysics 2020: Large Space Missions Beyond the Next Decade." I encourage the European community to get engaged early in considering the potential new opportunities offered by the VSE.

Finally, I wish to congratulate and thank my colleague, Duccio Macchetto, as he approaches retirement from ESA. It has been my pleasure during my almost 12 years at STScI to work with Duccio, who has guided me, as he has guided many other individuals, ESA, NASA, and STScI, in the intricacies and policies of the *HST* observatory and program. I wish him every success in the future.

Acknowledgements I am indebted to many colleagues at STScI for contributions to the material presented here. These include M. Mountain, B. Blacker, D. Golombek, D. Hunter, M. Livio, K. Long, J. Madrid, K. Sembach, D. Soderblom and L. Willers. This work was supported by STScI, operated by AURA, Inc under NASA contracts NAS5-26555 and NAS5-03127.

References

1. Bahcall, J.N., Spitzer, L. Jr., 1982. Sci. Am. 247, 40.
2. Massey, R., et al., 2007. Nature 445, 286.
3. Scoville, N.Z., et al., 2007. Astrophys. J. Suppl. Ser. 172, 1.
4. Scoville, N.Z., et al., 2007. Astrophys. J. Suppl. Ser. 172, 38.
5. Williams, R.E., et al., 1996. Astron. J. 112, 1335.

Chapter 50
Enabling Science with the Hubble Legacy Archive

Helmut Jenkner, W. Warren Miller III, and Bradley C. Whitmore

Abstract Significant enhancements to the Hubble archive will prepare it for legacy use and provide major new science capabilities. The focus is on higher-level data products than are available from the classic Hubble archive, including fully calibrated exposures and enhanced image products as well as source lists. All data products are immediately available on disk, eliminating the latency of queued data delivery. This capability will make HST data easier to obtain, combine, and compare with data from other space missions and with ground-based data. Image headers will be updated with improved astrometric information, greatly reducing the 1–2 arcsecond errors that frequently affect current pipeline products. A generalized footprint service allows users to determine quickly whether, and for how long, a particular region of the sky has been observed with HST instruments. In addition, all Hubble data become accessible through standard Virtual Observatory interfaces. This improved archive, the Hubble Legacy Archive (HLA), is expected to provide an essential research resource for many years, if not decades.

50.1 Motivation

The Hubble Space Telescope (HST) has the potential to provide excellent scientific data for several more years after the planned Servicing Mission 4. But as the conclusion of the HST mission inevitably draws nearer, interest will turn increasingly to the analysis of archival data and to the establishment of the true legacy of this mission. The research community already retrieves three times more data from the HST archive than is put in, indicating that HST data are used for purposes beyond their original intent. This is also reflected by the fact that roughly two thirds of the refereed papers based on Hubble data use archival information.

Much information remains to be gleaned from the archive, especially as holdings grow and larger statistical and correlative studies can be carried out. Existing and

H. Jenkner (✉) · W.W. Miller III · B.C. Whitmore
Space Telescope Science Institute, 3700 San Martin Drive, Baltimore, MD 21218, USA
e-mail: jenkner@stsci.edu

H. Jenkner
European Space Agency, Research and Scientific Support Department, Noordwijk, The Netherlands

F.D. Macchetto (ed.), *The Impact of HST on European Astronomy*, Astrophysics and Space Science Proceedings,
DOI 10.1007/978-90-481-3400-7_50, © Springer Science+Business Media B.V. 2010

diverse science cases, spanning all areas of astrophysics, could be accomplished and enabled by improvements to the existing system.

50.2 New Capabilities for the Hubble Legacy Archive

The existing HST archive, anchored by the Data Archive and Distribution System (DADS), serves well in an environment where the primary end-users are HST principal investigator teams. Archival researchers also currently benefit from the archive. To exploit the archive, they typically search for the relevant data in the form of exposures by using online tools and then issue requests for datasets. These requests typically require on the order of an hour to fulfill.

The organization of the current HST archive — by proposal, program, and exposure — supports *"vertical" data access* naturally (i.e., get all exposures for a certain PI in a certain program). The wealth of data in the HST archive, coupled with the increasing amount of complementary space- and ground-based data in online, immediate access databases, suggests that a more *"horizontal" organization* of the archive is now desirable in order to respond to broadly based queries, provide immediate access, and maximize data comparison possibilities. We need to provide a view of the HST archive that presents itself more like a sky atlas, albeit with sparse coverage owing to HST's comparatively small field of view.

To support effective and efficient browsing of data, a footprint service, quick-look data, preview images and cutouts are available. The current calibration system, On-The-Fly Reprocessing (OTFR), inserts a (sometimes substantial) delay in data access, but this has been acceptable for users who intend to download all associated exposures and perform in-depth analyses that rely on the best possible calibrations. The enhanced system provides nearly immediate access to calibrated science data, including combined images and mosaics. These images are of sufficient quality so that no further calibration is required. The new system presents the desired data in real-time, on the order of seconds rather than hours.

While some user-supplied high-level products are currently available on separate web pages in special cases, such as for the HDF, UDF, and GOODS, these high-level products will be made available as part of the Hubble Legacy Archive.

The set of new capabilities of the Hubble Legacy Archive, to become increasingly available over the next few years, is summarized below:

- *Enhanced image products* are available, such as combined calibrated drizzled images, mosaics, and color images.
- *Online access* is possible to all calibrated and enhanced image products in seconds instead of hours.
- *Improved astrometry* to about $0.1''$, and corrected World Coordinate System keywords in data headers are available.
- *A generalized "footprint" service* provides, given a position, a list or graphical representation of available exposures and image products, taking into account the aperture shape, telescope pointing, roll angle, etc.
- *Image cutouts* are available quickly to facilitate efficient browsing.

- *Source lists* based on HST images give coordinates, magnitudes and morphological information.
- *Community-provided catalogs and other high-level products* can be incorporated and made available.
- Access that is *compatible with Virtual Observatory standards* is provided for all data.

These capabilities, largely developed as independent prototypes, are unified and presented to the user by a modern, web browser-based interface. Emphasis is being placed on an interface that is as intuitive and simple as possible, while providing comprehensive and easily understood information.

50.3 Implementation and Plans

The Hubble Legacy Archive is developed in the framework of an international collaboration that includes the Space Telescope Science Institute (STScI), the Space Telescope — European Coordinating Facility (ST-ECF), and the Canadian Astronomy Data Centre (CADC). The CADC focuses on data associations and source lists, while the ST-ECF concentrates on spectroscopic data and Near-Infrared Camera and Multi-Object Spectrometer (NICMOS) grism extractions. STScI also works on associations and source lists, provides services such as cutouts and footprints, and performs the overall integration and the development of the interfaces.

The implementation of the Hubble Legacy Archive focuses on making new capabilities and enhanced data products available as early as possible. Feedback from the community is being solicited, so that the most desirable features can be implemented as early as possible.

An Early Data Release in the summer of 2007 incorporates the following release goals:

- Data products:
  - Calibrated Advanced Camera for Surveys (ACS) exposure-level images.
  - ACS combined images.
  - Point-source and extended-source, multi-wavelength source lists.
  - Improved astrometry.
  - Completeness level of 50% (minimum) to 80% (goal).
- Services:
  - Basic footprints, cutouts, and data download capabilities.
  - Simple web-based user interface demonstrator.
  - Simple Virtual Observatory access to data.

Following extensive field-testing and solicitation of user feedback, the Hubble Legacy Archive release in the winter of 2007/2008 is expected to have the following additional data products and capabilities:

- Data products:
  - ACS mosaics.

  - NICMOS grism extractions.
  - ACS-like products for Wide-Field and Planetary Camera 2 (WFPC2) data, including calibrated single exposures, combined images and mosaics.
- Services:
  - Improved basic services from the Early Data Release.
  - Advanced search capabilities.
  - Improved user interface.

The addition of remaining and even unforeseen features will continue in the following years.

50.4 Summary

The scientific community is becoming increasingly sophisticated in the use of online archive resources. Thus, the Hubble archive must modernize so that future "Hubble users" can maximally exploit this new research paradigm. The Hubble Legacy Archive achieves this by providing new services and essentially instantaneous access, paired with a modern web-based interface, making browsing of the archival data sets more comprehensive and intuitive.

Over the next years, archived data are expected to multiplicatively increase the scientific value of "original" Hubble data. Archived Hubble data are already a research and educational resource in their own right. To further this trend, the Hubble Legacy Archive makes available new enhanced science data products that have the potential to dramatically increase and optimize the total science output from HST, both now and in the future.

In the years immediately following the launch of HST, the Hubble archive and the data processing pipeline led to a new way astronomers worked and provided a substantial improvement in the ability to use observations for science. The Hubble Legacy Archive and the Virtual Observatory have the potential to lead to a further significant improvement in this area, enabling the community and the public to continue to explore this valuable scientific asset — the Hubble data set.

The Hubble Legacy Archive is available at hla.stsci.edu.

Chapter 51
Advanced Calibration Using Physical Instrument Models: HST, VLT and Beyond

Michael R. Rosa, Paul Bristow, and Florian Kerber

Abstract The Space Telescope European Co-ordinating Facility (ST-ECF), in the context of space-astronomy (ESA) and in close collaboration with ground-based astronomy (ESO), has pioneered the use of physical instrument models for instrument calibration and forward data analysis. The methodology provides real insight into the physical properties of the signal chain and therefore offers a high degree of predictability for the behavior of instruments and the measurements generated.

51.1 Physical Model Based Calibration

For more than a decade the Space Telescope European Co-ordinating Facility (ST-ECF) has pioneered the use of physical instrument models in instrument calibration and data analysis applications. The traditional approach to calibration consists of the linear application of empirically determined relations between input and output signals in the detection chain. Typical examples are the fitting of low-order polynomials to irregularly and scarcely distributed data points of wavelength calibration lines or the subsequent sequential, linear analysis of dispersion and distortion solutions in 2D spectroscopy. In contrast, a physical model based approach makes full use of physical principles and the engineering information that went into the design and construction of instruments. Such a model resembles an unsophisticated ray-trace embellished with additional functionalities, e.g. mechanical, thermal, or electrostatic, that provide a reasonably accurate description of the instrumental function from first principles. Instead of painstakingly deriving approximate

M.R. Rosa (✉)
Space Telescope European Co-ordinating Facility, Karl-Schwarzschild-Str. 2, 85748 Garching, Germany
e-mail: mrosa@stecf.org

P. Bristow · F. Kerber
ESO, Karl-Schwarzschild-Str. 2, 85748 Garching, Germany

P. Bristow
e-mail: bristowp@eso.org

F. Kerber
e-mail: fkerber@eso.org

F.D. Macchetto (ed.), *The Impact of HST on European Astronomy*, Astrophysics and Space Science Proceedings,
DOI 10.1007/978-90-481-3400-7_51, © Springer Science+Business Media B.V. 2010

empirical relations from noisy and insufficient experimental calibration data in a repetitive manner, the physics based modeling of measurements obtained with a given instrument provides deep insight into the actual physical properties and therefore offers a high degree of predictability for the behaviour of the instrument under various conditions of operation.

51.2 Instrument Physical Modeling at ST-ECF

Initiated at a 1995 "HST and VLT" calibration workshop [4], the idea was demonstrated to yield superior results for difficult calibration items of HSTs FOS and earned quickly the support of ESO [2]. Subsequently ESA, as part of the MoU between NASA and ESA on HST, established the Instrument Physical Modeling Group (IPMG) at the ST-ECF to apply the methodology in earnest, starting in 1999 with the Post-Operational Archive for the FOS, followed in 2002 by calibration enhancements for HST's STIS UV Echelle dispersion solutions and CCD charge transfer efficiency [1].

The high predictive power of the model based dispersion solutions for FOS and STIS very early on demonstrated the inadequacy of the line lists in use for the calibration lamps flown. In a fruitful collaboration with the Atomic Spectroscopy Group at the US National Institute of Standards and Technology (NIST) we have been able to produce very much improved wavelength standards for these Pt/Cr–Ne hollow cathode lamps. The superior performance of the new dispersion solutions, based on instrument models verified against the new Pt/Cr–Ne line list [5], have been validated on demanding scientific data [3]. The ST-ECF/NIST team in 2006 earned a Group Achievement Award from NASA for this coherent application of advanced calibration methodology in combination with custom generated superior laboratory standards.

51.3 Ongoing Activities: Ground and Space

The formal IPMG has now expired, but the methodology, physical instrument models in combination with superb laboratory assessment of components and standards, continues to be applied and further refined by former members of the IPMG, now both in the ESO Instrumentation Division and the ST-ECF. Current activities include supporting the CRIRES and X-shooter instruments for the VLT and laboratory work in support of the wavelengths calibration operations of the Cosmic Origins Spectrograph (COS) for HST.

Acknowledgements We are indebted to the continued support of ESO, in particular to the Director Generals Riccardo Giacconi and Catherine Cesarsky for taking personal interest in the development and application of the concept for ground and space astronomy; equally well to ESA, in particular Roger Bonnet and Brian Taylor for enabling the establishment of the IPMG. Duccio

Macchetto (ESA, HST) has been a steady patron for the work, especially in the face of bureaucratic barriers in a transatlantic multi-agency project. It is a special pleasure to thank him for his support. This is also the place to thank our two former team colleagues, Anastasia Alexov and Mauro Fiorentino, for the spirit and effort invested.

References

1. Bristow, P., Kerber, F., Rosa, M.R., 2006. In: Koekemoer, A.M., Goudfrooij, P., Dressel, L.L. (Eds.), The 2005 HST Calibration Workshop, p. 299.
2. Giacconi, R., 2005. Annu. Rev. Astron. Astrophys., 43, 1.
3. Kerber, F., Bristow, P., Rosa, M.R., 2006. In: Koekemoer, A.M., Goudfrooij, P., Dressel, L.L. (Eds.), The 2005 HST Calibration Workshop, p. 309.
4. Rosa, M.R., 1995. In: Benvenuti, P. (Ed.), Calibrating and Understanding HST and ESO Instruments, ESO Conf. & Workshop Proceedings, vol. 53, p. 43.
5. Sansonetti, C.J., Kerber, F., Reader, J., Rosa, M.R., 2004. Astrophys. J. Suppl. Ser. 153, 555.

Chapter 52
The Hubble Constant and HST

G.A. Tammann and A. Sandage

Abstract The Hubble diagram of distant type Ia supernovae, zero-pointed by nearby SNe Ia with Cepheid distances from HST observations yields $H_0 = 62.3$. The basis of this value has been weakened because the period–luminosity relation of Cepheids is not universal, but depends on their metallicity and possibly other parameters. An independent (Pop. II) distance scale is needed, as provided by RR Lyrae stars and the tip of the red-giant branch, to buttress the value of H_0. Present results show good agreement.

52.1 Introduction

Present determinations of the Hubble constant H_0 peak bimodially around 62 and 72 km s^{-1} Mpc^{-1}. One could conclude that a mean value of ~65 were satisfactory for all practical purposes. However, the interpretation of the CMB fluctuation spectrum requires a well determined value of H_0 as a prior. For instance the WMAP data are consistent with a Λ Universe if $H_0 \sim 72$ [22], whereas a value of $H_0 < 65$ questions an equation of state of the dark energy $w = p/\rho = -1$ and rather requires $w > -1$ and hence a quintessence model. The hunt for a precise value of H_0 therefore continues.

52.2 The HST Supernova Ia Project

The obvious way to determine the *large-scale* value of H_0 is to use the *tight* Hubble diagram of normal type Ia supernovae (proving SNe Ia to be excellent standard candles) and to zero-point it with the mean absolute magnitude of some nearby SNe Ia. The first attempt to follow this route, using brightest stars as distance indicators [18], was marred by the fact that one of the two calibrators (SN 1954A) turned later out to be of the overluminous variety like SN 1991T.

G.A. Tammann (✉)
Astronomisches Institut der Universität Basel, Venusstrasse 7, 4102 Binningen, Switzerland
e-mail: G-A.Tammann@unibas.ch

A. Sandage
Observatories of the Carnegie Institution of Washington, 813 Santa Barbara Street, Pasadena, CA 91101, USA

F.D. Macchetto (ed.), *The Impact of HST on European Astronomy*, Astrophysics and Space Science Proceedings,
DOI 10.1007/978-90-481-3400-7_52, © Springer Science+Business Media B.V. 2010

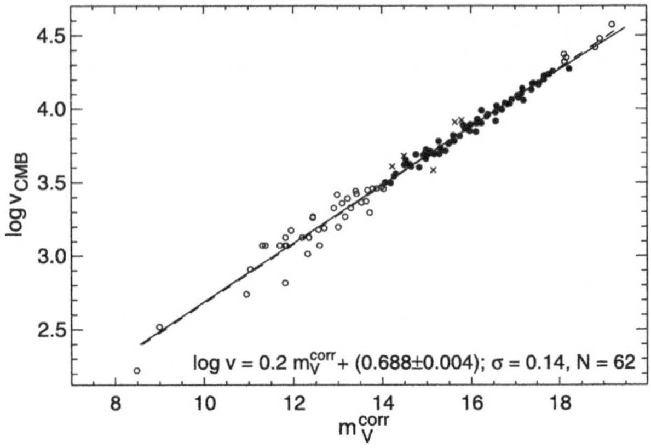

Fig. 52.1 The Hubble diagram in V for 108 normal SNe Ia. The *line* is the best fit to the 62 SNe Ia with $3{,}000 < v < 20{,}000\ \mathrm{km\,s^{-1}}$ (*black points*) for an $\Omega_M = 0.3$, $\Omega_\Lambda = 0.7$ Universe. The scatter is 0.14 mag in V as well as in B and I

With the advent of HST a special project was mounted by A. Sandage, A. Saha, F.D. Macchetto, N. Panagia, and G.A. Tammann. They were supported at times (chronologically) by L. Labhardt, H. Schwengeler, F. Thim, and B. Reindl. The aim was to determine *Cepheid* distances of host galaxies of well observed SNe Ia. The distances of eight such galaxies were published. This resulted in seven calibrators, because the spectroscopically distinct, overluminous SN 1991T cannot be used for the calibration. Three more Cepheid distances of normal SNe Ia became available in the literature. A total of ten calibrators is therefore available.

The analysis was presented in five papers; they are [13, 16, 20, 21, 23]:

Papers I & II: They define the different period–luminosity (P–L) relations in the Galaxy and in LMC. See Sect. 52.3.

Paper III: An up-to-date Hubble diagram is constructed using all well observed normal SNe Ia with $v < 30{,}000\ \mathrm{km\,s^{-1}}$. They are corrected for Galactic and internal absorption. For the internal reddening the slightly decline rate-dependent colors $(B-V)$ and $(V-I)$ at B maximum and 35 days thereafter of (nearly) absorption-free SNe Ia in E and S0 galaxies are used as a reference point. The SN magnitudes are reduced to a common decline rate of $\Delta m_{15} = 1.1$, requiring corrections in V from -0.18 to $+0.36$ mag. (Instead the SNe Ia can be reduced to a common Hubble type of the host galaxy; this results, however, in a somewhat larger scatter of the Hubble diagram.) It is essential that the calibrators and the distant SNe Ia be reduced the same way. (The independent reduction of [28] makes the calibrators 0.02 mag redder than the distant SNe Ia, which alone results in an 3% overestimate of H_0.) The 62 SNe Ia with $3{,}000 < v < 20{,}000\ \mathrm{km\,s^{-1}}$ define a Hubble diagram with a scatter of only 0.14 mag. The intrinsic scatter is even smaller as shown by the SNe Ia in E/S0 galaxies which have a scatter in I of only 0.11 mag.

Paper IV: Cepheid distances are derived, after a revision of the photometric WFPC2 calibration, for the eight program galaxies as well as for 27 galaxies from the literature using the P–L relations in V and I of the Galaxy *and* of LMC (see Sect. 52.3). Several tests are performed that the distances are independent of the metallicity of the Cepheids. The Cepheid distance scale has since been extended to still lower metallicities [24].

Paper V: The Cepheid distances of the ten calibrating galaxies yield mean absolute magnitudes of SNe Ia of $M_B = -19.49$, $M_V = -19.46$, and $M_I = -19.22$ (with random errors of 0.04 mag), which when combined with the appropriate Hubble diagrams in [13] give $H_0 = 62.3 \pm 1.3$ with an estimated systematic error of less than 10%. The random error is almost entirely determined by the adopted SNe Ia luminosities. The agreement of H_0 from B, V, and I data to within 0.3 units confirms the adopted absorption corrections. The *local* value of H_0, as determined from Cepheid distances of 25 galaxies outside the Local Group, from Cepheid-calibrated 21cm-line width distances of a complete, distance-limited sample, and from 20 SNe Ia within 2,000 km s^{-1}, is indistinguishable from the large-scale value. The invariance of H_0 with scale length is remarkable.

52.3 The P–L Relations of Cepheids

Two separate and independent P–L relations can be derived:

(1) The form and the zero-point of the Galactic P–L relation in B, V, and I is defined by 33 reddening-corrected Cepheids in Galactic clusters, fitted to the Pleiades main sequence at $\mu^0 = 5.61$ (the value of 5.64 from a recent review by [25] is sufficiently close), and by 36 reddening-corrected Cepheids with physical moving-atmosphere (BBW) distances. For details see [20, 23]. The two *independent* data sets are in good agreement. The slope of the combined P–L_{BVI} relations of these metal-rich stars ($[\mathrm{O/H}]_{\mathrm{Te}} = 8.60$) is quite steep ($a_V = -3.087 \pm 0.085$; see Fig. 52.2a). The equally metal-rich Cepheids in NGC 4321 [4] and NGC 3351 [6] have closely the same slope.

Claims of a flatter Galactic P–L relation by [5] and [1] have been contended in [24].

(2) Extensive B, V, I photometry of a large number of LMC Cepheids has been provided by [26], who have also determined their reddenings from surrounding red-clump stars. The data have been augmented by long-period Cepheids from the literature. For details see [20]. The metal-poor ($[\mathrm{O/H}]_{\mathrm{Te}} = 8.26$) LMC Cepheids cannot be fitted by a single-slope P–L relation, but the relation has a break at $\log P = 1.0$ and becomes significantly flatter for longer periods ($a_V = -2.507 \pm 0.069$; Fig. 52.2b). This behavior agrees with the pulsation models of [10]. For the zero-point of the LMC P–L-relation $\mu^0_{\mathrm{LMC}} = 18.54 \pm 0.02$ is adopted, which is the mean of 28 determinations in the literature from 1997 to 2006.

The differences between Galactic and LMC Cepheids are important. LMC Cepheids are brighter in V by 0.37 mag at $\log P = 0.4$ than their Galactic counterparts, and fainter by 0.13 mag at $\log P = 1.6$. The difference is also readily visible in the period–color (P–C) relations, Galactic Cepheids being redder. This is not only an effect of increased line blanketing, but also a temperature effect [7, 20], where also their position and different slopes in the $\log L - \log T_{\mathrm{e}}$ diagram are shown).

The non-uniqueness of the P–L relation poses a dilemma for the determination of Cepheid distances, particularly since the HST V and I magnitudes are not only used to determine the distance, but also the internal reddening. In [16] it was assumed

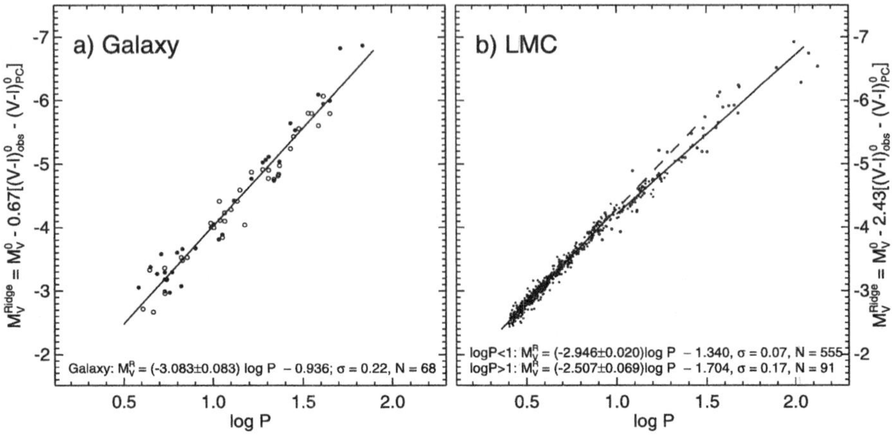

Fig. 52.2 The P–L relations in V of Cepheids in the Galaxy and LMC. The magnitudes are corrected for Galactic and internal absorption and shifted — in order to reduce the scatter — along constant-period lines with slope $\beta_{V,V-I}$ to the ridge line of the P–L relations ($\beta_{V,V-I}(\text{Galaxy}) = 0.67$, $\beta_{V,V-I}(\text{LMC}) = 2.43$, see [20]). The flattening of the LMC P–L relation for $\log P > 1.0$ is readily visible (6σ); it is also confirmed by pulsation models [10]

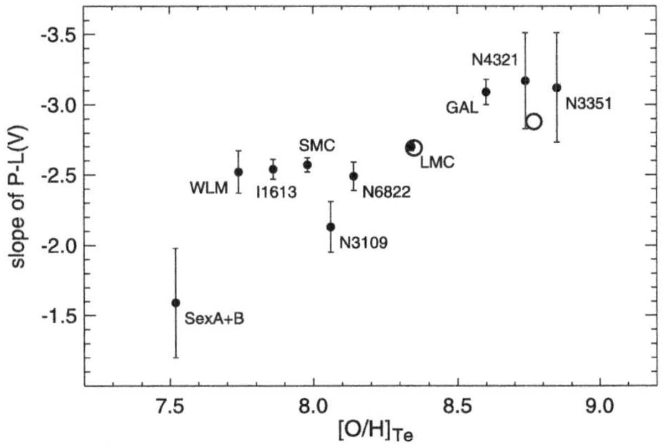

Fig. 52.3 The slope of various P–L relations in V in function of the metallicity $[\text{O/H}]_{\text{Te}}$ of the Cepheids. *Open circles* are the means of seven galaxies each, from [16]. The Cepheids of Sex A and B are combined because they have the same metallicity and TRGB distance

that the differences are a pure metallicity effect. Accordingly the distances from the Galactic and LMC P–L relations were interpolated according to metallicity. Since the LMC P–L relation yields larger distances at short periods and shorter distances at long periods (whereas the distances agree at $\log P \sim 0.9$), the metallicity corrections become necessarily period-dependent. The identification of the metallicity being the culprit is in line with the general trend of the P–L relation to become flatter as the metallicity decreases (Fig. 52.3). Several tests also justify the adopted procedure by showing that the final distances do not have a noticeable remaining metallicity error [16, 21, 24]. The weight of their zero-point is divided between the Pleiades, the BBW distances, and the adopted LMC modulus.

However, the full story is still more complex. Several metal-rich galaxies are known to have *flat* P–L relations similar to the LMC relation. A particularly well documented case are the Cepheids in the inner field of NGC 4258 [9]. The flatness appears to be due to an overluminosity of the shorter-period Cepheids, while those with longer periods have nearly the "normal" luminosity of Galactic Cepheids. This would be fortunate because most Cepheids known outside the Local Group have $\log P > 1.0$. But in case of the metal-rich Cepheids with a flat P–L relation one still depends on the *assumption* of a standard Galactic P–C relation for metal-rich Cepheids in order to obtain a correct value of the distance and reddening.

The existence of metal-rich Cepheids with flat P–L relations calls for a second parameter. This may well be Y. The pulsation models of [10] do indeed suggest that the slope of the P–L relations does not only depend on Z, but also on Y.

52.4 RR Lyrae Stars and the Tip of the Red-Giant Branch (TRGB)

A review of the absolute magnitude of RR Lyr stars has given $M_V = 0.52$ at $[\mathrm{Fe/H}] = -1.5$ with a non-linear dependence of M_V on [Fe/H] [19]. The value depends on fitting globular clusters to a main-sequence of appropriate metallicity as defined by SX Phe stars [11] and Hipparcos parallaxes, and on the theory of the ZAHB [2, 27] with a 0.09 mag correction to account for the average evolution of RR Lyr stars away from the ZAHB. The result is in excellent agreement with the RR Lyr stars in LMC as observed by [3] after correction for absorption and with $\mu^0_{\mathrm{LMC}} = 18.54$ from above, but the calibration does not depend on this agreement. The distance scales from Cepheids and RR Lyr stars are therefore *independent*.

The calibration of the RR Lyr stars can be applied to 24 galaxies, whose variables have been observed by numerous authors (see [24], Table 1). For twenty-two of these galaxies reliable apparent I-band magnitudes of the tip of the red-giant branch (TRGB) have been published. A combination of these magnitudes with the corresponding RR Lyr distances yields values of M_I^{TRGB}, the mean of which is -4.05 ± 0.02 with an impressively small scatter of 0.08 mag. The value holds for a mean metallicity of $[\mathrm{Fe/H}] = -1.5$ of the galaxies involved (see [24], Table 2). The same value was derived independently by [14].

The use of the TRGB as a standard candle and distance indicator has a long history [17, 24]. The TRGB is reached during the helium flash in the fully degenerate cores of old, metal-poor red giants, as modelled by [8, 12, 15], and others. While $M_{\mathrm{bol}}^{\mathrm{TRGB}}$ increases somewhat with increasing metallicity, the effect is partially offset in the I-band by line blanketing, such that M_I^{TRGB} becomes quite stable. In fact the RR Lyr calibrated values of M_I^{TRGB} from above reveal no dependence on metallicity. The question was also investigated by a number of authors, cited in [24], using theoretical and observational data. The compromise conclusion is that M_I^{TRGB} deviates from the mean value by only ± 0.05 for $-2.2 < [\mathrm{Fe/H}] < -1.2$; only for the unfrequent interval $-1.2 < [\mathrm{Fe/H}] < -0.5$ the TRGB becomes fainter by up to 0.2 mag.

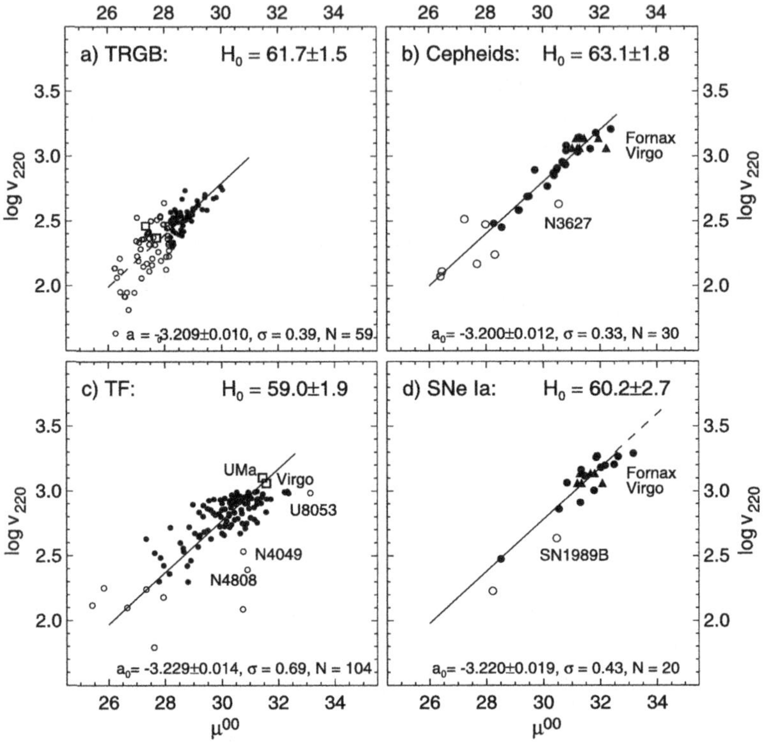

Fig. 52.4 Distance-calibrated Hubble diagrams for **(a)** TRGB distances; the M 81, Cen A, and IC 342 groups are shown as *squares* at their mean position; **(b)** Cepheid distances (the Virgo und Fornax cluster members are plotted at $v_{220} = 1{,}152$ and $1{,}371$ km s^{-1}, respectively); **(c)** 21cm-line width distances of a complete sample of field galaxies with $v_{220} < 1{,}000$ km s^{-1}; the Virgo cluster and the UMa cluster (at $v_{220} = 1{,}236$ km s^{-1}) are also shown; **(d)** SN Ia distances with $v_{220} < 2{,}000$ km s^{-1}; the *dashed line* is the downward extension of the Hubble line from [13] defined by 62 SNe Ia with $3{,}000 < v_{\rm CMB} < 20{,}000$ km s^{-1}, reflecting the large-scale value of H_0 (from [21]). *Triangles* denote cluster members. *Open circles* are objects with $\mu^0 < 28.2$ or in **(c)** with $v_{220} < 200$ km s^{-1}, and a few deviating objects (identified); *open symbols* are not considered for the solution. All distances are plotted as seen from the barycenter of the Local Group

TRGB distances are relevant for the determination of H_0 in three ways:

(1) A comparison of Cepheid with TRGB distances shows no metallicity dependence of the adopted Cepheid distances.
(2) TRGB magnitudes have been published in the literature for over 200 galaxies, many of which are observed with the WFPC2 or ACS of HST. Their distances have been derived using a constant value of $M_I^{\rm TRGB} = -4.05$. Fifty-nine of these galaxies lie outside of 4.5 Mpc (i.e. well outside the gravitational deceleration of the Local Group) and reach out to ∼10 Mpc. They define a local value of $H_0 = 61.7 \pm 1.5$ (Fig. 52.4a).
(3) For two host galaxies of SNe Ia TRGB distances are available. Two more SNe Ia are in groups for which at least two group members have a TRGB distance.

One obtains in this way a tentative mean luminosity of the four SNe Ia of $M_V = -19.47 \pm 0.13$, which inserted into the Hubble diagram of distant SNe Ia (Fig. 52.1) gives $H_0 = 62 \pm 4$.

The TRGB distance scale will eventually carry the same weight as the one based on Cepheids.

52.5 Summary

In place of a summary, four Hubble diagrams are shown in Fig. 52.4. Here the *independent* TRGB diagram gives, within the errors, the same value of H_0 as the Cepheids as well as the Cepheid-calibrated 21cm-line width distances and the equally calibrated *local* SNe Ia. All four values agree well with the cosmic value of $H_0 = 62.3$ from distant SNe Ia.

Admittedly the range of useful TRGB distances is still restricted to 4.5–10 Mpc (at shorter distances the pull of the Local Group decelerates the nearest galaxies), but already some TRGB information is available for four host galaxies of SNe Ia [24]. If inserted into the Hubble diagram of distant SNe Ia they yield a preliminary large-scale value of ~62. Good TRGB distances for at least seven SNe Ia will be accessible in the foreseeable future. They will provide an important test of H_0 as determined from Cepheids.

Acknowledgements G.A.T. acknowledges that the HST Supernova Program was co-sponsored by the Swiss National Science Foundation and subsequently by the Swiss Prodex Program of ESA. He thanks Dres. M. Rejkuba and A. Renzini for helpful discussions. A.S. thanks the Observatories of the Carnegie Institution for post-retirement facilities.

References

1. Benedict, G.F., McArthur, B.E., Feast, M.W., et al., 2007. Astron. J. 133, 1810.
2. Catelan, M., Pritzl, B.J., Smith, H.A., 2004. Astrophys. J. Suppl. Ser. 154, 633.
3. Clementini, G., Gratton, R., Bragaglia, A., et al., 2003. Astron. J. 125, 1309.
4. Ferrarese, L., Freedman, W.L., Hill, R.J., et al., 1996. Astrophys. J. 464, 568.
5. Gieren, W., Storm, J., Barnes, T.G., Fouqué, P., Pietrzynski, G., Kienzle, F., 2005. Astrophys. J. 627, 224.
6. Graham, J.A., Phelps, R.L., Freedman, W.L., et al., 1997. Astrophys. J. 477, 535.
7. Laney, C.D., Stobie, R.S., 1986. Mon. Not. R. Astron. Soc. 222, 449.
8. Lee, M.G., Freedman, W.L., Madore, B.F., 1993. Astrophys. J. 417, 553.
9. Macri, L.M., Stanek, K.Z., Bersier, D., Greenhill, L.J., Reid, M.J., 2006. Astrophys. J. 652, 1133.
10. Marconi, M., Musella, I., Fiorentino, G., 2005. Astrophys. J. 632, 590.
11. McNamara, D.H., 1997. Publ. Astron. Soc. Pac. 109, 1221.
12. Pietrinferni, A., Cassisi, S., Salaris, M., Castelli, F., 2004. Astrophys. J. 612, 168.
13. Reindl, B., Tammann, G.A., Sandage, A., Saha, A., 2005. Astrophys. J. 624, 532 (Paper III).
14. Rizzi, L., Tully, R.B., Makarov, D., et al., 2007. Astrophys. J. 661, 815.
15. Salaris, M., Cassisi, S., 1998. Mon. Not. R. Astron. Soc. 298, 166.

16. Saha, A., Thim, F., Tammann, G.A., Reindl, B., Sandage, A., 2006. Astrophys. J. Suppl. Ser. 165, 108 (Paper IV).
17. Sandage, A., 1971. In: O'Connell, D.J.K. (Ed.), Nuclei of Galaxies, Pont. Acad. Scient. Scripta Varia, vol. 35, p. 601. North Holland, Amsterdam.
18. Sandage, A., Tammann, G.A., 1982. Astrophys. J. 256, 339.
19. Sandage, A., Tammann, G.A., 2006. Annu. Rev. Astron. Astrophys. 44, 93.
20. Sandage, A., Tammann, G.A., Reindl, B., 2004. Astron. Astrophys. 424, 43 (Paper II).
21. Sandage, A., Tammann, G.A., Saha, A., Reindl, B., Macchetto, F.D., Panagia, N., 2006. Astrophys. J. 653, 843 (Paper V).
22. Spergel, D.N., Bean, R., Doré, O., et al., 2007. Astrophys. J. Suppl. Ser. 170, 377.
23. Tammann, G.A., Sandage, A., Reindl, B., 2003. Astron. Astrophys. 404, 423 (Paper I).
24. Tammann, G.A., Sandage, A., Reindl, B., 2008. Astrophys. J. 679, 52 (TSR 07).
25. Terndrup, An.D., Pinsonneault, D.M., et al., 2007. Astrophys. J. 655, 233.
26. Udalski, A., Soszynski, I., Szymanski, M., et al., 1999. Acta Astron. 49, 223.
27. VandenBerg, D.A., Swenson, E.J., Rogers, F.J., Iglesias, C.A., Aexander, D.R., 2000. Astrophys. J. 532, 430.
28. Wang, X., Wang, L., Pain, R., Zhou, X., Li, Z., 2006. Astrophys. J. 645, 488.

Chapter 53
Recent Progress on the Cepheid Distance Scale with *HST*

Lucas Macri

Abstract I present some results and ongoing work on several projects using *HST* to improve the precision and accuracy of the Cepheid Distance Scale. The ultimate goal of these projects is to obtain a measurement of H_0 with a total uncertainty of 5% or less. A measurement with such precision and accuracy would have a significant impact on the determination of other cosmological parameters, such as the equation of state of dark energy.

53.1 Introduction

The impact of the *Hubble Space Telescope* on the decades-long question of the age of the Universe was one of its greatest achievements and a key driver of its original development. Two independent teams used the Wide Field and Planetary Camera 2 to discover $\sim 10^3$ Cepheid variables in ~ 20 galaxies. The Cepheid-based distances to these galaxies served to calibrate several secondary distance indicators (type Ia supernovae, the Tully–Fisher relation, etc.) which were applied to objects in the Hubble flow to measure the value of the Hubble constant, H_0.

Two major teams undertook this challenging project. The *HST* Key Project on the Extragalactic Distance Scale, obtained a value of $H_0 = 72 \pm 8$ km s^{-1} Mpc^{-1} [1]. The *HST* Program for the Luminosity Calibration of Type Ia Supernovae (which includes our friend Duccio Maccheto), obtained a value of $H_0 = 62 \pm 5$ km s^{-1} Mpc^{-1} [2]. The 14% disagreement between these results is a far better situation than the pre-*HST* dichotomy in H_0 values of "50 or 100" km s^{-1} Mpc^{-1}. However, the past decade has seen outstanding advances in other areas of cosmology, such as the discovery of dark energy from observations of high-z type Ia supernovae [3, 4], the accurate determination of many cosmological parameters from observations of the Cosmic Microwave Background by the WMAP satellite [5], and the measurement of baryon-acoustic oscillations from SDSS observations [6].

In this new era of "precision cosmology", a Cepheid-based measurement of H_0 which could resolve the disagreement between [1] and [2] *and* have a total error budget of 5% would have a significant effect in reducing the degeneracy between Ω_m and w that is present in CMB-based determinations of the properties of dark energy [7].

L. Macri (✉)
National Optical Astronomy Observatory, 950 N Cherry Ave, Tucson, AZ, USA
e-mail: lmacri@noao.edu

F.D. Macchetto (ed.), *The Impact of HST on European Astronomy*, Astrophysics and Space Science Proceedings,
DOI 10.1007/978-90-481-3400-7_53, © Springer Science+Business Media B.V. 2010

53.2 Geometric Anchors for the Cepheid Distance Scale

One of the leading sources of uncertainty in the measurements of H_0 by [1] and [2] is due to the broad dispersion in different distance estimates to the current "anchor" of the Extragalactic Distance Scale: the Large Magellanic Cloud. Both studies adopt $\mu_{\mathrm{LMC}} = 18.5 \pm 0.1$ mag as a representative value of the distance modulus to our neighbor and its uncertainty. One can avoid this problem by relying on geometrical techniques to calibrate the Cepheid P–L relation; I would like to mention three recent developments on this area:

1. The determination of parallaxes to nine Galactic Cepheids by [8] using FGS on *HST*. The average parallax uncertainties are $\sigma_\pi/\pi = 8\%$, which yield a final uncertainty in the Cepheid P–L zeropoint of 2%.
2. The discovery of a large sample of Cepheids in NGC 4258 [7] using ACS on *HST*. This galaxy has a geometric distance estimate with an uncertainty of 8% based on VLBI observations of water masers orbiting an accretion disk around the central massive black hole [9]. This uncertainty will soon be reduced to a value as low as 3%.
3. The determination of a distance to the LMC based on detached eclipsing binaries by [10]. While this is a "pseudo-geometric" technique (due to its reliance on model stellar atmospheres), it is a well-characterized problem that can yield very precise and accurate distances, specially when far-UV spectra of these systems is obtained with *HST*. This team has studied four systems, which can be combined to obtain a distance to the LMC with an uncertainty of 3%.

These three results give remarkably similar results for the distance modulus to the Large Magellanic Cloud: 18.40 ± 0.05 mag [8], 18.41 ± 0.06 mag [10], 18.41 ± 0.16 mag [7]. Therefore, this long-lasting source of systematic uncertainty in H_0 may have been finally put to rest.

53.3 The Metallicity Dependence of the P–L Zeropoint

The metallicity dependence of the zero-point of the Cepheid Period–Luminosity relation is due to the fact that given two stars of equal masses but different metal abundances (in this case, LMC vs. solar), the evolutionary tracks of the latter will be shifted to lower effective surface temperature [11]. Hence, Cepheids populations with different abundances follow different period-color relations. Neglecting this effect while relying on the intrinsic P–C relation of LMC Cepheids to correct solar-abundance variables for extinction will result in distances that are biased towards smaller values.

Following the initial work of [12], two recent studies have obtained a 3σ determination of the metallicity dependence of the Cepheid P–L zeropoint. [13] relied on the Tip of the Red Giant Branch as a metallicity-insensitive distance indicator and compared the differences between Cepheid and TRGB distances as a function of abundance. Reference [7] carried out a differential analysis of the Cepheids in

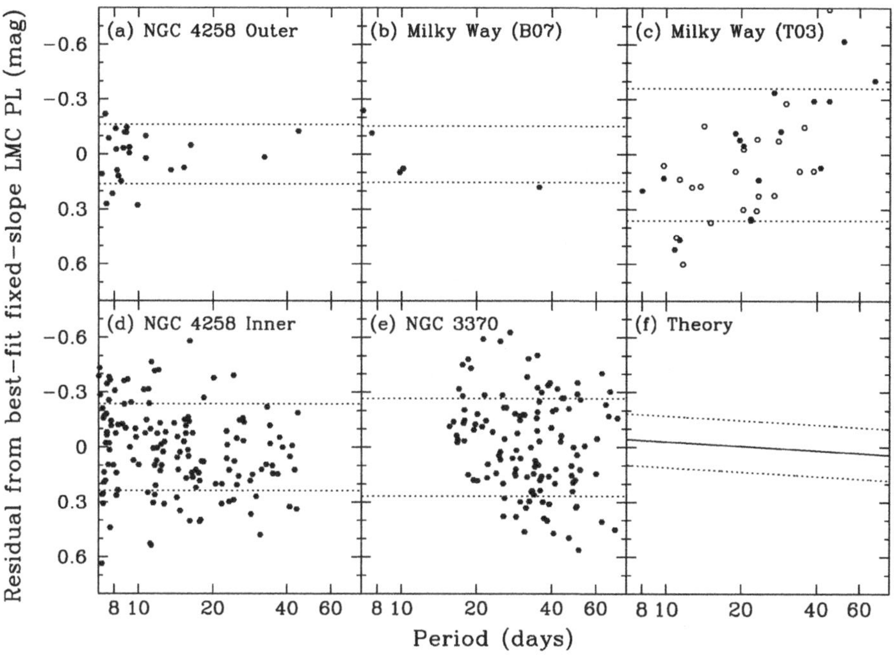

Fig. 53.1 Residuals from the best-fit LMC P–L relation (with a fixed slope) as a function of period, for different samples. The *four left-most panels* (**a**, **b**, **d**, **e**) are entirely consistent with a universal slope, which is expected from theoretical models (**f**). The samples from [15] are shown in panel (**c**). *Dashed lines* indicate the dispersion of each sample

NGC 4258. Now that the effect has been characterized with greater statistical significance, the metallicity correction is no longer a dominant source of uncertainty in H_0. Nevertheless, it is desirable to continue work on this area, specially given theoretical predictions [14] that are at odds with the observational results.

53.4 A Universal Slope for the Cepheid P–L Relation

Recently, [15] have postulated that solar abundance Cepheids exhibit a P–L relation with a significantly steeper slope that the relation derived by [16] from LMC Cepheids. We disagree with this conclusion for several reasons. First, [15] use a calibration of the Baade–Wesselink technique that has been shown by [17] to yield the same abnormally steeper slope for LMC Cepheids. Second, the recent parallaxes to Milky Way Cepheids by [8] yield a slope that is entirely consistent with the LMC value. Third, the two sub-samples of Cepheids in NGC 4258 studied by [7] are consistent with a common slope, even though they cover a range of abundance of ~0.5 dex.

Figure 53.1 shows our argument in a graphical manner. We take the observed I-band magnitudes of Cepheids in a variety of galaxies and fit them with the LMC

P–L relation derived by [16], keeping the slope fixed. We plot the *residuals* of the sample as a function of period, which should show no correlation if the P–L relation has a universal slope.

The samples in panels (a), (b), (d) and (e) are entirely consistent with a universal slope, even though they are drawn from very different populations: (a) the outer field of NGC 4258; (b) the Milky Way (with *HST* parallaxes from [8]); (d) the inner field of NGC 4258; (e) a field in NGC 3370. The first one has LMC metallicity, while the other three have solar abundances. By comparison, the Milky Way samples of [15] (panel c) show the largest scatter and the residual slope appears to be driven by a few outlying points. Lastly, panel (f) shows the theoretical expectations for slope variations due to differences in abundance.

The on-going SHOES project (PI: A. Riess) is using NICMOS on board *HST* to obtain accurate Cepheid distances to hosts of type Ia supernovae using the H-band P–L relation. Since these Cepheid samples have abundances near solar, it is imperative to use a similar sample as the "anchor" of the Distance Scale. We have chosen the inner field of NGC 4258 from [7] for this task, in anticipation of the very precise forthcoming maser distance to that galaxy. By bypassing the LMC, we will avoid the problems associated the metallicity dependence in the zero-point of the P–L relation, and turn the issue of a non-universal slope into a moot point, since all samples will have the same abundance. In this way, we hope to obtain a value of H_0 that is accurate to 5%, with a consequent improvement in the determination of other parameters of cosmological significance.

References

1. Freedman, W., et al., 2001. Astrophys. J. 553, 47.
2. Sandage, A., et al., 2006. Astrophys. J. 653, 843.
3. Riess, A., et al., 1998. Astron. J. 116, 1009.
4. Perlmutter, S., et al., 1999. Astrophys. J. 517, 565.
5. Spergel, D., et al., 2007. Astrophys. J. Suppl. Ser. 170, 377.
6. Eisenstein, D., et al., 2005. Astrophys. J. 633, 560.
7. Macri, L., et al., 2006. Astrophys. J. 652, 1133.
8. Benedict, G., et al., 2007. Astron. J. 133, 1810.
9. Herrnstein, J., et al., 1999. Nature 400, 539.
10. Fitzpatrick, E., et al., 2003. Astrophys. J. 587, 685.
11. Pietrinferni, A., et al., 2004. Astrophys. J. 612, 168.
12. Kennicutt, R., et al., 1998. Astrophys. J. 498, 181.
13. Sakai, S., et al., 2004. Astrophys. J. 608, 42.
14. Fiorentino, G., et al., 2002. Astrophys. J. 576, 402.
15. Tammann, G., et al., 2003. Astron. Astrophys. 404, 423.
16. Udalski, A., et al., 1999. Acta Astron. 49, 201.
17. Gieren, W., et al., 2005. Astrophys. J. 628, 695.

Chapter 54
Seeing Dark Energy

Adam G. Riess

Abstract This talk is largely based on the work now published in (Riess et al. in Astrophys. J. 659:98, 2007).

This talk is largely based on the work now published in [31].

The accelerating cosmic expansion first inferred from observations of distant type Ia supernovae (SNe Ia; [25, 28] indicates unexpected gravitational physics, frequently attributed to the dominating presence of a "dark energy" with negative pressure).

(For students of the history of science, a copy of my lab notebook page from 1997 showing where I recorded the first indication that the data pointed to present acceleration and dark energy can be found in Symmetry Magazine, Fall 2007, http://www.symmetrymagazine.org/cms/?pid=1000557.)

Increasingly incisive samples of SNe Ia at $z < 1$ have reinforced the significance of this result [5, 6, 12, 21, 34]. Using the new Advanced Camera for Surveys (ACS) and refurbished NICMOS camera on the *Hubble Space Telescope* (*HST*), our collaboration secured observations of a sample of the most-distant known SNe Ia. These half-dozen SNe Ia, all at $z > 1.25$, helped confirm the reality of cosmic acceleration by delineating the transition from preceding cosmic deceleration during the matter-dominated phase and by ruling out simple sources of astrophysical dimming [30]. The expanded sample of 23 SNe Ia at $z \geq 1$ presented here are now used to begin characterizing the early behavior of dark energy.

Other studies independent of SNe Ia now strongly favor something like dark energy as the dominant component in the mass-energy budget of the Universe. Perhaps most convincingly, observations of large-scale structure and the cosmic microwave background radiation provide indirect evidence for a dark-energy component. Measurements of the integrated Sachs–Wolfe effect (e.g., [1, 9, 23]) more directly suggest the presence of dark energy with a negative pressure. Additional, albeit more tentative, evidence is provided by observations of X-ray clusters [4] and baryon oscillations (e.g., [15]).

The unexplained existence of a dominant, dark-energy-like phenomenon presents a stiff challenge to the standard model of cosmology and particle physics. The apparent acceleration may result from exotic physics such as the repulsive gravity predicted for a medium with negative pressure or from entirely new physics. The explanation of strongest pedigree is Einstein's famous "cosmological constant" Λ (i.e.,

A.G. Riess (✉)
Space Telescope Science Institute, Baltimore, MD, USA
e-mail: ariess@stsci.edu

F.D. Macchetto (ed.), *The Impact of HST on European Astronomy*,
Astrophysics and Space Science Proceedings,
DOI 10.1007/978-90-481-3400-7_54, © Springer Science+Business Media B.V. 2010

vacuum energy; Einstein, 1917), followed by a decaying scalar field similar to that already invoked for many inflation models (i.e., quintessence — [10, 24, 36]). Competitors include the Chaplygin gas [7], topological defects, and a massless scalar field at low temperature. Alternatively, alterations to General Relativity may be required as occurs from the higher-dimensional transport of gravitons in string theory models [13] and braneworlds, or by finely-tuned, long-range modifications (e.g., Cardassian type, [18]; or [11]; see [33] for a review). Empirical clues are critical for testing hypotheses and narrowing the allowed range of possible models.

SNe Ia remain one of our best tools for unraveling the properties of dark energy because their individual measurement precision is unparalleled and they are readily attainable in sample sizes of order 10^2, statistically sufficient to measure dark-energy-induced changes to the expansion rate of $\sim$1%. Specifically, the equation-of-state parameter of dark energy, w (where $P = w\rho c^2$) determines both the evolution of the density of dark energy,

$$\rho_{DE} = \rho_{DE,0} \exp\left\{3 \int_a^1 \frac{da}{a}(1 + w(a))\right\},$$

and its gravitational effect on expansion,

$$\frac{\ddot{a}}{a} = -\frac{4\pi G}{3}(1 + 3w(a)),$$

where $\rho_{DE,0}$ is the present dark-energy density. Measuring changes in the scale factor, a, with time from the distance and redshift measurements of SNe Ia,

$$\frac{d_l(z)}{c(1+z)} = \int_t^{t0} \frac{dt'}{a(t')} = \int_0^z \frac{dz'}{H(z')},$$

constrains the behavior of $w(a)$ or $w(z)$ and is most easily accomplished at $z < 2$ during the epoch of dark-energy dominance.

Ideally, we seek to extract the function $w(z)$ for dark energy or its mean value at a wide range of epochs. Alternatively, we might constrain its recent value $w_0 \equiv w$ $(z = 0)$ and a derivative, $dw/dz \equiv w'$, which are exactly specified for a cosmological constant to be $(-1, 0)$. Most other models make less precise predictions. For example, the presence of a "tracker" dark-energy field whose evolution is coupled to the (decreasing) dark matter or radiation density may be detected by a measured value of $w' > 0$. In truth, we know almost nothing of what to expect for $w(z)$, so the safest approach is to assume nothing and measure $w(z)$ across the redshift range of interest. SN Ia at $z > 1$ are crucial to constrain variations of w with redshift. These measurements can only be made from space, and we report here on that endeavor. We have discovered and measured 21 new SN Ia with HST and used them to constrain the properties of the dark energy.

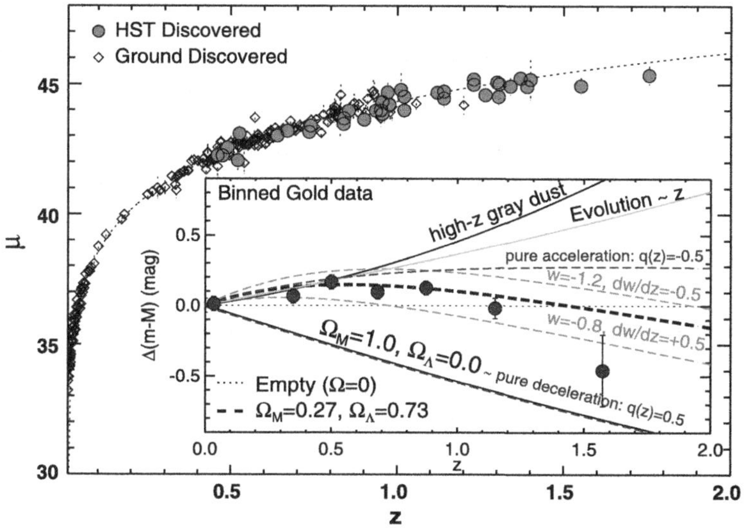

Fig. 54.1 (Color online) Hubble diagram of distance moduli and redshifts for all of the *HST*-discovered SNe Ia in the Gold and Silver sets from our program

54.1 Higher-z SNe Ia

In Fig. 54.1 we show the Hubble diagram of distance moduli and redshifts for all of the *HST*-discovered SNe Ia in the Gold and Silver sets from our program. The new SNe Ia span a wide range of redshift ($0.21 < z < 1.55$), but their most valuable contribution to the SN Ia Hubble diagram remains in the highest-redshift region where they now well delineate the range at $z \geq 1$ with 23 SNe Ia, 13 new objects since [30]. This territory remains uniquely accessible to *HST*, which has discovered the dozen highest-redshift SNe Ia known, and its exploration is the focus of the rest of this paper.

In the inset to Fig. 54.1 we show the residual Hubble diagram (from an empty Universe) with the Gold data uniformly binned. Here and elsewhere, we will utilize uniform, unbiased binning achieved with a fixed value of $n\Delta z$, where Δz is the bin width in redshift and n is the number of SNe in the bin.[1] In Fig. 54.1 we use $n\Delta z = 6$ which yields seven bins for our sample. Although binning is for illustrative purposes in the Hubble diagram, there are some specific advantages of binning such as the removal of lensing-induced asymmetrical residuals by flux averaging [35] and the ease of accounting for systematic uncertainties introduced by zeropoint errors in sets of photometric passbands used at similar redshifts.

The distance-redshift relation of SNe Ia is one of few powerful tools available in observational cosmology. A number of different hypotheses and models can be

[1] The last bin ends abruptly with the highest-redshift SN; thus, its $n\Delta z \leq$ value is smaller than the rest.

tested with it, including kinematic descriptions of the expansion history of the Universe, the existence of mass-energy terms on the right-hand side of the Friedman equation, and the presence of astrophysical sources of contamination. Testing all interesting hypotheses is well beyond the scope of this paper and is best left for future work. Instead, we now undertake a few narrowly posed investigations.

54.2 Alternatives to Dark Energy

After the detection of the apparent acceleration of cosmic expansion (and dark energy) by [28] and [25], alternative hypotheses for the apparent faintness of high-redshift SNe Ia were posed. These included extragalactic gray dust with negligible tell-tale reddening or additional dispersion [2, 3, 26, 27], and pure luminosity evolution [14].

In [30] we found that the first significant sample of SNe Ia at $z > 1$ from *HST* rejected with high confidence the simplest model of gray dust by [19], in which a smooth background of dust is present (presumably ejected from galaxies) at a redshift greater than the SN sample (i.e., $z > 2$) and diluted as the Universe expands. This model and its opacity was invented to match the 1998 evidence for dimming of supernovae at z 0.5 without invoking dark energy in a universe with $\Omega_m = 1$. This model is shown in the inset of Fig. 54.1. The present Gold sample (at the best fitting value of H_0) rejects this model at even higher confidence ($\Delta\chi^2 = 194$, i.e., 14σ), beyond a level worthy of further consideration.

54.2.1 Dark Energy

Strong evidence suggests that high-redshift SNe Ia provide accurate distance measurements and that the source of the apparent acceleration they reveal lies in the negative pressure of a "dark energy" component. Proceeding from this conclusion, our hard-earned sample of SNe Ia at $z > 1.0$ can provide unique constraints on its properties. Strong motivation for this investigation comes from thorough studies of high-redshift and low-redshift SNe Ia, yielding a consensus that there is no evidence for evolution or intergalactic gray dust at or below the current statistical constraints on the average high-redshift apparent brightness of SN Ia (see [16, 17] for recent reviews). We summarize the key findings here. (1) Empirically, analyses of SN Ia distances versus host stellar age, chemical abundance, morphology, and dust content indicate that SN Ia distances are relatively indifferent to the evolution of the Universe [5, 29, 32, 35]. (2) Detailed examinations of the distance-independent properties of SNe Ia (including the far-UV flux, e.g., as presented in the last section) provide strong evidence for uniformity across redshift and no indication (thus far) of redshift-dependent differences (e.g., [8, 20, 32]). (3) SNe Ia are uniquely qualified as standard candles because a well understood, physical limit (the Chandrasekhar limit) provides the source of their homogeneity. Based on these studies, we adopt a limit on redshift-dependent systematics is to be 5% per $\Delta z = 1$ at $z > 0.1$ and make quantitative use of this here.

Many have studied the constraint placed by the redshift-magnitude relation of SNe Ia on the parameter combination Ω_M–w, where w (assumed to be constant) is the dark energy equation-of-state parameter. There are few models for dark energy that predict an equation of state that is *constant*, different from the cosmological constant, and not already ruled out by the data. On the other hand, a prominent class of models does exist whose defining feature is a time-dependent dark energy (i.e., quintessence). While the rejection of $w = -1$ for an assumed constant value of w would invalidate a cosmological constant, it is also possible that apparent consistency with $w = -1$ in such an analysis would incorrectly imply a cosmological constant. For example, if $w(z)$ is rising, declining, or even sinusoidal, a measured derivative could be inconsistent with zero while the average value remains near -1. Therefore, when using $w(z)$ to discriminate between dark-energy models, it is important to allow for time-varying behavior, or else valuable information may be lost. Here, we seek to constrain the value of $w(z > 1)$ and bound its derivative across the range $0.2 < z < 1.3$. This is unique information afforded by the *HST*-discovered SN Ia sample.

Finally, we may consider whether three additional parameters to describe $w(z)$ are actually needed to improve upon a flat, Λ-cold-dark-matter (ΛCDM) model fit to the data. To determine this we can calculate the improvement to the fit,

$$\chi^2_{\mathrm{eff}} \equiv -\Delta(2\ln\mathcal{L}) = 2\ln\mathcal{L}(w = -1) - 2\ln\mathcal{L}(w_i = \mathcal{W}_i), \qquad (54.1)$$

with i additional free parameters. For the weak, strong, and strongest priors we find an improvement of $\chi^2_{\mathrm{eff}} = 4$, 5.5, and 5.5, respectively, for the three additional degrees of freedom, in no case requiring the additional complexity in dark energy (improvements of >14 would be noteworthy). Likewise, there is no improvement at all for the Akaike Information Criterion (i.e., $\Delta\mathrm{AIC} = \Delta\chi^2 - 2i$; [22] with changes of -2, -0.5, and -0.5, respectively, which fail to overcome the penalty of increased complexity in the model.

54.3 Conclusions

(1) We present 21 new *HST*-discovered SNe Ia and an improved calibration of the previous sample from [30]. Together this sample contains 23 SNe Ia at $z \geq 1$, extending the Hubble diagram over 10 Gyr.

(2) We derive uncorrelated, model-independent estimates of $H(z)$ which well-delineate current acceleration and preceding deceleration. The *HST*-discovered SNe Ia measure $H(z > 1)$ to slightly better than 20% precision.

(3) The full *HST*-discovered SN Ia sample, presented here, provides a factor of two improvement over our present ability to constrain simple parameterizations of the equation-of-state parameter of dark energy (w) and its evolution.

(4) Stronger priors and tighter constraints on the preferred cosmological model can be extracted from independent measurements tied to the surface of last scattering, but the use of these requires assumptions about the behavior of dark energy across a wide range of redshift ($1.8 < z < 1{,}089$). The strongest of these priors, like the simplest dark energy parameterizations, appears unjustified in the presence of our current ignorance about dark energy. Assuming the effect of dark energy at $z > 1.8$ is minimal, we derive meaningful constraints on the early properties of dark energy: $w(z > 1) = -0.8^{+0.6}_{-1.0}$ and $w(z > 1) < 0$, i.e., negative pressure, at 98% confidence.

(5) At present, we find that the use of additional parameters to describe $w(z)$ does not provide a statistically significant improvement to the fit of the redshift-magnitude relation over the use of a simple cosmological constant.

(6) An analysis of the $z > 1$ sample-averaged spectrum shows it to be consistent with the mean spectrum of SNe Ia over the last 10 Gyr, failing to reveal direct evidence for SN Ia evolution.

References

1. Afshordi, N., Loh, Y.-S., Strauss, M.S., 2004. Phys. Rev. D 69, 083524.
2. Aguirre, A.N., 1999. Astrophys. J. 512, L19.
3. Aguirre, A.N., 1999. Astrophys. J. 525, 583.
4. Allen, S.W., Schmidt, R.W., Ebeling, H., Fabian, A.C., van Speybroeck, L., 2004. Mon. Not. R. Astron. Soc. 353, 457.
5. Astier, P., et al., 2006. Astron. Astrophys. 447, 31.
6. Barris, B., et al., 2004. Astrophys. J. 602, 571.
7. Benot, M.C., Bertolami, O., Sen, A.A., 2002. Phys. Rev. D 66, 4, 3507.
8. Blondin, S., et al., 2006. Astron. J. 131, 1648.
9. Boughn, S., Crittenden, R., 2004. Nature 427, 45.
10. Caldwell, R.R., Davé, R., Steinhardt, P.J., 1998. Astrophys. Space Sci. 261, 303.
11. Carroll, S.M., Duvvuri, V., Trodden, M., Turner, M.S., 2004. Phys. Rev. D 70, 043528.
12. Conley, A., et al., 2006. Astrophys. J. 644, 1.
13. Deffayet, C., et al., 2002. Phys. Rev. D 66, 4019.
14. Drell, P.S., Loredo, T.J., Wasserman, I., 2000. Astrophys. J. 530, 593.
15. Eisenstein, D.J., et al., 2005. Astrophys. J. 633, 560.
16. Filippenko, A.V., 2004. In: Freedman, W.L. (Ed.), Measuring and Modeling the Universe, Carnegie Observatories Astrophysics Series, vol. 2, p. 270. Cambridge Univ. Press, Cambridge.
17. Filippenko, A.V., 2005. In: Sion, E.M., Vennes, S., Shipman, H.L. (Eds.), White Dwarfs: Cosmological and Galactic Probes, p. 97. Springer, Dordrecht.
18. Freese, K., 2005. New Astron. Rev. 49, 103.
19. Goobar, A., Bergstrom, L., Mörtsell, E., 2002. Astron. Astrophys. 384, 1.
20. Howell, D.A., et al., 2005. Astrophys. J. 634, 1190.
21. Knop, R., et al., 2003. Astrophys. J. 598, 102.
22. Liddle, A.R., 2004. Mon. Not. R. Astron. Soc. 351, 49.
23. Nolta, M.R., et al., 2004. Astrophys. J. 608, 10.
24. Peebles, P.J., Ratra, B., 2003. Rev. Mod. Phys. 75, 559.
25. Perlmutter, S., 1999. Astrophys. J. 517, 565.
26. Rana, N.C., 1979. Astrophys. Space Sci. 66, 173.
27. Rana, N.C., 1980. Astrophys. Space Sci. 71, 123.

28. Riess, A.G., et al., 1998. Astron. J. 116, 1009.
29. Riess, A.G., et al., 1999. Astron. J. 118, 2668.
30. Riess, A.G., et al., 2004. Astrophys. J. 607, 665.
31. Riess, A.G., et al., 2007. Astrophys. J. 659, 98.
32. Sullivan, M., et al., 2003. Mon. Not. R. Astron. Soc. 340, 1057.
33. Szydlowski, M., Kurek, A., Krawiec, A., 2006. Phys. Lett. B 642, 171.
34. Tonry, J.T., et al., 2003. Astrophys. J. 594, 1.
35. Wang, Y., Mukherjee, P., 2006. Astrophys. J. 650, 1.
36. Wetterich, C., 1995. Astron. Astrophys. 301, 321.

Part V
Closing Remarks

Chapter 55
Closing Remarks

Duccio Macchetto

Abstract The primary aim of the 41st ESLAB Symposium was to review the key contributions that HST has made in all areas of astronomy and emphasize their impact on European astronomical research. I believe that goal was fully met. The presence of many US scientists added to the success of the Symposium and emphasised the fact that astronomy is truly an international endeavour.

I have spent most of my professional life working for the Hubble Space Telescope Project. First as the ESA Study Scientist, in 1974 and 1975, I was charged with the studies that led to the selection of the Faint Object Camera as the European scientific instrument that was fly on HST. I participated in the discussions between NASA and ESA that led to the Memorandum of Understanding (MOU) between the two agencies, an MOU whose key terms open the door to the participation by all European astronomers to the scientific promise of HST. After the HST Project was approved both in Europe and the United States, I became the ESA Project Scientist and the Principal Investigator for the Faint Object Camera.

When I left Europe in 1983, it was to take up a position at the Space Telescope Science Institute (STScI), directed at that time by Riccardo Giacconi. While I held a number of different positions at the STScI, as the ESA Senior Astronomer I was also responsible for all aspects of ESA's participation in HST. I made it my responsibility to ensure that access to HST was open to everyone on an equal basis without favouritism or bias; the process of the Time Allocation Committee has become a model for all major observatories space- or ground-based. I am very proud that European astronomers have consistently been awarded a considerable fraction of observing time on HST on a purely competitive basis.

Since its launch in 1990 the Hubble Space Telescope has made major contributions to all areas of astronomy and astrophysics, from the study of nearby planets, the processes of star and planetary formation, the stellar and interstellar components of galaxies, the discovery that most if not all galaxies harbor a massive black hole in their centres, which has profound implications for our understanding of the process of galaxy formation feed back and evolution, to the realization that the universe as a whole is undergoing acceleration as a result of a yet unknown form of "dark energy".

D. Macchetto (✉)
Space Science Department of the European Space Agency, Space Telescope Science Institute, Baltimore, MD 21218, USA
e-mail: macchetto@stsci.edu

F.D. Macchetto (ed.), *The Impact of HST on European Astronomy*, Astrophysics and Space Science Proceedings,
DOI 10.1007/978-90-481-3400-7_55, © Springer Science+Business Media B.V. 2010

European astronomers have participated in many of these observations and have made significant contributions to the progress in our understanding. However, the impact of HST goes beyond these direct contributions and continues to shape the key questions that need to be addressed not only with HST but with other space and ground based observatories. In addition Hubble has had an important role on the performance and productivity of several European facilities, most notably on the VLT. It can be argued that the Hubble–VLT synergy has been even more effective than the synergy between Hubble and Keck in the early years. Programmes such as GOODS and COSMOS are a direct result of this. Other important missions that have benefited from the HST are XMM-Newton and in the future, Herschel and ALMA.

The incredible success of the HST mission is the result of the efforts from many dedicated individuals, and I cannot forget the key contributions made by my ESA colleagues, engineers and scientists alike, that have built and operated the Solar Arrays, the Faint Object Camera and have participated in the scientific operations of HST both at the STScI in Baltimore and at the Space Telescope European Coordinating Facility in Garching.

The primary aim of the 41st ESLAB Symposium was to review the key contributions that HST has made in all areas of astronomy and emphasize their impact on European astronomical research. I believe that goal was fully met. The presence of many US scientists added to the success of the Symposium and emphasised the fact that astronomy is truly an international endeavour.

Life in the HST programme has been exciting to say the least. I always felt as if I was riding a roller-coaster! At the time of writing this summary, after many delays, instrument failures and recoveries, the astronauts have just completed Servicing Mission 4 and what an incredible feat that was! The HST has been fully restored and upgraded and we believe that it will continue to make its unique mark on astronomy for the next five to ten years: **PER ASPERA AD ASTRA!**

The manufacturer's authorised representative in the EU is Springer Nature Customer Service Centre GmbH, Europaplatz 3, 69115 Heidelberg, Germany. If you have any concerns regarding our products, please contact ProductSafety@springernature.com

Printed and bound by CPI Group (UK) Ltd, Croydon, CR0 4YY

30/09/2026

02225447-0011